Search for a Rational Ethic

George D. Snell

Search for
a Rational Ethic

Springer-Verlag
New York Berlin Heidelberg
London Paris Tokyo

George D. Snell
The Jackson Laboratory
Bar Harbor, Maine 04609, USA

Library of Congress Cataloging-in-Publication Data
Snell, George D. (George Davis), 1903–
    Search for a rational ethic.
    Bibliography: p.
    1. Ethics. 2. Social ethics. I. Title.
BJ1012.S55   1988   170   88-4898

Typeset by Coghill Book Typesetting Co., Richmond, Virginia.
Printed and bound by Edwards Brothers, Inc., Ann Arbor, Michigan.
Printed in the United States of America.

9 8 7 6 5 4 3 2 1

ISBN 0-387-96767-2 Springer-Verlag New York Berlin Heidelberg
ISBN 3-540-96767-2 Springer-Verlag Berlin Heidelberg New York

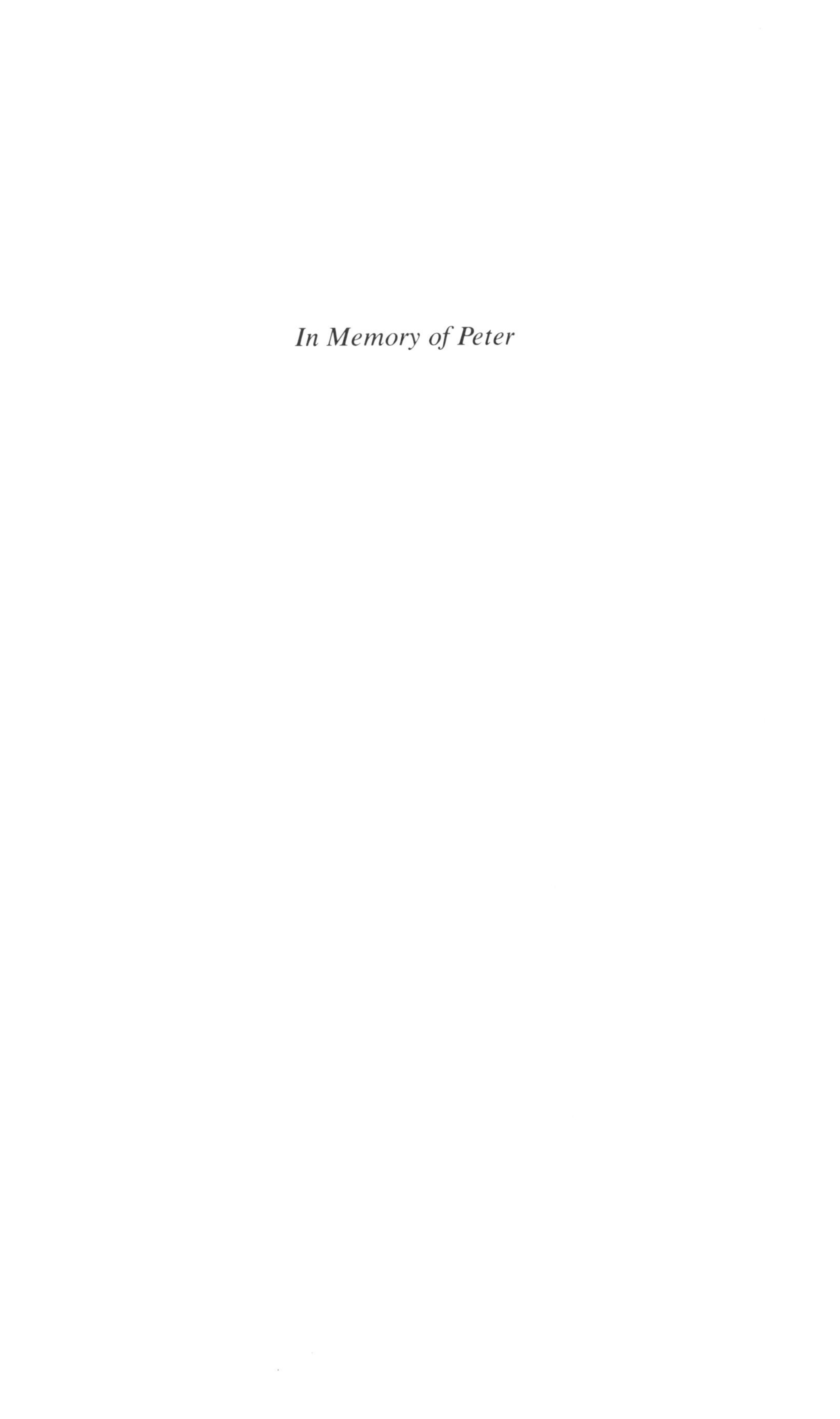

In Memory of Peter

Foreword

Knowledge we have in great abundance, and enough exists if wisely used to solve many of the most threatening problems of humanity. The key word is wisely; wisdom we sorely lack. There is a special role to be played by distinguished scholars who, having passed the most challenging tests of their specialized fields, are willing to confront the central questions of human existence. What is life (where is the boundary between life and non-life)? Why do we behave as we do? What is the meaning of human existence? Where do ethical precepts come from? What should be the goals of civilization, beyond mere survival and hedonic reward? These are the kinds of topics George Snell boldly addresses in *Search for a Rational Ethic*.

Scientific knowledge is especially important in any such endeavor, because we are in the golden age of science, and scientific research increasingly impinges on the domain of philosophy. Indeed, it is not too much to say that philosophy has consisted to a large extent of failed neurological models. Much of its investigation pivots on how the mind works, that is, to what extent the mind can perceive reality, how concepts are formed, what is the source of moral reasoning, and so forth. Increasingly, scientific research is leading us to the physical basis of mind. If we are ever to create the correct neurological model, it will be through science. But the account of physical basis and physiological process will be only half the story. The other half, as Snell notes in the book before you, is history. Not just cultural history, about which we know a great deal, but also genetic history, about which we know extremely little. The mind is the product in each generation of the interaction of biological features, which evolved over tens or hundreds of thousands of years, and the cultural environment, which evolved over mere decades or centuries. It is far from being solely a product of culture.

Science is also important in its ability to control and change culture. Through technology it offers power and energetic efficiency billions of times greater than that available to preliterate societies. It has opened the way to alter the genes and hence the nature of life itself. It has inadvertently generated problems in ethics undreamed of by traditional religion. In short, science is the source of both titanic new moral dilemmas and the only reliable source of the knowledge required to solve them.

In *Search for a Rational Ethic,* Snell gives a penetrating analysis of the special role of science in the new moral journey on which humanity has been launched. He believes he has discovered new strength in the pragmatic principle of "self interest rightly understood," as de Toqueville once phrased it, in other words behavior shaped to serve both individual happiness and the common good. Snell sees little profit in a return to the unexamined strictures of traditional religion or the expectation of unrequited sacrifice. He places emphasis on knowledge of human biology made as exact as possible, but applied by emotion-guided consensus. He insists that because our understanding of human biology and social behavior is so incomplete, the better part of wisdom is humility. Finally, in the best tradition of humanism, Snell concludes that ethics is indeed the binding fabric of society. We have tended to play down that venerable principle in recent years. Some have gone so far as to suppose that the rise of scientific materialism serves as a rationale for increased selfishness and relativism. This book shows that the opposite is true.

Harvard University   Edward O. Wilson
April 6, 1988

Preface

This book has a long history. My interest in ethics goes back to the 1920s when, as an undergraduate at Dartmouth College, I took courses in logic and ethics with Professor James MacKaye. Professor MacKaye's training was in engineering rather than philosophy, and he brought some of the exactness of the engineering method into his philosophical approach. In particular, he emphasized the importance of careful definition of terms. He also gave a lucid analysis of the methods of definition. I have made consistent use of his ideas on definition in writing this volume. I have also drawn on his ideas concerning ethics, though I no longer fully accept some aspects of his approach.

Another experience that contributed to my concern with ethics occurred in 1933–1934 when I taught a course in genetics and evolution at Washington University in St. Louis. Teaching genetics was easy—that was my specialty—but while I had the familiarity with evolution expected of any graduate student in biology, my reading in that area was limited. I understood the basic principles of the theory of natural selection but had never gone deeply into that subject. When I came to teach it, it seemed to me that I was teaching an anti-ethical doctrine. The implications of the survival of the fittest, to use the expression coined by Herbert Spencer and adopted by Darwin, did not seem compatible with my New England upbringing. I found this conflict quite disturbing. When subsequently I had an opportunity for more extensive reading, I discovered that many biologists, including Darwin himself, have felt this same concern and that a great deal of thought and effort have gone into reconciling evolution and ethics.

While through most of my career I have concentrated on certain specific areas of genetic research, I have consistently maintained my interest in ethics. In 1953–1954, when I was due a sabbatical leave from The Jackson Laboratory, I decided that I would like to devote it to gathering material for a book on ethics. With the generous consent of Dr. Little, founder and original director of the laboratory, and with the support of a Guggenheim Fellowship, I spent a year reading in the libraries at the University of

Texas in Austin and at Dartmouth College. Seven years later, when another sabbatical was due, and again with the generous consent of the director of the laboratory, Dr. Earl Green, who had succeeded Dr. Little, I spent a year at home writing a book on ethics. At the end of the year, I found a publisher who would accept it, but who stipulated certain changes. I was not entirely happy with the suggested changes, but also and more compelling, I had to get back to the laboratory and had no time to make them.

By hindsight, it is very fortunate that the book was not published at that time. The scientific background that I see as essential for a thoroughly rational ethic has developed enormously since 1950. Also my understanding of the philosophy of ethics, I think, has improved. And perhaps because of what appears to be a growing concern with ethics in this country, the timing is better. The present volume, although directed to the same end using many of the same ideas as the earlier volume, is entirely reorganized and rewritten. The last 8 or 10 years of my retirement have been largely dedicated to its completion.

Because, in my view, an appropriate factual background is the necessary foundation for a rational ethic and because relevant facts can come from many areas of knowledge, this book ranges over a wide field. Although I believe that this broad approach is justified, it carries obvious dangers. I have had to make considerable use of secondary sources and I have ventured into areas where I have no established competence. It is possible that some errors have resulted. If so, however, I am confident that the errors are minor and that they do not invalidate my basic conclusions.

Many people have contributed in many ways to the writing. I have mentioned my debt to Prof. James MacKaye and Dr. Clarence Little who, years ago, did much to make this book possible, and also to Dr. Earl Green and the Guggenheim Foundation. Miss Sheila Counce (now Dr. Sheila Counce) kept my experiments going during the year I was away on sabbatical and the late Prof. T. S. Painter acted as my host at the University of Texas. Mrs. Helen Bunker kept my work at the laboratory going through 1960–1961, when I was at home writing. Dr. John Fuller read the early version of this book and offered many constructive suggestions.

I have made use of the libraries at the University of Texas, Dartmouth College, the University of Maine, the College of the Atlantic, The Jackson Laboratory, and of the Bangor Public Library and Maine State Library in Augusta, and I have received extensive assistance from librarians at all of them. Particular thanks are due to the Misses Joan Staats, Alison Baker, and Ann Jordan, librarians at The Jackson Laboratory, for helping with various questions and for ordering many books and journals on interlibrary loan, and to Mrs. Mary Smith and Mr. Edward Knight for locating material for me in the Library of Congress. My thanks also to Stan Short for helping with the preparation of figures. Dr. Barbara Sanford, Director of The Jackson Laboratory, kindly made the typing facilities of The Jackson Laboratory available to me, and Miss Barbara Dillon has patiently deciphered my handwriting and seen the manuscript through to completion, retyping and retyping again some of the more difficult portions that required several rewritings. Mrs. Ann Bingham has drawn on her editing experience to help me get the manuscript and list of references in final form before the final typing. Drs. Donald W. Bailey, Peter Demant, and Herbert C. Morse, and Profs. John Buell, James F. Crow, Richard R. Hernstein, George C. Lodge, Mark A. Lutz, Arnold R. Sanderson, Morris Storer, and Edward O. Wilson have read all or parts of the manuscript. I have profited greatly from their suggestions. If errors remain, it is through no fault of theirs. My thanks are due to Springer-Verlag and to its editors who saw the manuscript through the final stages of editing and printing. Finally, thanks are due to my beloved wife for her patience with the life-style imposed by having an author in the house and for her continued interest in the work.

Bar Harbor, Maine   George D. Snell
December 1987

Contents

1

Introduction

In his classic, *Democracy in America,* Alexis de Tocqueville wrote "The Americans . . . are fond of explaining almost all the actions of their lives by the principle of self-interest rightly understood" (2). De Tocqueville was speaking of the early 1830s, approximately 150 years ago. At this time, according to his observations, the principle was widely accepted and often explicitly acknowledged and was seen by the Americans to whom he talked as guiding them toward unselfish and civicly oriented conduct. "American moralists," he wrote, "do not profess that men ought to sacrifice themselves for their fellow creatures *because* it is noble to make such sacrifices, but they boldly aver that such sacrifices are as necessary to him who imposes them upon himself as to him for whose sake they are made." In the same chapter he states his own judgment: "I am not afraid to say that the principle of self-interest rightly understood appears to me the best suited of all philosophical theories to the wants of the men of our time, and that I regard it as their chief remaining security against themselves. Towards it, therefore, the minds of the moralists of our age should turn; even though they judge it to be incomplete, it must nevertheless be adopted as necessary."

As de Tocqueville points out, the concept was already old at the time of his travels in America. Some 250 years earlier, Montaigne had written: "Were I not to follow the straight road for its straightness, I should follow it for having found by experience that in the end it is commonly the happiest and most useful track."

Today, under the name enlightened self-interest, the principle probably is still viewed as valid by many Americans, though is perhaps less apt to be publicly noted. Walter Lippmann, philosopher and journalist, was one American who, nearly 100 years after de Tocqueville, specifically defended the principle. "The difference between good and evil," he wrote, "must be a difference which men themselves recognize and understand. Happiness cannot be the reward of virtue; it must be the intelligible consequence of it. It follows, too, that virtue cannot be commanded; it must be willed out of personal conviction and desire. Such a morality may properly be called humanism, for it is centered not in superhuman but in human nature. . . . (Civilized men) must live by the premise that whatever is righteous is inherently desirable because experience demonstrates its desirability" (3).

Following these earlier writers, I shall adopt the principle of self-interest rightly understood as the starting point for our search for a rational ethic. Our problem then becomes to truly understand the dictates of self-interest. How do we recognize the truly enlightened choice?

If, when faced with a choice between alternative courses of action, we are to base our choice on enlightened self-interest, we must have some means of determining, in the specific situation at hand, what self-interest dictates. We must be able to evaluate the consequences of the available choices, and to do this we must have a standard of measurement and an instrument wherewith to measure.

A standard of measurement implies some set of values. A value system may or may not be an acknowledged part of the judgment and if acknowledged may or may not conform to the values actually applied. This is a subject as old as philosophy and one on which philosophers still do not agree.

The instrument for measuring enlightened self-interest is reason adequately informed. To choose intelligently between alternative acts, we must know, or have a reasonable estimate of, their consequences, and the only way we can determine consequences is to gather and analyze appropriate facts.

Choices can run an enormous gamut in their significance. They can range from whether or not to water one's lawn, to the selection of a site for building a major factory, to the decision on whether to loose or not to loose an atomic bomb. They can affect virtually no one except the decision maker; they can affect a few or many other people; or they can affect the whole world. The effect or effects on those impacted can be slight or profound. We can call these aspects of choice *the circle of impact* and *the magnitude of impact*.

Choices can also differ in their complexity. It may be easy or it may be extraordinarily difficult to know their consequences. It is easier to gauge the effect of watering or not watering one's lawn than it is to determine the consequences of building or not building a contemplated factory.

In de Tocqueville's day, business and national leaders, on occasion, were faced with difficult and momentous decisions. I think few people would question, however, that leaders today are sometimes faced with decisions whose magnitude and circle of impact are enormously greater than any within the purview of leaders of the past. Technology has given us awesome powers, and the significance of choice has been expanded correspondingly.

Technology has also increased the complexity of the world we live in. In today's world, the hurtful consequences of acts can often be remote and easily overlooked rather than clear and immediate. The complexity of choice has increased and so has the volume of facts necessary to choose rationally. In de Tocqueville's day, many people apparently felt they could apply successfully the principle of self-interest rightly understood. They had the facts they needed. Today, it is harder to apply intelligently. Perhaps that is one reason it is less frequently mentioned, and why public opinion on many major issues has become fragmented, often into single-issue pressure groups. It may also be why there are many wrongs that are widely condoned.

Another problem that we face in the United States today in decision-making is the increased pluralism of our society. There has been a growth in ethnic, religious, and philosophical diversity. If self-interest is to guide us toward the common good we must, despite this diversity, achieve compromise and accommodation at many levels. It will not, in the long run, serve our interest to attempt this through the suppression of diversity. If we are to bring people together, it must be by assembling and disseminating the facts relevant to the problems we face.

For ethical group choice, then, a shared factual basis is essential. In today's world, assembling and getting agreement on the facts relevant to a single major issue can be a formidable undertaking. Virtually the whole body of human knowledge has relevance, somewhere and sometime, to one or another of the sum total of decisions we have to make. Also, there may be important questions of fact that are unresolved or inadequately resolved, leaving gaps that only time and effort can remedy. But there is also, it seems to me, a body of facts with broad relevance, facts that can help many individuals in many situations to apply the principle of self-interest rightly understood and, through its application, to come up with the same basic conclusions.

Ethics, as a subject of study, usually is approached either from the philosophical viewpoint or through an examination of the ethical implications of relatively restricted issues. What we have been saying, however, implies that ethics can also be approached by an orderly assembling of facts relevant to a broad variety of decision-making situations. Perhaps even the foundations of ethics can be illuminated by way of this route. Opinions would differ as to what facts are most relevant, but in their sum they should be facts that provide the background for a working ethical philosophy.

It is the purpose of this book to assemble at least some of this body of essential information. This requires ranging over a wide field. I think anyone must face such an undertaking with trepidation. Nevertheless, it would seem that the value of such a broad or holistic approach to dealing with our problems should warrant the effort.

The reader can easily gather from the table of contents something of the specific areas covered in this volume. It would be pointless merely to elaborate on this. It may, however, be relevant to list some of the reasons for my choices.

In his comments on the propensity of Americans to explain their unselfish actions in terms of self-interest rightly understood, de Tocqueville also said, "In this respect I think they frequently fail to do themselves justice; for in the United States as well as elsewhere people are sometimes seen to give way to those disinterested and spontaneous impulses that are natural to man" (2).

This perceptive comment by de Tocqueville raises questions as to the origin and properties of man's social nature. Are our characteristics as social animals shaped by nature or nurture, by our genes or by our upbringing, or by some mixture of the two? Are we good or bad by nature, or both? And if and insofar as our social conduct comes from within, are we all formed in the same mold or in many different molds? Answers to those questions, it seems to me, certainly belong within that fundamental body of knowledge basic to many ethical choices that we are seeking.

Sociobiology is a relatively new name for an old, interdisciplinary area of study concerned with the genetically determined component of social behavior in animals. Two of its major domains are the genetics of social behavior and the evolution of social behavior. We shall devote a chapter to each of these, drawing information from animal studies, but concentrating primarily on man. With respect to our three questions, we shall conclude that, within the limits of our present capacity for measurement, our social nature is determined about equally by heredity and environment, that we are naturally capable of helpfulness and love, but also of selfishness and aggression, and that we display great individual diversity in our social tendencies.

Sociobiology and its contributory sciences have added enormously to our understanding of the properties and evolution of man's sociality. As developed by its most competent exponents, it is a thoroughly sound branch of science. It is also true, however, that because the ultimate value of sociobiological thinking is biological survival, the subject often is viewed as leading to a basically selfish ethic. Richard Dawkins's generally excellent book, *The Selfish Gene,* illustrates the point. The title in itself is indicative. In his text, also, Dawkins says, "Be warned that if you wish, as I do, to build a society in which individuals cooperate generously and unselfishly towards a common good, you can expect little help from nature. Let us try to *teach* generosity and altruism, because we are born selfish" (1). It seems to me that there is a contradiction in this statement. While it clearly indicates that Dawkins's particular approach to sociobiology has led him to a dim view of human nature, it also implies, in his stated wish for a cooperative society, a streak of goodness that his reason has denied. Dawkins also fails to tell us how he arrives at his moral imperative that we "teach generosity and altruism."

Actually, in the hands of some of its exponents, sociobiology has gone far towards solving a problem inherent in Darwinism, the problem of explaining through natural selection the gentler aspects of human nature. Nevertheless, a rounded view of ethics requires that we broaden our perspective beyond that provided by sociobiology alone. In particular, we need a rational ground on which to base our moral imperatives.

One philosopher, Peter Singer, has clearly recognized the importance of sociobiology for ethics, though he disagrees with many of the philosophical conclusions that have been drawn (4). I did not learn of his book until I had essentially completed this volume, and it is, I think, significant that he, a philosopher by training, and I, a biologist, are in general agreement as to the premises appropriate for a modern moral philosophy.

The social sciences are complementary to sociobiology. They deal with the other half, the environmental half, of those forces that shape man's sociality. Ethics is also complementary to sociobiology in that if it is to work, it must be through nurture not nature, but it is distinct from the social sciences in that it deals with values, asking not merely what human social conduct is, but what it ought to be.

While, as we have indicated, our emphasis is on the factual rather than the theoretical background necessary for rational choice, a book of this sort would be incomplete without some consideration of the more typical subject matter of ethics. Two chapters, therefore, will be devoted to this. One perhaps unconventional area of concern that I emphasize derives from my answers to my three questions, and particularly the answer to the question about the natural diversity of man's sociality. If we are formed from many different mixtures of the whole range of mankind's capacity for good and evil, certainly this is important for any planning we do concerning moral choice. We shall have a good deal to say about this. Here I offer only a few comments.

We have referred to a circle of impact. Another circle important in ethical thinking is the *circle of participation*. Most important decisions are the result of some degree of collaboration. Collaborating groups in our complex society are numerous, diverse, and in varying degrees fluid or fixed. In big business, small groups of management personnel or boards of directors make major decisions. Voluntary associations such as the League of Women Voters make decisions that can carry considerable weight even though the power of such organizations is limited. Some appointed commissions have the power to make major decisions. In a democracy, all voters form a participating group.

If there is diversity in man's moral nature, this is important in connection with the circle of participation because it must bear on decisions about who should participate. Are there people we must exclude? Viewed in a quite different context, variability in sociality must be important in our interpretation of Christ's injunction: "Love thy neighbor as thyself." It is sufficient here to point out these problems; any attempt at answers will be left to subsequent chapters.

This volume, insofar as it deals with our ethical shortcomings and their causes, is confined primarily to the current situation. We live, however, in a changing world, and ethical problems change accordingly. Chapter 2, "A Revolutionary Age," examines the changes taking place and, in the light of historical parallels, suggests the course they may take in the future and urges the need for preparedness and planning. Three potential problem areas are (1) our growing powers for modification of the environment—and this can include our socioeconomic as well as our natural environment; (2) extraordinary developments in medicine, including a rapidly expanding capacity to modify the human genotype; and (3) alterations in the political process resulting from new technologies and the growing complexity of our society. An adequate appreciation of threats to our political system, in turn, requires a clear understanding of the workings of, and ethical contrast between, democratic and absolute governments. The creators of our American democracy believed that only an ethical society can be a free society. If this is so, it is a major reason why all men of good will, for their own good, must strive for an ethical world.

Information concerning all of these issues is an appropriate part of the factual basis for a rational ethic. We do not deal with any of these issues here, though we do mention some of them. They are appropriate material for a second volume.

Ethical understanding is meaningless unless it is put to use. Human ethical behavior, as we have indicated, is determined by both environment and heredity. Since the hereditary component in our behavior is, for all practical purposes, immutable, any practical routes intended for the modification of our conduct must be directed at the environment. Probably the two most important such routes are education and institutional change. There are already signs of a growing public will to modify ethical education. Some progress here is likely in the near future. In practice, major institutional change does not take place except in an atmosphere of crisis. In Chapters 2 and 3 of this volume, I suggest that conditions may be on the way that will make such change possible. In a crisis, many things can happen. We in this country could easily turn to authoritarianism, thereby risking a slide into absolute rule. Before any crisis does develop, it is essential that we have alternatives to a blind acceptance of authoritarianism, and plans for assuring that we seek and effectuate from among these alternatives such institutional changes as will best assure the optimum combination of freedom and personal accountability, and the selection of wise and responsible leaders.

One basic conclusion will emerge from this volume. Whatever the ultimate value sought—whether it be biological survival, happiness, self-realization, or some other end—those who cherish freedom and good-will cannot afford to do no less than to choose, according to the best of their ability, a course of cooperation, considerateness, and personal integrity. Some choices will be easy and obvious, others extremely difficult, but self-interest rightly understood does dictate that we seek a basically moral world.

2

A Revolutionary Age

Introduction

There are periods in history when troubles multiply, change accelerates, and nations enter an era of revolution. Signs abound in this last quarter of the twentieth century that we are nearing such an era. Our economy is in deep trouble; there is a widespread sense of disaffection; intellectuals find fault with our institutions; dishonesty and crime are rampant; and yet, through it all, technological marvels with a potential for both good and evil are being generated at an unprecedented rate. A revolution is in the making. Whether it moves us toward the realization of our hopes or of our fears depends on the wisdom with which we guide it.

I shall be writing in this chapter primarily about the United States, but our land is not unique. Situations with similar revolutionary potential are found throughout the industrialized democracies. Communist nations have their own full share of problems, but their consequences under communist dictatorship could be quite different. In the less technically developed parts of the world, problems are both deep and often seemingly insoluble.

Here in the United States, the growing public disaffection and distrust has been documented in numerous polls. There have been ups and downs depending on economic conditions, on political circumstances such as President Reagan's skilled use of the media, especially television, and on events such as the Vietnam War and the revelation of the secret sale of arms to Iran to raise funds for the Contra revolutionaries in communist Nicaragua. The general trend, however, reflects a worsening situation. A Harris Poll that has been repeated over the past 15 years shows a steady increase in the public's sense of alienation. People increasingly distrust their leaders in both business and in government.

It is encouraging to note accounts in the news media of leading Americans who see major changes ahead and urge positive steps to cope with them. Television newsman Walter Cronkite, in an address to the 1980 Harvard University graduating class, referred to the "impending revolution" and listed issues that require the urgent attention of the rising generation. Richard Strout, dean of the Washington press corps, in an interview with a leading news magazine, likewise expressed the belief that change is inevitable. He urged the adoption in this country of a parliamentary system. Many leaders have recommended significant but smaller governmental changes. A popular proposal advocated by former President Carter among others, is a single presidential term of six years. Irving Shapiro, former president of the DuPont corporation, has written that "the 1980s will be a period of building new institutions in government and society." In addition to the suggestions concerning possible changes in government, there have been numerous proposals aimed at enhancing the societal benefits of business. It is a hopeful sign that leaders who recognize the magnitude of our problems are also concerned with ways to cope with them. The more one examines the suggestions, however, the more apparent it is that they are either vague or so diverse that an early consensus on solutions is unlikely. It is also reasonable to ask: are they of a scale adequate to cope with the magnitude of the problems we now face?

If, indeed, we are living in a revolutionary age, it may be profitable to seek lessons from an earlier, comparable period in history.

The second half of the eighteenth century was a period of extraordinary change in the western world. It was then that the Industrial Revolution took place in Britain and that political or Democratic Revolutions occurred in America and in France. Major changes in religious thought were going on at the same time, though most historians would place the high point of the Reformation, or Protestant Revolution, in the first half of the sixteenth century when Martin Luther and John Calvin broke openly with the Catholic Church. Thus during this rather broad period of time the western world underwent a triple revolution: religious or ethical, industrial, and political.

I suggest that we face today a similar triple revolution that is also likely to extend over a broad period, though with the peaks perhaps less separated. Events move rapidly in today's world. To designate the ethical, industrial, and political components of this new revolutionary wave I shall use, respectively, the terms *Rational, Technological,* and *Environmental.*

This chapter will deal with these six revolutions, three in the past and three just beginning.

To avoid ambiguity, it will be well to define the meanings I attach to the term *revolution*. It is, and here will be, used in three related senses: (1) any major industrial or social change of lasting significance occurring in a short period of time; (2) the overthrow of a government or change of a political system by force; (3) a major change of leadership by whatever means achieved.

While the major changes brought about by each of the three earlier revolutions occurred within a relatively short period of time, each revolution was part of a trend that extended over a much larger period. Following the practice of some writers on these matters, I shall distinguish the longer periods from the shorter, peak periods by the omission of initial capital letters in their designations.

Revolutions in sense one (some very major change) are not only historically important but also philosophically fascinating. What brings about these rare occasions when great change is condensed into a short period of history? The evidence, I believe, points to an unusual concentration of mutually reinforcing circumstances, any one of which acting alone is not only capable of moving forward an ongoing process, but capable also of reinforcing one or more of its companion circumstances. The result is a complex system of positive feedbacks. This phenomenon of mutual reinforcement probably is now or before many years will be at work in all the waves of transformation with which we are concerned.

We turn now to an examination, in companion pairs, of the six revolutions.

The Industrial Revolution

The Industrial Revolution (written with initial capitals), while identified by definition with a short period in the late 1700s, is still with us today in a less intense form and has roots that go far back into history. A few critical developments may be singled out.

In the eleventh century there was a great improvement in conditions through western Europe (41). Creative people no longer had to struggle merely to live; they could create. The growth of towns that began about this time produced other conditions necessary for creativity. To a considerable extent, the towns escaped the domination of the feudal lords (19) and the resulting freedom engendered an atmosphere congenial to innovation. The towns also were centers of trade and industry, and this increased contacts and the flow of information. These three factors—communication and travel, leisure, and freedom—produced conditions favorable for the growth of knowledge, and knowledge is the most important single element in the soils that nourish the growth of change.

Industry, more than most human endeavors, thrives on the expansion of knowledge. During the era of industrialization, information and understanding grew through the rediscovery of the literature of classical Greece, through many fundamental scientific developments, and through technical innovation itself. Its diffusion was hastened by the development in Germany of printing with movable type, by the expenditure of much effort, especially in England, on books and lectures aimed at the artisan and inventor, and thorugh the movement between countries of men skilled in various industries, most notable being the emigration of the Huguenots during their persecution in France. French chemists contributed to the textile industry through the development of chlorine bleaches and new dyes. The Dutch made important contributions to optics. The studies of mathematicians—especially Napier, the inventor of logarithms—and of astronomers advanced the arts of navigation and surveying. In many other areas of industrial advance, particularly those resulting from mechanical inventions, science made little obvious contribution, but knowledge of mechanics was certainly important and the belief in progress and a faith in experimentation and science helped to engender an atmosphere in which technical innovation could thrive.

Besides the expanding intellectual horizons, there were other factors producing an atmosphere favorable to entrepreneurship. The growth of Protestantism and, somewhat later in England, of Puritanism, led to a gradual increase in individual freedom. In the fifteenth and sixteenth centuries many small but cumulatively important developments in mining, iron working, weaving, and other industrial processes occurred. Late in the sixteenth century, during the reign of Queen Elizabeth, the concept of patents was developed (44). This stimulated invention and was one of the contributing factors that led to substantial progress in all branches of industry during the following century. During this time, too, the use of coal increased, but its successful application to iron smelting, though attempted, was not achieved. Webster has referred to the interval from 1626 to 1660 as the Great Instauration (44); Nef has called essentially the same time span an early industrial revolution (32).

Some thinkers of the sixteenth and seventeenth centuries, most notably Francis Bacon and his followers, foresaw an enormous growth in technology and industry. The time for the great expansion, however, had not yet come. James I corrupted the patent laws to give his favorites monopolies in various sections of the economy; there were years of poor harvests; the Thirty Years' War in Europe reduced exports; the Puritan Revolution in England further disrupted economic progress; the all-important woolen industry declined sharply; prices rose, the population outran production, and there was widespread poverty, unemployment, and discontent (44). Business could not thrive under these conditions.

In the eighteenth century, economic conditions improved and there was a renewed burgeoning, centered in England, of invention, industrialization, and trade. The list of inventions is long. One authority has singled out as particularly significant James Watt's steam engine, patented in 1769, the spinning machines of Arkwright and Crampton, patented in 1770 and 1779, and Henry Cort's reverberatory furnace, invented in 1784 (10). These and related developments made possible fabrics produced in factories by machines which in turn were driven by steam engines made from cheap and abundant iron, the iron smelters and steam engines both being fueled by coal. This group of developments constituted a self-reinforcing system.

The resulting growth was reflected in England by a rapidly expanding importation of cotton: two and one half million pounds in 1760, 22 million 1787, and 366 million in 1837 (24). Iron production showed a comparable increase (45).

While the Industrial Revolution originated in England, other countries contributed to and, in due course, copied it. Before the Revolution, France was the leader in fabric technology and Germany in mining and metallurgy. China, as Needham has shown, was for centuries distinguished in pure and applied science, but fell short of the industrial take-off point (31). The United States, though first affected by the Industrial Revolution in the demand for Southern cotton, soon adopted the new industrial techniques and added its own contributions, such as Eli Whitney's cotton gin developed in 1793. During the nineteenth century, industrialization spread throughout the western world and by 1900, Japan's industry was well on the way to its meteoric rise (15).

As the revolution spread, it also progressed. Whether or not it is now ending is largely a matter of definition, but there can be no doubt that industrial developments such as the railroad, electric light, airplane, and mass-produced auto have transformed our lives.

A hallmark of the industrial age was the energy source that fueled it. In the preindustrial era, waterpower and wood were the staple sources of energy. Broadly speaking, coal and then oil and gas have powered transportation and industry and have heated our homes and offices since the beginning of the eighteenth century. We are now threatened with the ultimate exhaustion of these energy sources. Adapting to this exhaustion will in itself bring major upheavals.

One of the most important outcomes of many revolutions is a major change in leadership. Such a change was partly but not completely accomplished by the Industrial Revolution. Business men, whether in trade, manufacturing, or finance, were the guiding force behind industrialization. This was the age of the entrepreneur, and the entrepreneur was correspondingly rewarded with an increase in influence, power, and wealth. Only following the Democratic Revolution, however, was the change in leadership completed.

Because of the importance of business leaders in determining the course and social consequences of the industrial era, we examine this subject here. At the risk of some oversimplification, we can say that the leadership has gone through three stages. Our characterizations are based primarily on four studies (10, 24, 25, 28).

In the eighteenth century, the typical leader was the entrepreneur who was both inventor and manufacturer. Often the business started by the entrepreneur became a family business. Well before the eighteenth century, bankers and merchants had occupied an important place in the business community, and they, of course, continued to do so, but it was the entrepreneur who typified the new era. Entrepreneurs have continued to provide much of the spark for industrial progress. Henry Ford, who both created and mass produced the model-T Ford, and Edwin Land, founder of the Polaroid Corporation and, with the aid of a research team, developer of the instant-picture camera, are twentieth century examples. In the founding of exciting small businesses, entrepreneurs still play a major role, but they play a minor role in big business.

During the nineteenth century, the dominant business type was the empire builder. It was men of this type who built the railroads, steel mills, and the oil trusts. Their ruthlessness earned them such names as "robber baron" and "jungle fighter."

In the twentieth century, the increasing size and complexity of the major corporations necessitated a new change in leadership. In what has been referred to as the managerial revolution, ownership and management became separated with operating responsibility passing to salaried professionals. In the larger businesses, several layers of management were introduced. Even in family-owned businesses, operation generally was transferred to hired specialists. A new and powerful professional elite had come into being.

Maccoby, in a 1976 study based on in-depth interviews with 250 managers of 12 major corporations, has analyzed the personalities of present-day business leaders. He distinguishes four types, with overlap between them of course, but each well-defined and recognizable. We need mention here only the dominant type whom he calls the *gamesman*. These men are highly competitive; winning is of great importance to them. They are hard working, risk taking, good at team leadership. They view their job as a game, but nevertheless are subject to continual anxiety (28).

The corporations selected by Maccoby for study were "elite" corporations, in the forefront of new developments. It is therefore not surprising that the gamesmen in Maccoby's study showed a considerable degree of both knowledge of and competence in the new technologies. These particular business leaders may foreshadow the wave of the future.

In their technological competence, the gamesmen that Maccoby studied were not typical. In the majority of industries, the typical leader has specialized in finance or marketing, not technology. He has two major goals, growth and profit. Growth is seen as a source of security. Additional managers are required and the existing management can move up in the hierarchy. The struggle for profits leads to a concentration on short-term rather than long-term goals. The resulting pursuit by subordinates of financial returns may lead them to the use of shady methods that their superiors would not accept in their private lives. Products are designed to maximize sales, not serviceability. Salesmanship becomes more important than engineering. The financial power of the corporation is used to insure the passage of legislation favorable to management's goals. Leaders such as these are not fully serving the general welfare.

The Industrial Revolution brought great wealth and an outpouring of goods from which the whole population ultimately benefited, but it also brought grinding poverty to factory workers, child labor, slums, and pollution. Workers were separated from the means of production and became "hands" tending machines they did not own. Many Puritan intellectuals were distressed by these consequences of the industrial system they had helped create, and made conscientious efforts to remedy them. Improvement was slow in coming, but wages and working conditions are certainly now far better. Serious problems, however, remain. And they are problems that, without vigilance, in all probability will be duplicated in the forthcoming Technological Revolution.

These problems result primarily from the greatly expanded circle of impact of decisions made by our leaders. In primitive society, decisions seldom affected more than a few dozen people; now they can affect millions. The damaging effects can conveniently be considered under two headings.

Modern industries, as environmentalists emphasize, generate a diversity of waste products, many of them poisonous, and these by-products have often been discarded without adequate protection against potential damage. Also such vast quantities of useful products ultimately end up as trash that their disposal has become a major problem. The resulting damage to the environment may not always have been easy to foresee, but it is also true that many businesses and some of our political leaders have chosen not to see them because remedial action would have reduced profits.

The second problem associated with modern, large-scale industry is more inclusive than and overlaps the first. It derives from a trait of human nature. People seldom want to cheat or deceive friends or acquaintances, even in small ways, but it is easy to take advantage of the gullibility of strangers. And, as we shall see in the next chapter, big business is often guilty of just such practices even though our competitive economy provides people who use it intelligently with a considerable degree of protection.

The Technological Revolution

The era of the Industrial Revolution was recognized by people then living as a time of remarkable change. Indeed, in the early seventeenth century, more than a century before the true Revolution, British scientists and inventors were convinced that their country was entering a new era of social and industrial development. Today, in the United States and in other industrialized nations, a similar sense of anticipation is widespread in the technological community. Basic and applied research have produced such marvels as space exploration, the computer, the laser, and genetic engineering, and more developments seem inevitable. The concatenation of circumstances necessary to produce an expansion comparable to the Industrial Revolution is easily imaginable.

What will be the nature of this new revolution and when and where will it occur? These are fascinating questions, but predicting the future is such an uncertain business that I shall offer only a few suggestions.

Information will proliferate and will be generally and easily available through small computers connected to central data banks. Computers will play an increasing role in the diagnosing of disease. Automation of factories will be greatly expanded (8). Autos and many household items in common use will be made more efficient and more durable. New substances such as ceramics with remarkable properties will benefit industrial processes and household products. New energy sources and new ways of conserving energy will be developed. Our capacity for genetic engineering, already considerable, will expand enormously and, if we use it wisely, bring enormous benefits to mankind. Undoubtedly there also will be many technological marvels we cannot now even imagine.

It is still too early to say definitely when the concatenation of circumstances competent to produce a truly revolutionary acceleration in the rate of technological change will occur. Perhaps we are already entering this phase, but it may also be that the present stage of the technological revolution is closer to the early or mini-industrial revolutoin of the seventeenth century than it is to the ultimate stage—the Industrial Revolution that occurred a hundred years later. Most of the methodology necessary for a technological take-off may be at hand, but just as seventeenth century industrial growth was hampered by an unfavorable economic, political, and international environment, so today, worldwide problems may delay the ultimate fruition.

The Industrial Revolution was clearly localized in one country, Britain. The United States has the potential to lead in the Technological Revolution, but only if we first solve our economic and educational problems. Japan, at the moment, is the country that seems to be forging ahead. Perhaps we may hope that a number of countries will be major participants.

While the Technological Revolution, strictly speaking, will be a revolution in goods—in the material aspects of our society—it inevitably also will have profound effects on the economic and social aspects of our lives. If we are to achieve, in this new era, a beneficent world, it will be through wise planning, and a willingness to innovate.

We need, first, to develop a clear picture of the society we want. This is an undertaking that deserves extensive thought and study. I offer only a few personal thoughts.

We want a safe, uncontaminated environment, not one poisoned by industrial wastes. We want a diverse environment, one in which the variety of tastes characteristic of humankind can be satisfied. We want an uncrowded and beautiful environment, at least to the degree that persons to whom these attributes are important will not be deprived. We want interesting work and a congenial workplace. We want a sense of community and a sense of belonging. We want freedom, but only to the extent that it cannot be used to the detriment of others.

The extent to which we achieve a truly desirable society will depend, in large part, on our leaders. Some of the qualities we must seek derive directly from the technological revolution. Technological competence will be increasingly necessary, at least in business. Ability in finance and other competences also will be needed. Certainly we do not want practitioners of financial legerdemain, but while the technological revolution alone probably can insure the choice of leaders with technical skills, it cannot insure the necessary emphasis on cooperativeness and the other qualities of mind that the long haul requires.

If we are to insure these qualities, it will be through our institutions and through a public dedication to a well-reasoned morality. These in turn will depend on the two companion revolutions of the technological revolution. To these we now turn.

The Democratic Revolution

The Democratic Revolution, like the Industrial Revolution, had an extended history and an interlude of accelerated change. The longer period lasted at least from 1642 to 1942, a period of three centuries or more; the shorter period has been defined as the 40-year interval from 1760 to 1800 (34).

The Puritan Revolution in England, one of the milestones in the early stages of the struggle for democracy, pitted the forces of business and democracy, led by Oliver Cromwell, against the aristocracy. The first battle of the Revolution, fought in 1642, is an appropriate marker for the beginning of the new era. The Puritans won the war but lost the subsequent struggle for democracy. During the seventeenth century, Parliament achieved a permanent increase in influence as a result of the war and its own firm stand against James I and Charles I. Despite a few subsequent concessions from the ruling forces, however, it did not gain an approximation to its present powers until 1884 (17, 20).

The next great era of struggle was the Democratic Revolution itself. It was marked by the American and French Revolutions (34). Both revolutions succeeded, but democracy in France fell prey to internal dissension and corruption and external pressures. There were uprisings in England and abortive revolts in Hungary, Holland, and Italy during the same period. Democratic unrest continued into the nineteenth century and peaked in revolutions throughout much of Europe in 1848. These were generally unsuccessful, but the forces of democracy had demonstrated their strength and major reforms occurred in subsequent years.

Another major stage in the democratic movement was the establishment of effective and stable parliamentary governments in Germany and Japan at the end of World War II. In 1889, Japan had adopted a constitution and a parliamentary system, but the emperor retained extensive powers. The parliament was replaced by a military dictatorship in the 1931–33 depression. No change comparable to the post-war shifts in Germany and Japan has occurred since that time. In many countries, the battle for democracy has yet to be won.

In all the revolutions (in the military sense of the term), except possibly the American, two driving forces can be distinguished. The first was the mass of underprivileged peasants, workers, and slum dwellers, the second the leaders who were typically businessmen, professionals, and intellectuals. In Colonial America, unlike Europe at that time, there was very little true poverty. There were some 400,000 Negro slaves in the South (36), but these contributed little to the motivation of the American Revolution. Hence it was not primarily poverty but the widespread love of freedom that, in America, sparked the Revolution. In Europe, the masses rebelled simply to improve their lot; the intellectuals injected the elements of direction and idealism. These elements were then to a considerable extent taken over by the businessmen and professionals, who also sought greater personal power and freedom.

The timing of the French Revolution and of accompanying uprisings in other European countries was in considerable part determined by an accumulation of economic problems. All of western Europe in 1789 was in the grip of an economic depression. In both the towns and the countryside there was much unemployment. Prices of agricultural products had been declining for more than a decade, adding to the hardship in rural areas. The problems of the farming community were brought to a climax in 1788–89 by a major drought, devastating hailstorms, the most severe winter in 80 years, and spring floods. Food shortages resulted. The influx of gold and silver from the New World and the repeated issuance of paper money brought on inflation which, though modest in comparative terms, worked hardship because it was not accompanied by a comparable increase in wages. In France over a period of 48 years, the rise in prices was triple the rise in income of the laboring classes.

Many other economic problems were particularly acute in France. The distribution of wealth was extremely uneven. The nobility and the clergy evaded payment of taxes and opposed tax increases needed to bring the budget into balance. The church had extensive landholdings and a correspondingly large income. Much of this was turned into gold and silver ornaments, nominally to honor the church, but in fact as a hedge against inflation. Faced by these intractable economic problems, Louis XVI turned to economic conservatives who advocated economic policies curiously similar to those favored by President Reagan in the United States today. The forces of the marketplace were to be given free play. Competition and the law of supply and demand were to regulate both prices and wages. Strikes were forbidden but were frequent nevertheless. The people of Paris generally opposed these moves, favoring a policy of greater government intervention.

The economic problems were compounded by violent crime and extensive corruption. Marauding bands roamed the countryside, robbing and terrorizing the populace. Judicial and administrative posts were sold to provide income for the royal treasury. It was common practice for the purchasers of these posts to recoup their costs by accepting bribes. Another common form of graft was the diversion by wealthy landowners of money intended for the upkeep of public roads to the needs of their own chateaux. These conditions produced a public mood of distrust, cynicism, and fear similar to that which we are experiencing today.

These historical parallels involve conditions of undoubted import, but there were other less serious but nevertheless revealing parallels that accompanied the Revolution itself. Parents adopted a permissive approach toward the upbringing of their children. A French gardener told a visiting English lady, "During the Revolution we dared not scold our children for their faults. Those who called themselves patriots regarded it as against the fundamental principles of liberty to correct children. This made them so unruly that very often, when a parent presumed to scold his child, the latter would tell him to mind his own business." Pornographic literature abounded and was easily available to youngsters. These conditions ultimately led parents to turn to priest-run schools (14). It is interesting to note that the Durants, from whose history I take most of these items about the conditions that preceded and accompanied the French Revolution, were writing in 1975 when any parallel to Reagan's policies could not be known and when the other similarities to today were less obvious then they are now.

The economic problems in Europe in the late eighteenth century provided a widely felt motive for revolt, but rebelliousness alone could generate little leadership or sense of direction. Hate of the ruling classes was the driving force; leadership came from the business and professional classes, and the success of the revolution in different countries was directly related to the size and strength of these classes. In a number of countries, wealthy businessmen were incorporated into the aristocracy and, in varying degrees, lost their ties to the business community. Insofar as this occurred, the potential for revolutionary leadership was decreased. In France, the Revolution was largely led by lawyers who did not gain acceptance into the aristocracy. Colonial America presented a somewhat unique situation in that the prosperous, land-owning farmers as well as the businessmen were intellectually committed to revolution. The United Empire Royalists were the one notable exception. The early settlers also were experienced in self-government, except for the slaves and a small minority of propertyless poor. Although women could not vote, some had considerable influence in local or national affairs. Abigail Adams is the most famous example.

A final and essential factor in the guidance of the revolutionary movement was a background of philosophical ideas. Hobbes, Locke, and Hume in England and Voltaire and Rousseau in France, as well as other now less well-known but at the time very influential writers, provided a clear intellectual foundation for the democratic movement. The concepts of liberty, equality, and the right of people to choose their own form of government were widely understood and accepted. The myth of the divine right of kings had been effectively demolished. The British and French writings were well known in Colonial America, and de Tocqueville could say with considerable justification, "The Americans seemed only to be putting into practice ideas which had been sponsored by our writers" (13).

When it came to the precise form of government, there was a widespread and lively interest but little unanimity. The principle of representation was well known and generally accepted, but in such matters as the structure of the representative body there was more disagreement. Bailyn, in an interesting study, has shown that many pamphlets were published in Colonial America in which the duties of government were debated. The authors were men with a variety of backgrounds, but generally not professional students or writers (3).

There were many reasons for the success of the American Revolution and the failure of the democratic revolutions in Europe in this age of revolution. A majority of settlers in Colonial America probably had left Europe to gain greater freedom. In America, they were able, despite some intervention from England, to create and gain experience with democratic forms of government. At the time of the American Revolution, a significant number of loyalists supported the Crown, but they did not represent a power structure comparable to the European aristocracies. Moreover, after they fled the country, they could not rally any substantial support other than the already dedicated forces of England. The thousands of French royalists who fled to neighboring European countries, on the other hand, were able within a few years to enlist the aid of Prussia and Austria. The resulting military threat was a major reason for the replacement of constitutional government by the Napoleonic dictatorship.

There were also more subtle reasons for the failure of the European revolution. Polarization was greater in Europe. There was radicalism of both the right and the left. Class distinctions had been deeply ingrained and class feelings did not die easily. The separation of church and state, which had become an accepted principle in America by the 1770s, was certainly not recognized in Cromwell's time in England and was incorporated in an imperfect form in the first French constitution. Perhaps most damaging of all, corruption crept into the revolutionary government in

France, with its inevitable erosive effect on the mutual confidence so essential to the successful practice of democracy.

In the nations of Europe before the establishment of democracy, power was unmistakably centered in the monarch and the aristocracy. The hierarchy of the Catholic Church was a lesser power center. The Industrial Revolution conferred the power of wealth and industrial know-how on the business community, but of itself did little to erode the power and privileges of the Crown. It was the Democratic Revolution that began, and ultimately completed, the final transfer of power. By design, however, its architects created a multiplicity of power centers. This was very specifically an objective of the men who drafted the Constitution of the United States. The power of the business community was increased, but it was shared with elected and appointed officials and with the voting public.

The Environmental Revolution

The modern equivalent of the Democratic Revolution has only recently begun to take form, but its general outline is clear. It has three components. The first is a demand for the protection of the environment from the hazards of industrial and population growth, technological innovation, and the misuse of land. The second is the desire of consumers for better, more durable products, honestly and understandably represented so that intelligent choices can be made. The third is the dissatisfaction of workers with being simply a cog in an industrial machine and their demand for more individual input into the industrial process. All of these components can appropriately be grouped under environmentalism, if we include in the environment, the environment of the marketplace and the environment of the workplace. The *Environmental Revolution* seeks a more livable world in all its aspects.

These desires for a more livable world constitute of themselves a motive force, but, as in the case of the Democratic Revolution, the urgency required for major change will come only from economic decline and widespread hardship. In keeping with the greater diffusion of power under democracy as compared with autocracy, the parties targeted in the current revolution are less defined than they were when the king and the aristocracy were the clear objects of disaffection. However, dissatisfaction generally centers on big business, with the federal government a lesser target. Neither, it is felt, serves the public interest as it should.

The consumer and environmental movements have extensive roots, but there was a substantial increase in public awareness of our problems in the 1960s sparked by the publication of such books as Vance Packard's *The Wastemakers,* Rachel Carson's *Silent Spring,* and Ralph Nader's *Unsafe at Any Speed.* In terms of membership in consumer and environmentalist organizations, the numbers of participants is still small, but active groups are widespread and influential and have much public support. In Canada alone, a 1977 count showed some 500 volunteer groups with environmentalist interests (25).

As to the nature of the reforms that are needed, there is nothing approaching unanimity with respect to either business or government. Views range from communism or socialism to the conservative businessman's counter proposal of greater freedom of enterprise and less bureaucratic interference. The Democratic Revolution appeared to have clearer goals, but whether this represents simplification by hindsight or a simpler world is not clear.

Very substantial changes, in my personal judgment, will be needed to achieve the goals of the Environmental Revolution. It is not the purpose of this volume to go into this subject in detail, but some trends deserve noting.

The one most outstanding trend is the growth of regulatory agencies. These grew in number from very few before 1900 to 179 in 1978. Growth accelerated following the 1929–35 depression and again in the 1970s. The agencies have almost become a fourth arm of government. Businessmen complain about the huge volume and sometimes frivolous detail of the regulations issued. The agencies serve a necessary function, but we may reasonably ask if their design might be improved.

Other recent trends, largely borrowed from Japan, seek improved "quality of work life" through more flexible work hours, more involvement of workers in job planning, and greater government-business-labor cooperation (27).

The Democratic Revolution was fought on the battlefields of Europe and America. In France, one of its consequences was the series of executions by guillotine commonly known as the Reign of Terror. In the current age, the democratic machinery which those earlier struggles created should make similar violence unnecessary. If, however, we permit polarization, corruption, and greed to grow and spread, some of the tragedies of the earlier era could be repeated.

The leadership created by the Environmental Revolution will be substantially influenced by the institutions which this revolution engenders. As to the qualities that should be sought over and above the inevitable high order of technological competence, I can do no better than to quote Thomas Jefferson. In one of his many letters to John Adams, Jefferson states very clearly what these qualities should be. "I agree with you," he wrote, "that there is a natural aristocracy among men. The grounds for this are virtue and talent. . . . There is also an artificial aristocracy, founded on wealth and birth, without either virtue or talents; for with these it would belong to the first class. The natural aristocracy I consider the most precious gift of nature, for the instruction, the trusts, and the government of society. . . . May we not even say, that that form of government is the best, which provides most effectively for a pure selection of these natural aristocrats into the offices of government?" (1).

Jefferson clearly hoped that the democratic government that he and Adams had helped to create in the United States would result in a leadership approximating this ideal. Successful democratic governments everywhere have produced remarkable leaders, but the mean is far short of the aristocracy that Jefferson envisaged. The ideal is unattainable, but sometime in the century ahead of us we will have an opportunity to bring it closer.

The Protestant Revolution

The Reformation, or Protestant Revolution, antedated the Industrial and Democratic Revolution by about two centuries. It, in turn, was preceded and made possible by the growth of knowledge and the intellectual ferment that characterized the Renaissance.

A number of factors combined to produce this period of awakening from the darkness of the Middle Ages. Rapid growth in trade brought new contacts and fresh experiences. In the search for a shorter trade route to the East Indies, the New World was discovered. Refugees fleeing to western Europe following the collapse of the Byzantine Empire introduced to the West the long-forgotten literature of classical Greece. The Moors, when they conquered Spain in the eighth century, brought with them a rich treasure of Greek thought and knowledge that the Arab nations almost alone had kept alive for some centuries. Spanish Jews later carried this to other parts of Europe (37). The invention of the printing press in Germany in 1454 and the contemporaneous development of methods for manufacturing paper from linen provided the means for the wide dissemination of much of the new knowledge (16).

An important aspect of the Renaissance was the appearance and growth of the intellectual movement known as *humanism* (16). This movement started in Italy in the fourteenth century but soon spread over all Europe. The humanists were scholars who gathered and interpreted the newly discovered classics and added their own contributions in art and science.

Erasmus, a Dutch scholar who ultimately settled in Basel, is generally recognized as the leading figure in this movement. One of his most important contributions was a translation of the Bible into Latin, the universal language of scholars of that day, directly from the Greek. This translation, which revealed significant errors in the Catholic version, was printed and widely circulated. Erasmus was also known for his attacks, often in the form of satire, on the corruption and authoritarianism of the period. These attacks increased his influence in the intellectual community, but also made many enemies even though he carefully avoided any open break with the Catholic Church. It fell primarily on two other men, Luther and Calvin, to attack the Church and lead the way to the Reformation.

It is interesting to note the differences in intellect and personality of Erasmus as compared with these two reformers and the resulting differences in their influence on this period of history. While Martin Luther is the acknowledged initiator of the Reformation, I shall emphasize Calvin in my discussion because the Protestantism of England and Colonial America traced back primarily to his teachings.

Erasmus was a true liberal, tolerant of the views of others, undogmatic, and capable of doubt. Calvin, like Erasmus, was a brilliant scholar, but his mental set was different. After a brief period in a monastery, which he entered at age 22, his subsequent training was mostly in law. At the French university he attended, he was exposed to humanist thinking, and the new ferment in theology which this had created came more and more to dominate his own thinking. Over a period of years he developed his own theological doctrine, which he thereafter clung to with an absolute and often intolerant conviction and expounded with brilliance. His dogmatism won converts but led to standards which we can only view as needlessly severe.

The views on religion and ethics that Luther and Calvin developed differed in detail but shared essential common principles. This essential core of their doctrines is well summed up by Morely, who described the details of the Reformation as "God, the Bible, the conscience of the individual man, and nothing more nor beyond. The substitution of the book for the church was the essence of the Protestant revolt" (20). Both Luther and Calvin also attacked the Catholic Church on specific moral grounds as well as on the more general grounds of the source of moral authority. The open break with the Church was first made by Luther on All Saints' Day in 1517 when he posted his Theses on the door of the church in Wittenberg, home of a leading university. The Theses attacked the sale of indulgences by the priesthood, a practice that amounted essentially to the exchange of presumed safe conduct to Heaven in return for money to finance the high living of the Papal hierarchy.

Luther and Calvin, unlike Erasmus, both openly broke with the Catholic Church, but Calvin went further and became actively involved in the creation of a reformed community. In France, where he was born, Protestants were subject to persecution, so Calvin ultimatley settled in Geneva. The business leaders of this city had seized control from the old aristocracy and established a considerable degree of freedom, but they had not succeeded in eliminating factionalism. This situation provided an opportunity for Calvin, already widely known through his writings, to assume an active role in the life of the city and eventually to become its accepted leader. Once in power, Calvin set out to create a community founded on the teachings of the Bible. In his view, the relatively harsh morality of the Old Testament was quite as authoritative as the ethics of love taught by

Jesus. His doctrines thus led him to set up a theocracy that did indeed enforce high standards, but also was Puritanical in the worse sense of that word. Calvin himself scrupulously observed his own laws, and in view of earlier corruption and moral debasement and of even more oppressive rule elsewhere, they were welcomed by many of the city's leaders. It is nevertheless true that the enforcement of these laws, using as guidance the early books of the Old Testament, was severe in the extreme. Fisher, in his review of this period, cites as examples the execution of a child for striking its parents and the burning for heresy of Servetus, Calvin's most outspoken opponent (16).

Calvin's place in history has been the subject of extensive debate. He did indeed do much to deserve the opprobrium of his more severe critics, but many aspects of his teachings and of the government he created in Geneva were progressive, and could be used by his followers to gradually raise the levels of freedom and tolerance (11, 16, 17, 43).

Calvin aligned himself with business leaders and craftsmen, not the old aristocracy. Calvinism thus became an important element in the subsequent political revolution. He gave an important role to education and scholarship, and to strengthen them, he brought to Geneva one of his former humanist teachers. Under their joint guidance, Geneva became a center of learning. The city's growing reputation attracted many Protestant immigrants, refugees from persecution in their native countries. Despite the harsh elements in Calvin's rule, he did create a generally attractive environment. These immigrants, as they returned home, tended to take the best elements of Calvinism back with them.

In founding ethics on the Bible rather than on the leadership of the Catholic Church, Calvin opened the way to a decrease in the authority of human rulers and an increase in the authority of intellectual judgments. In theory, at least, the government Calvin set up created separate centers of authority for church and state. True separation of church and state was not achieved until some two centuries later, but Calvin did leave a foundation on which his heirs could build.

Parishioners were given a limited role in the choice of their pastors. This was an advance over the Catholic practice and was enlarged by the Pilgrims, whose Protestantism was founded on Calvinism, into a truly democratic church government.

The high ethical standards Calvin set, even though they continued for a long time to be tainted with intolerance, did provide a necessary background for successful democracy and industrial enterprise. Together with Luther and the other leaders of Protestantism, he accomplished what the humanists may have hoped for but did not seriously attempt: the creation of an alternative to the monolithic power of the Catholic Church.

Luther and Calvin and most of their immediate followers, no less than the Catholic hierarchy, were intolerant of views contrary to their own. The concept of religious liberty and of the separation of church and state in the modern sense of that expression were foreign to their thinking. An almost inevitable consequence was persecution and war, both in the name of religion. In Germany, the Lutherans fought the Calvinists and the Catholic rulers attacked both. The Huguenots, followers of Calvin, were persecuted in France. The Thirty Years' War that decimated Europe between 1618 and 1648 was in substantial part of religious origin. In England, the Anglican Church was tied to the King, and under Charles I and Laud (appointed by Charles as Archbishop of Canterbury), became increasingly aristocratic and oppressive.

Although intolerance dominated the early Protestant churches, voices of moderation appeared here and there from the very beginning, and in the course of time these gained the ascendancy. Zwingli, a contemporary of Luther who set up a Protestant church in Zurich that antedated Calvin's rise to influence in Geneva, was more liberal and more in the humanist tradition than either Luther or Calvin. Unfortunately, although his influence did spread in Switzerland and ultimately to England, the tradition of the early Protestant churches was derived from his two more famous contemporaries (16).

In England, early in the seventeenth century, the Puritans rebelled against the autocratic rule exercised by the Anglican Church leadership of James I and Charles I. One expression of this was the flight to Lyden and subsequently to Plymouth of the Pilgrims. John Robinson, the spiritual guide and intellectual leader of this small group, was notable for his tolerance and his denial of any absolute certainty in revelation (2). There were other notable Puritans, among them the poet Milton, who spoke out for religious liberty.

Despite the relative liberality of the Pilgrims, and the democracy they practiced in the choice of pastors, Calvinist thinking dominated the early days of Colonial America. Gradually, however, the voices of moderation gained more and more influence. A proliferation of sects in the eighteenth century, largely as a result of new immigration to America from Europe, created an economic motive for tolerance. Businessmen found it merely common sense to be on good terms with people of differing religious views. While the early American settlements were essentially theocracies, the American leaders, including the leaders of the American Revolution, came more and more to accept religious liberty and the separation of church and state as essential ingredients in a democratic society (36). Thus protestantism ultimately achieved the liberal position common to most of its denominations today.

A final question that we need to ask concerning the Protestant Revolution is the effect that it had on the Industrial and Democratic Revolutions. The evidence indicates that it contributed positively to both.

Calvin drew his principal support from merchants and artisans and gave a high place in his writings to creative work. This link between industry and Protestantism continued in England and Colonial America. As it had in Geneva, Protestantism in its new homes also tended to encourage education. According to Weber, Protestantism was favorable to the work ethic necessary for a successful society (43). This view has been disputed, but it may have some validity. Protestant teaching also accepted the right to private property. Finally, the emphasis placed by both Luther and Calvin on personal integrity was calculated to produce an atmosphere of mutual confidence favorable to trade and cooperation. All these factors contributed to the emerging Industrial Revolution.

Protestantism also was favorable to the movement towards democracy. Rossiter, in particular, has emphasized this linkage in his *Seedtime of the Republic*. The struggle for religious liberty and for political liberty went hand in hand. The English Parliaments called into session by James I and Charles I, held firm in their struggles with the monarchs and thereby contributed to the cause of both religious and political freedom (20). The Pilgrim's search for democracy in church affairs naturally spread to a similar search in the affairs of the community. And just as a high level of education and high ethical standards were favorable to industry, so also they were favorable to the self-rule of democracy.

The Rational Revolution

The ethical component of the triple revolution of the twentieth and twenty-first centuries might appropriately be called either the *Humanist Revolution* or *Rational Revolution*. I shall use the name Rational Revolution because the essence of the new ethics is the basing of morality on knowledge and reason rather than on "revelation." This requires a knowledge of man's moral nature, a search for a broad understanding of the influence of man's conduct on man, and the rational use of all such knowledge to advance the welfare of mankind. The fundamental principles of morality are timeless and remain essentially constant, but there is a need to adapt them in a rational manner to the complex problems of the modern world.

The Rational Revolution, like the Protestant Revolution before it, has its roots in the expansion of knowledge. Like that Revolution, too, its growth has been marked by a proliferation of sects or factions in conflict both with unaltered Protestantism and with one another. It has not yet, in any direct way, influenced the Technological or Environmental Revolutions, but it has the potential, if developed appropriately, to interact substantially and beneficiently. A recent surge in interest in ethics and its practical applications justifies the hope that this potential will be realized.

The knowledge that nourished the Protestant Revolution was the many-faceted surge of information that characterized the Renaissance. The Rational Revolution has its roots in a more restricted but very fertile intellectual soil: the study of the biological origins of social behavior.

By all odds, the most important single sources are Darwin's theories of evolution and natural selection and the enormous body of information that has grown up around them. These inturn were based on prior advances in geology, paleontology, and biology. Particularly influential was the demonstration that the earth has undergone and has been substantially modified by a series of ice ages extending over thousands of years. Because the evolutionary origin of life makes a literal interpretation of Genesis untenable, it seems less likely that the Bible as a whole is divinely inspired. A fundamental tenet of the old Protestantism is undermined. An alternative, if we cannot rely on a supernatural foundation for ethics, is to rely on a humanist or rational one.

Other important knowledge sources of the Rational Revolution can be found in genetics and its various ramifications, in studies of animal societies (especially of primate societies in their natural habitats), in observations of social behavior in the laboratory, in anthropology, and in the social sciences. All such sources, together with the factual and theoretical foundations of evolution and natural selection, have been woven in recent years into a domain of thought known as sociobiology.

Humanism, a major contributor to Protestant beliefs, is still very much alive today as a broad body of thought that gives man himself and his welfare a central place in the search for guides to human affairs. Most humanists do not link humanism and sociobiology. However, Tarkunde, distinguished both through his career as a jurist in India and as a humanist, has clearly recognized in a recent article the importance of the association (though he does not use the term *sociobiology*) (42), and Singer, a philosopher by training, has explicitly linked sociobiology to an ethical system that is clearly humanist (39). Darwin himself foresaw the ethical implications of his theories. In the *Descent of Man,* published in 1871, he devotes considerable space to discussing the evolution of "the social and moral faculties" (12).

Biologists generally agree on the basic facts of evolution, variation, and preferential survival through natural selection of the best adapted variants. There are, however, some peripheral areas where there is still substantial disagreement. Two of these have broad ethical implications and are therefore relevant in the present context. The first area is the age-old nature-nurtue debate. The second is the question of whether or not the principle of natural selection or "the survival of the fittest" justifies a predatory attitude toward life.

Darwinism reawakened the nature-nurture debate. Sir Francis Galton, a contemporary and interpreter of Darwin and the founder of the eugenics movement, took a strongly hereditarian position. Philosophers and scientists with other backgrounds have argued for the importance of environment and they have gained a wide following. In general, geneticists have tended to be hereditarian and sociologists and psychologists proenvironment, a division that suggests a certain amount of prejudice. Popular opinion has swung back and forth between the two viewpoints. As long as the debate has remained an intellectual exercise, it has been harmless, but when leaders have taken extreme positions, damage has resulted.

Karl Marx was familiar with Darwin's *Origin of Species* and regarded it "as a basis in natural science for the class struggle in history" (23). Unlike Darwin, however, he was an extreme environmentalist. Much of the bad in human nature, he contended, was the result of capitalist oppression. In the coming communist society, he therefore believed, crime would largely disappear and cooperative behavior would predominate. Needless to say, these outcomes have not been realized (7).

Adolph Hitler's views on nature-nurture were the exact opposite of Marx's. His extreme hereditarianism led to policies of mandated "eugenics" and genocide.

Such extreme views were made possible by the lack of clear evidence as to the relative roles of nature and nurture in shaping personality. Fortunately, evidence on this subject has greatly expanded, and both social scientists and geneticists are finding a common middle ground. We shall examine this subject in Chapter 4.

The second area of controversy with ties to both evolution and ethics concerns the potential of natural selection to endow humankind with a moral nature such that we can reasonably hope for and demand socially responsible behavior. Darwin himself held this to be possible (12). As we shall show in Chapter 5, we now have firm evidence on this point, but for years rigorous proof was lacking, and as a result some evolutionists, in at least some of their writings, stressed the aggressive and self-seeking aspects of human nature, creating in the process a school of thought that has come to be known as Social Darwinism (23). This cult has been seized on by individuals who want to justify self-seeking behavior on their own part. Of the various scholars who contributed to the various schools of Social Darwinist thought, none was more influential than Herbert Spencer. In many respects, Spencer was a brilliant interpreter of Darwinism and its implications. He coined the expression "survival of the fittest" that evolutionists still find useful. But in the application of this concept to human affairs, he arrived at conclusions about the permissible treatment of the unfortunate that many people find repulsive. His ideas nevertheless were adopted by less critical thinkers and widely applied to human affairs.

The application of Social Darwinism to economics led to the advocacy of an extreme form of laissez-faire. The American empire builders or "robber barons" of the nineteenth century justified their acts as the normal conduct of man in an economic arena where survival of the fittest governed the rules of the game.

Spencer's version of Darwinism was also used to justify war. This school of thought is typified by Emperor William II and the German militarists at the time of World War I. As I know from a friend who lived in

Germany at the time, school children were asked to write themes in which the virtues of war were extolled.

By the last half of the twentieth century, Social Darwinism had lost much of its popular appeal. Ethical thinkers had rejected some forms of extremism but had produced no guide with an authority comparable to Calvinism. Humanists preached an ethics that contained much idealism but that tended to be nebulous.* In the schools of the United States, the volume of moral teaching began to decline in about 1850 and dropped almost to zero by 1930. This has been well documented by Benson and Engeman. Ethical teaching in American schools of necessity has been affected by the constitutional requirement of the separation of church and state, but a general change in public attitude, substantially influenced, Benson and Engeman suggest, by Freud and John Dewey, has also played a role (6).

Along with the decline in ethical instruction has gone an increase in dishonesty and crime. Disciplinary problems in many schools have reached crisis proportions. Benson and Engeman, while not urging any single, simplistic explanation of antisocial behavior, believe there is a connection.

In recent years, inspired in part by doubts about our leadership and the health of our economy, many middle class Americans have begun to sense a need for some kind of moral reform. Yankelovich, who has been following social trends and public attitudes in the United States for some 20 years, depicts in his book *New Rules,* sweeping changes in our outlook that are now taking place. He sees these as "no less than a search for a new American philosophy of life." A search for "self-realization" in the 1960s and 1970s proved unsatisfying. No well-formulated alternative has emerged, but Yankelovich sees an implied need for "new rules," in effect, a new ethic (46).

The Rational Revolution will have failed unless it can raise our ethical standards. Some parents are trying to deal with the needs of their children by sending them to sectarian schools, but much more than this is needed. The rising concern with ethical issues that has marked the last decades of the twentieth century is an encouraging sign, but we have a long way to go. No other single search is more important to human affairs.

*One clear and excellent "Statement of Principles and Values" has been made by Paul Kurtz, a leading secular humanist (*Free Inquiry,* Spring 1987, pp. 4–5).

3

What's Gone Wrong?

A. Introduction

What's gone wrong? In searching for an answer to this question, we shall find evidence that many of the problems that beset us today have ethical roots. If this indeed be the case, the evidence will suggest also that self-interest is served by high ethical standards. A moral society benefits us all.

The democratic, free enterprise system that prevails in the United States and the other industrialized democracies, has provided a wonderful abundance of the goods and services that most people desire. Few of us would choose to go back to the more primitive life-style that prevailed a century or two ago. Yet despite the success of our system, we appear to be entering an era of mounting problems. For most of us, these problems appear only as dark clouds on the horizon—we have not, up to now, been hit by the storm. Many farmers, however, and workers in steel and some other industries where foreign competition has forced plant closings have been overwhelmed by the deluge. In 1986, 5% of the nations's farmers, already deeply in debt because of falling income and rising costs, were unable to finance the spring planting and were forced off the land they had worked so diligently and so long. Many banks in farming areas are in trouble. Some commentators are comparing farm conditions today with those in 1929 at the beginning of the Great Depression (11).

Besides these problems, which are now seriously affecting the lives of our citizens, there are other, very major problems that are less of a threat to our present life-styles than they are to our future. It is with these problems that the remainder of this chapter is concerned.

Not only are there financial problems; there are also problems of public attitude. We noted these in Chapter 2. People questioned in a May 1987 poll as to their views of American leaders, expressed "disappointment, cynicism and concern" (12).

If indeed we have major problems, it is important to seek to understand their causes and to find cures. The problems are complex and can be viewed in more than one way. A great deal of skilled effort has been devoted to examining them from the economic point of view. I suggest that they can also be profitably examined from the ethical point of view. Such an examination is the major purpose of this chapter. Before we begin the examination, however, it will be useful to discuss and compare the nature of the two approaches. A brief treatment may carry the risk of over-simplification, but I do not think that I will seriously mislead the reader.

From the point of view of standard economic theory, the function of our economic system is to satisfy people's wants. The term *preferences* is sometimes substituted for *wants* but the essential meaning is the same. Wants are a flawed guide. Clear evidence of this is found in our ban on the sale of marijuana, cocaine, and heroin. There are addicts who want these desperately, but we decree that if we can prevent it, their wants shall not be satisfied. From the theoretical point of view, the problem with wants is that they are subject to conflicts, both intrapersonal and interpersonal. We may want rich desserts but also a svelte figure, wants that are almost inevitably incompatible. The gratification of the wants of some people now, particularly if combined with poor planning, may deny the wants of other people in other places or other times.

Wants are also subject to manipulation by companies with goods or services to sell. Advertisements often appeal to our desire for immediate gratification to the detriment of the present or future needs of others. In a similar vein, George Lodge, Professor of Business Administration at the Harvard Business School, notes that in some circumstances "the sum of consumer desires in the marketplace is not sufficient to define community need" (3).

The virtue of wants as the basic value system in economic theory is the parallelism between wants and expenditures. Purchases can be assumed to measure preferences. A numerical value can be assigned to both. This potential for quantitation gives conventional economics a precision generally lacking in the social sciences. And a major goal of economists of the dominant, "neoclassical" school is to give economics the precision of an exact science. They exclude all value judgments and they strive for mathematical formulations which, when used with appropriate economic data, will predict economic trends.

There are other schools of economics which start with a broader factual or theoretical base (2,4,8), and also, critics, some of them economists, who see flaws or deficiencies in the neoclassical approach.

One objection to their approach is that it is too theoretical and insufficiently empirical (1). Certainly no science can be considered exact unless it is based on an adequate body of facts. Economics is no exception, but I might note that unlike physics, chemistry, and biology, whose exactness derives from experimentation, this source of well-proven facts is unavailable as a foundation for economics. Also the consistency and repeatability of the sort of observations needed for building sound theory are uncharacteristic of much human behavior (6).

Of all the criticisms of neoclassical economics, the ones most relevant in the context of this volume are those concerning its exclusion of value judgments and ethical considerations. There are a few economists who have voiced such criticisms, but the most explicit statements have generally come from people trained in other disciplines.

Derek Bok, president of Harvard University and a lawyer by training, is one of the more notable exponents of this viewpoint:

> Universities have gone much too far in trying to produce value-free teaching and research. This happened primarily because scholars within universities were deluded by the thought that only completely objective, 'scientific' inquiry was respectable and that values were not amenable to rigorous scholarship. . . . Many scholars even distorted the reality they studied in order to make it susceptible to their 'rigorous' methods of analysis. Now we are beginning to understand that while scientific methods have utility in softer fields such as the social sciences, they are not adequate to understanding all questions (9).

Hirsch, a Briton, is one economist who shares this viewpoint. He writes: "Truth, trust, acceptance, restraint, obligation—these are among the social virtues grounded in religious belief which are also now seeen to play a central role in the functioning of an individualistic, contractual economy" (2).

Before this century, leading economists did not hesitate to show their esteem for morally responsible behavior. Adam Smith, whose *The Wealth of Nations* laid the philosophical foundations for our free enterprise system, was not a stranger to moral analysis. At one time he taught moral philosophy and before writing his economic opus, he published a book on ethics, *The Theory of Moral Sentiments*. His personal emphasis on moral standards is clearly indicated by his admiration for the Greek Stoics who contended, in his words, that "Man . . . ought to regard himself, not as something separate and detached, but as a citizen of the world, a member of the vast commonwealth of nature. To the interest of this great community, he ought at all times to be willing that his own little interests be sacrificed" (7). This quotation from Smith would certainly seem to warrant the assumption that his advocacy of free enterprise presupposed a widespread integrity in the business world. Amasa Walker, whose book *The Science of Wealth,* was first published in 1866, went through a total of seven editions, and was reprinted in 1969, did not hesitate to introduce ethical considerations. Thus he wrote that anyone in a position of responsibility should have "moral power . . . which gives such a control over appetites, passions, and propensities as affords assurance that under no circumstances of trial or temptation will he ever depart from the strictest line of duty" (10). Before his career in economics, Walker had operated a wholesale shoe and leather business from which he retired with a comfortable fortune, so his words are those of a practical man as well as a theorist. It was not until the 1930s that the urge to make economics an exact science led to the rejection of all value judgments and the exclusion of subjective elements other than human wants (2,5). Now there are increasing signs of a swing in the other direction.

My own approach to the problems, many of them economic in nature, that we examine in the remainder of this chapter is specifically and intentionally ethical. This does not mean in any sense that I reject the economic approach. The two are complementary, not mutually exclusive. The same act can be examined both with respect to its economic and its ethical properties. The moral factors in economic events are perhaps in some cases further back in the causal chain than the economic factors, and they are also more difficult to quantitate, but they are no less amenable to observation and analysis.

What are the properties of the ethical judgments we form concerning conduct? We examine ethics in detail in Chapters 6 and 7. A bare outline must suffice here.

Ethics is based on a system of values other than wants. We may, for present purposes, select the general welfare as our standard. A suitable standard should enable us, in theory at least, to choose between competing needs. General welfare meets this requirement. Another property of our ethical judgments is that they are usually guided by an extensive and ancient body of moral laws. These laws require that we be truthful, that we live up to our promises, that we be considerate of the welfare of others, that we be fair and just. While the ultimate test is the general welfare, we usually do not need to go back to this source; moral laws are a sufficient guide. Of course, these laws have themselves to meet the test of furthering welfare.

I think we can confirm from our own observations that the great majority of people are aware of this body of law, and that most of them render it at least lip service. However, as we shall show in this chapter, many people evade these laws in many ways. Evasions tend to be dismissed as minor; they usually affect people remote from us, not our family or friends; and they often are condoned by the general public.

The types of wrongdoing that I consider can be grouped under seven headings: waste, misguidance in the market place, debt encouraged, unneeded wealth misused, security sought through power or privilege, dishonesty condoned, and major crime unpublished. The list that I offer is not complete—prejudice, avarice, malice, and deceit come in many forms. As we shall show, however, the perpetration of the wrongs listed, in its totality, seriously harms us all. Self-interest is not well served in a society that tolerates wrongdoing, even wrongdoing of a seemingly minor sort. It can only be truly served if we strive for and in substantial measure achieve a society where moral standards are widely respected and generally observed.

B. Waste

Americans today are often unmindful of the old adage, "Waste not, want not." We have become the most wasteful nation on earth. Waste, however, is as universal as man, at least as old as civilization, and as diverse as our patterns of production and consumption.

Waste is a manifestation of shortsightedness often abetted by sloth, greed, and false pride. Its ill effects are typically delayed. Sometimes they catch up with the wasteful person, but at least as often nowadays the harm falls on others. In today's affluent and highly industrialized society, indeed, the greatest harm from waste may well fall not on the living but on the unborn. Whether the damage from any inappropriate or unnecessary use of resources comes soon or only after many years, people with a conscience can only judge the use as wasteful and wrong.

The varieties of waste can be treated conveniently under eight headings: water and soil, energy, minerals, food, forest products, manufactured products, and armaments. Sections under these headings are followed by a section on waste disposal; industrial waste and one on the economic consequences of waste.

Water and Soil

Of all natural resources, water imposes the most unalterable limit on population and carries the greatest threat of discontent and ultimately hardship if overtaxed. In parts of the western United States, the threshold of discontent is rapidly being reached.

The Colorado River, source of water for a number of states, is now drained to the last permissible drop (2). Aquifers, a very major source of western water, are being used 26% faster than nature can refill them. Antiquated water mains in our cities lose by leakage millions of gallons of water a day. Many water sources have suffered severe damage from discharged wastes. The demands of agriculture for irrigation water have almost tripled in the last three decades and now account for 83% of all the water used in the United States (8, 35, 40).

The loss of land, soil, and soil fertility owing to human intervention, is one of the oldest forms of waste and one of the most disastrous in its consequences. It goes back at least to the beginning of intensive agriculture and has been the cause of the decline or disappearance of civilizations in both the old world and the new (6). In the United States at the present time it takes three major forms: the loss of farmland due to urban development, the loss of soil from erosion by water and wind, and the loss of the fertility of irrigated land owing to the rise in the level of underlying salt water.

Land loss in the United States due to the construction of roads, houses, shopping centers, factories, and reservoirs amounts to three million acres per year, of which one million is prime farmland (13, 31, 32, 39). At present rates of conversion, Florida, New Hampshire, and Rhode Island will lose virtually all their prime farmland in 20 years (13).

Soil erosion is an even more serious problem. Soil is eroding faster than it is being replaced on half of our crop land. A billion tons of suspended solids are discharged into U.S. surface waters each year. The Mississippi alone discharges an average of 15 tons of topsoil per second into the Gulf of Mexico. Iowa, which had 16 inches of topsoil a century ago, now has only 8 (10, 39). Erosion of forested land, particularly on hillsides where lumbering is done by heavy machinery, is also significant.

Irrigation is another source of soil damage. In dry areas, the subsurface water often contains salt, and irrigation, by raising the water level can bring salt into the topsoil. Worldwide, it is estimated that since the earliest use of irrigation centuries ago, some 25% of the Earth's irrigated cropland has become too salty to farm (27).

If land abuse is a serious problem in this country, it is far worse in many Third World countries. In Africa, it has contributed to the recent famines (30).

Waste of water and soil is atypical of waste in general, in that to a major extent the problem originates in and must be solved by the government and society as a whole. Thus population growth and poor government planning are major factors. We certainly use more water than we need, but as long as population goes up, so will consumption until demand outruns supply.

The extensive use of water for irrigation has been encouraged by the government, often through pork barrel projects originating in Congress, and in the long run has contributed to the overproduction problem on the farms, and the destruction of land value by salting (36).

The loss of soil on the farms could be avoided, to a considerable extent, by the use of improved methods, but under present economic conditions the costs are prohibitive (5). Only major government intervention can alter this. The loss of farm land through development is due, to some extent at least, to poor planning at various government levels.

ENERGY

Most of our energy for transportation, lighting, heating, and the running of machines comes from fossil fuels. We also use fossil fuels for making fertilizers, plastics, and synthetic fibers. At the moment, an abundance of coal and oil is available for these uses. Because of the world's generous coal reserves, including large deposits in the United States and Canada, and the huge oil reserves in the Middle East, the ultimate crunch may not come for a century or more. But energy consumption started to grow very rapidly about 30 years ago and is still increasing. At the 1984 extraction rate, the known oil reserves of the United States will last only 9 years. The Soviet Union and Great Britain will fare only somewhat better. The oil crunch caused by the 1973 oil embargo imposed by the Organization of Petroleum Exporting Countries (OPEC) could be repeated again in the 1990s (16, 44). While coal reserves in the U.S. and Canada are extensive, the bulk of them are in forms that will be much more costly to use.

A major constraint on our use of coal and oil as energy sources that we must begin to take seriously is the resulting release of vast quantities of carbon dioxide (CO_2). This, when incorporated in the atmosphere, produces a greenhouse effect that will increasingly alter the world's climate (33).

In the face of these threats to our future, we always have and still continue to use energy wastefully. In the early days of the coal and oil industries in this country, the rush to exploit these potential sources of huge profits resulted in inefficiences of extraction. The goal was quick extraction of the most accessible and most easily marketable sources. Too many mines were opened and only the best seams were worked, the difficulty of mining poorer sources being increased in the process. The same sort of inefficiencies prevailed in the oil industry. Also, huge quantities of natural gas were flared off at the wellhead because at the time there was no way to get it to market (19, 43).

These conditions in both coal mining and oil extraction have been greatly reduced; the major problems now are in the areas of consumption. Shifting some of our freight transportation from trucks to trains, which use one-fourth as much energy per ton mile, could produce substantial savings. A 1973 study indicated a potential of 42% energy saving in home heating and another study 6 years later placed the potential saving in all categories of space heating at 67% (18, 42).

There is some good news. The oil crisis of 1973 made the American public suddenly energy conscious. Many people insulated and otherwise weatherized their houses, and improvements are still being made, some of them with the encouragement of electric utilities. There was a big increase in the sale of cars capable of achieving good gas mileage, and American auto makers were forced to redesign their autos to compete with the more efficient European and Japanese models. Some corporations, among them Dow Chemical (whose founder was an energy hobbyist) and Du Pont, found ways to achieve substantial energy savings, demonstrating both the feasibility and practicality of such measures (11). Overall, we have made considerable progress—in 1983 the United States consumed 22%

less energy per dollar of gross national product than would have been the case had 1973 efficiency levels held (20).

There have, however, also been setbacks. When oil became plentiful again in the 1980s, the American auto companies saw a chance to increase their profit margins. Taking advantage of Americans' short memories in public matters, and playing on people's fondness for power and pep, they quickly undid much of the gasoline savings they had achieved only a few years earlier.

We can only guess as to our energy future, but we do have a few clues. We have already mentioned the possibility of another energy crunch in the 1990s, and the inevitability of the ultimate exhaustion of fossil fuel available at reasonable cost. There is a substantial chance that science will ultimately bring into being virtually inexhaustible energy sources. The two most likely are atomic fusion and the transformation of sunlight into electricity by means of relatively inexpensive solar cells. It has been estimated that an array of cells on an area of 30 square miles could generate the electricity equivalent to 20% of the total United States electrical generating capacity (46). Solar energy is a 100% renewable source, but it does have the major disadvantage of being available only during daylight hours in the absence of some means of storage. Less spectacular, but having substantial potential, are devices to increase the efficiency of home and office heating. Methods of reducing heat loss through windows and the use of cogeneration and pulse furnaces are examples of them (3, 26, 38).

Despite the potential of these energy safeguards that science may provide for us, the possibility of difficult and even disastrous energy shortages remains. The best safeguard of all would be more energy savings here and now. And any money saved from elimination of wasteful consumption could certainly find better uses.

MINERALS

Modern industry has created a demand, unimaginable to a few centuries ago, for a variety of minerals. The supply of most of these, at least in concentrations permitting easy mining, is limited, and their unequal distribution throughout the world adds a major strategic significance to the shortages (17). Estimates of the potential for exhaustion are subject to a large error, but critical shortages in the next 10 to 60 years are seen as possible for lead, zinc, tin, platinum, silver, copper, tungsten, mercury, and possibly several other metals (1, 15, 25).

Much of our waste of minerals is tied to our throw-away society. Millions of tons of iron and tens of thousands of tons of aluminum, zinc, copper, lead, and tin are thrown away in this country every year (17), mostly in the form of discarded products ranging from beer cans to junked automobiles. There has been some move towards recycling, but we need much greater effort in this direction and much more emphasis on durable products.

Food

Turning to food, we find again that Americans have a poor record of thrifty use. An unusual study provides some precise evidence. Professor William Rathje and his students at the University of Arizona have carried out an extensive investigation of the refuse from 7,000 homes in Tucson. The results indicate that 15% of the food purchased is discarded. An extrapolation to the whole city of Tucson suggests that 9,500 tons of food worth $11 to $13 million end up in landfills each year. Poor families discard less than middle income families and middle income families make greater use of the relatively expensive name brands than the rich (34).

Although millions are starving in Africa and a small fraction of Americans are going hungry at least part of the time, there is no food shortage in this country and no real evidence that one is in sight. In fact, there is a potential for further increases in crop yields through the application of genetic engineering.

These trends do not point to any early food shortage for the United States, but the more distant future remains uncertain. The only immediate social cost of our wasteful food habits is the diversion of money from better uses.

Forest Products

We are already close to a shortage of forest products. During the past decade, the prices of newsprint and lumber have soared. This has worked a hardship on both newspapers and homeowners or would-be homeowners. A study by the Council on Wage and Price Stability concluded that "over the long term the projected demand (for lumber) cannot be met by the supply of timber that will be forthcoming at current relative prices and under current management policies" (4). With better management policies, according to one optimistic estimate, production could be doubled or even tripled (41). Worldwide, it is estimated that one-third of all forest land has been lost since the beginning of intensive agriculture several thousand years ago, and since 1967 the per capita production of wood has declined.

In the United States we use approximately 1.4 tons of forest products per person per year (29). A large share is used as paper or one form or another of containerboard (15). It is these products that are the largest sources of waste. They make up almost 30% of the nation's trash (29), much of it unnecessary. In most homes in the United States, trash cans contain a considerable volume of unwanted mail, paper bags from grocery stores that are used only once (a practice rare in Europe), and newspapers and containerboard that could be but are not recycled. The bulk of discarded newspapers and magazines, moreover, is increased by pages and pages of advertisements that are completely passed over by most readers, that are often much larger than they need to be if their purpose simply were to convey information, and that in a few cases, such as cigarette advertisements, tout products that are harmful.

MANUFACTURED PRODUCTS

Some of the sources of waste so far considered involve products whose value has been increased by processing. Paper is an example. We now turn to products whose value derives much more from the manufacturing process than from the raw materials that enter into them. It is in the consumption of these products that wastefulness has indeed become a way of life.

In the days when products were made by hand, often with great labor and often in the household where they were to be used, there was every inducement to make them last as long as possible. Even when they changed hands before use, the maker in most cases took sufficient pride in his work so that he built them for long service. When the industrial revolution culminated in the development of the assembly line and mass production, all that was changed. The satisfaction of the owners and managers of industry derived more from profits than from quality. And profits were dependent on large volume and steady output. Success and selling became almost synonymous, and the using up of goods a virtue and not a sin. The culmination was an attitude well summarized by Victor Lubow, a marketing consultant, who wrote in *The Journal of Retailing* in 1955:

> "Our enormously productive economy . . . demands that we make consumption our way of life, that we convert the buying and using of goods into rituals, that we seek our spiritual, our ego satisfactions, in consumption. . . . We need things consumed, burned up, worn out, replaced, and discarded at an ever increasing rate" (23).

This is an unusually uninhibited advocacy of the gospel of unlimited consumption, but if anyone imagines that it is an exceptional viewpoint, a perusal of Vance Packard's *The Wastemakers* should convince him of the contrary.

The requirement for a mass market resulting from the development of mass production has generated several economic trends of dubious morality. We now turn to one of these, the design of products to encourage waste. In the following two sections of this chapter we shall examine two others.

American corporations have used three major methods to encourage the turnover of the goods they manufacture. The first is planned obsolescence, or the design of goods to wear out faster than efficient and available engineering practices would dictate. The second is the use of style changes to hasten turnover. The third is the development, in the name of convenience, of disposable or throwaway products.

I suspect that most Americans hate to believe that American corporations have conned them into the purchase of goods designed, not to last, but to wear out. Certainly this has been my own feeling. Unfortunately, the evidence is overwhelming that, beginning at least as early as the 1930s, there has been widespread though perhaps generally tempered use of this unethical practice. I cannot begin to give the evidence here—an excellent presentation can be found in Packard's volume. Suffice it to say that in some major corporations such as General Electric, planned obsolescence was discussed and accepted at the highest administration levels, and engineers, certainly in many cases greatly to their distaste, were induced to incorporate it in their designs (9, 28).

Style change as a means of product turnover was adopted much more openly. After all, it had been practiced in the design of women's clothes for generations, leading Oscar Wilde to remark, "Fashion is a form of ugliness so intolerable that we have to alter it every six months." It was adopted with particular openness and particular effectiveness in the auto industry, with General Motors leading the way. And with the aid of brainwashing by an intensive advertising campaign, the American public accepted it readily.

The costs were susbstantial. In the auto industry, even though the annual style changes were made as superficial as possible, they cost some 40 or 50 million dollars for a single high-volume line (24). The costs for the whole industry were a good many times this, all of them ultimately borne by the consumer. An even greater waste resulted from the rapid turnover of cars, which was of course the purpose of this whole misapplication of free enterprise.

The development of throwaway products is too familiar to need elaboration. These products range from beer cans to test tubes for the laboratory and from tins of squirtable whipped cream to deodorant pads. Many of them undeniably save labor and, especially in homes where both parents hold outside jobs, some of them fill a real need. But convenience has to be balanced against the littering of highways with disposable beer and soft drink cans, and the cost of materials that end up in the trash can and have to be disposed of at further additional cost.

ARMAMENTS

Armaments are one of the greatest and most tragic forms of waste, but under present world conditions also the most intractable. It is estimated that the total annual expenditure worldwide for arms and armies is about $500 billion, a sum approximately equivalent to the combined gross national products of all the developing nations (14).

Even undeveloped countries are spending heavily. The impoverished nations of Africa spend over $50 billion annually for arms. The exportation of arms is not limited to the major powers; in recent years Argentina, Brazil, India, Israel, and South Africa have become suppliers. Altogether, the annual value of major-weapons exports is estimated at nearly $60 billion. Not only do these arms expenditures represent an enormous waste of materials, but also of manpower, including the effort of approximately 400,000 skilled scientists and engineers (22).

The intractability of the arms problem resides in the need for international understanding and trust before any broad solution is possible. Unilateral disarmament or major unilateral arms reduction by the United States would be an invitation to disaster. In the present world climate, we can hope for limited arms agreements but not the ultimate solution—a United Nations with adequate power to enforce world peace.

Waste Disposal: Industrial Waste

So far in this section we have been talking about waste used as a verb—to waste or to be wasteful. We now turn to waste used as a noun, to trash, the unwanted leftovers of our society that have to be disposed of. Not all waste is the result of wastefulness. Ash from coal-burning furnaces has to be discarded and, with rare exceptions, has to end up in a dump.

The volume of trash generated in the United States is enormous. We have been called the disposable society. American communities are faced with the disposal of somewhere between a third of a ton to a ton of waste per person per year, the variation being largely a function of the amount of industrial waste included. The total is 133 million tons. This waste is composed (approximately in descending order) of paper, containerboard, food wastes, glass, yard and garden wastes, metals, and products such as plastics and textiles (29). Much of this ends up in landfills, but land suitable for this form of disposal is becoming increasingly hard to find. By the 1990s, in 27 states landfills will be just that—filled. A garbage barge loaded in New York City in the spring of 1987 travelled for months along the Atlantic coast, going even as far as Mexico, seeking a disposal site. There were no takers. The fly-swarmed scow became a national disgrace and an international joke (21).

A particularly devastating result of our throwaway habits is the death of hundreds of thousands of marine animals that have eaten or become entangled in discarded plastic. We produce enormous amounts of plastic—1.2 trillion cubic inches were manufactured in the United States in 1985. It is indestructible; and ships, since all their waste goes overboard, have dumped so much that all sorts of plastic objects are common in the world's waters. The victims range from sea birds to whales (45).

Considerable industrial waste is included in municipal waste, but there is additional corporate waste that has to be disposed of independently. Unlike most domestic waste, a considerable amount of corporate waste contains toxic substances, and much of this has been dumped in sites where people are endangered. This is a wrong of major proportions, and hence an appropriate subject for this chapter, but it is too vast a subject for even condensed treatment here. A book published by the Sierra Club deals with it very thoroughly (12).

The problem of waste disposal in the United States has reached the point where a major search for solutions has become necessary. This is another vast subject that I can only touch on. As European countries have clearly demonstrated, recycling of many items in our trash is both possible and practical (7). There have been notable successes in this country, but they are all too few in number. A report of recycling successes achieved by the Minnesota Mining and Manufacturing Company (3M), best known as the producer of Scotch Tape, shows what can be done. In 1976, the company initiated an antipollution program for its factories in 15 countries. Instead of buying costly pollution control devices, the 3M managers and engineers redesigned both products and equipment. Both water and waste products were recycled. The result, by 1979, was a savings of over $20 million while increasing production by 40% at the same time (37).

ECONOMIC CONSEQUENCES OF WASTE

How harmful are the varieties of waste we have described? Are they really that bad? Some people, including traditional economists, but not such institutionalists as Kapp, might argue that the types of economic behavior that I have described are merely the result of people satisfying their wants and are therefore not to be condemned. People want big autos, they want to be able to use water freely even if they live in relatively dry areas, they want highways and an abundance of shopping malls, they want inexpensive food even if this means the use of injudicious agricultural methods to produce it, they want a variety of catalogs and an abundance of newspaper advertisements to guide their shopping even if these end up in trash cans and increasingly expensive landfills.

The problem with this argument is that it is a short-range view. Because a major consequence of waste is the gradual depletion of natural resources, the harm that it does is particularly apt to be delayed. We have already experienced significant damage in two areas.

The oil crisis following the oil embargo of 1973, and the period of inflation that followed, could certainly have been lessened if we had been more thrifty in our use of oil. Because of our wasteful consumption, we had become excessively dependent on oil from the Middle East, and were therefore vulnerable.

A more serious consequence of our wasteful practice was the loss of much of our auto market to European and especially Japanese competition. The design of American cars to be big, showy, and short-lived may temporarily have maximized profits, but it created a ready-made opportunity for the more conservative, better-built models being developed abroad. The oil shortage and hence the demand for gas economy came at just the right time for our foreign competitors.

The greatest damage from our wastefulness is probably yet to come, and will be due to the depletion of resources. The future is very uncertain, but the first major crisis could well be due to a shortage of water. Unless technological developments intervene, shortages of minerals and fuel could bring us close to disaster. The greatest danger of all could come from loss of and damage to cultivateable land. Nations in the past have succumbed to this, and it is already causing serious problems in Africa and other less developed areas. We still have a chance to prevent this, but major new policy initiatives will be necessary.

Waste in all its forms is wrong because in the long run it hurts people everywhere.

C. Misguidance in the Marketplace

The need for a mass market that accompanied the rise of mass production created in turn the need for effective marketing. The goods produced had to be sold, and success tended to go to the company that sold most successfully. Because products were both diverse and complex and because most consumers were unskilled in rational choice, sales tended to reward the use of obfuscation and artifice rather than truth. Misguidance in the marketplace became a way of life.

In no country has industry exploited the power of advertising more effectively or more universally than in the United States. But wherever mass production has gone, mass advertising has followed. In South America, for example, the introduction of advanced industries was accompanied by major growth in advertising (6).

There obviously are legitimate functions for advertising; for example, its use to call attention to new products or services, or to price reductions, or to provide information that will be of genuine use to the consumer in making a rational choice between products. If a company truly has a superior and reasonably priced product, it has a right, not to say a duty, to proclaim it from the housetops. We shall be concerned here only with the abuses of marketing, and it is these, unfortunately, that fill the mass media.

CHARACTERISTICS OF ADVERTISING

Reeves, a leading executive in the advertising business, has likened advertisers to the Red Queen in *Through the Looking Glass*. They both have to run at top speed to stay in place (11). Competing products are often very similar. The different brands of cigarettes and the different asprin-containing pain-relievers are familiar examples. Sales are more likely to be determined by advertising success than by any real advantage of one product over another. If a mass-produced product is to hold no more than its own, it must be continually and effectively advertised. This puts great pressure on the manufacturer and the advertising agency to devise advertisements that, whatever their nature, will persuade people to buy.

Advertising executives have devoted enormous amounts of thought and study to devising guides to successful advertising. In the process, they have come up with some widely followed rules and practices, some of them sufficiently acceptable to talk and write about, others of a more dubious nature.

Reeves's *Reality in Advertising* offers a number of rules: (1) Keep advertisements simple; people are unlikely to remember more than one point, (2) repeat the same theme over and over, (3) don't try to make an advertisement a really attractive work of art, (4) use a "unique selling proposition" for each product, i.e., some claim to uniqueness, no matter how trivial. This "U.S.P." can well be used as the oft-repeated theme (11).

Among the practices less freely acknowledged by advertisers are the use of sex appeal, snob appeal, appeals to the urge to keep up with the Joneses, and half-truths. Some authorities believe that many advertisements contain "hidden persuaders" that act via the viewer's subconscious mind, but the effectiveness of any such approach is difficult to prove. Advertisements for cosmetics, lotions, and lingerie may appeal indirectly to sexual motives and in that sense perhaps do work through the unconscious (9).

Competent authorities attest to the wide use of half-truths in advertising. In an interview with *U.S. News and World Report,* Stephen Newman, a professor at the New York Law School and a specialist on consumer law, was asked, "How prevalent is deception in advertising?" His answer: "No one has ever taken the full measure of the problem, but I believe it is very widespread. Very rarely can an individual tell from the face of an ad whether or not it is true. Shading the truth goes on all the time, and material omissions are almost a tool of the advertisers' trade. Whether or not you define that as deception depends upon your perspective" (14).

Deception or not, there are many well-documented examples of advertisements that mislead the consumer. Since the American public became energy conscious, billions of dollars have been spent annually on products that, according to the claims of the manufacturers, offer energy savings. In a study by the General Accounting Office (GAO), Congress's watchdog agency, claims were tested against actual performance. Many did not stand up. The study concluded that "Consumers face difficulties in assessing product claims that are potentially inaccurate, misleading, and difficult to compare." The "GAO believes that continued government efforts are needed to protect consumers from being misled by advertisements making energy-saving claims" (1).

A ruling by a Federal Trade Commission judge provides another example. Television commercials that touted Bayer asprin, Bayer children's asprin, and Vanquish as better than other brands were judged false and misleading because there was no proof. Actual falsehood in advertising is limited by government regulations, but misinformation can be delivered in other ways. The message conveyed by the picture of the Marlboro he-man on horseback in Marlboro cigarette advertisements is quite as false in its implications and probably more effective with youngsters than any false statement could be.

There are two areas where, in recent years, advertisements have conveyed genuine information. One area is the tar and nicotine content of cigarettes, the other the gas mileage delivered by autos under specific test conditions. In both cases, the information has been required and provided by federal regulatory agencies. Cigarette advertisements have tended to confuse the meaning of the tar and nicotine data with their accompanying rhetoric, but the data probably still serve a useful purpose.

One reason that the misguidance in advertising works as intended is because of repetition. As Hitler, that master of propaganda, noted, a lie repeated often enough comes to be accepted as truth. The reader or viewer succumbs to saturation.

COSTS OF ADVERTISING

The total cost of advertising in the United States in 1978 was placed at $43.7 billion. This was approximately 2% of the gross national product. The cost as a percent of the gross national product was nearer to 3% in 1929 (5) but for the last three decades has run close to the 2% mark (6). The disposal of the paper used in advertising imposes a further cost of an uncertain but clearly substantial amount. Another cost is the time spent by consumers in cashing in on coupons, Green Stamps, and other selling devices. In 1980, 81 billion coupons were printed for distribution. The diligent user may realize slight personal savings, but the net result is higher costs that are passed on to the public at large.

MISGUIDANCE AT THE RETAIL LEVEL

Honest representation is much more common at the retail level than in the sales pitches brewed by Madison Avenue for the manufacturer. This is particularly true in locations where there is a sense of community, though even here distributors are stuck with much misleading packaging and labeling that they did not initiate. But at this level there are also glaring examples of misguidance. This is perhaps particularly true in the case of some automobile dealers.

Auto salesmanship is at its worst in poor, usually minority, districts where the unscrupulous dealer can use the high vulnerability of the residents to sell them cars often far beyond their means. MacDonald, in *Detroit 1985,* and Gibney, in *The Operators,* describe some of the more extreme methods used (3,8).

Almost as soon as a potential customer has parked near the dealership, he is approached by a salesman who skillfully whets the customer's desire to make a purchase, either of a new or a used vehicle. Any questions about the problems of financing are turned aside as unimportant. Instead, the customer's vanity is appealed to, arousing dreams of bigger and fancier cars. At this point, he is led into one of several cubicles, all of them bugged, and the first salesman gives way to a second, unbound by any previous promises. The customer is never left alone, never given a chance to think or make notes. Other salesmen that he meets have listened in via the bugs and know the degree of his vulnerability. The likely end result is the sale of a car not only more costly than the customer can afford, but also more costly than he has been led to understand. The dealer always is protected against default, sometimes by a chattel mortgage on the per- sonal property of the customer and his wife written into blanks in the contract. Misguidance has produced another victim.

Vendor misguidance is found in the sale of real estate and a great variety of products and services, and it can assume innumerable forms. In its extreme forms it merges with outright fraud. It includes artificial mark- downs from previously inflated prices, mislabeling, e.g., cashmere sweat- ers with little cashmere; misrepresentation of interest rates by lending institutions; and misrepresentation of work done in home remodeling and auto and appliance repair (3,13). Even Uncle Sam can be a victim. Many corporations win government contracts by underestimating their costs and then collecting payment of cost overruns. The General Accounting Office estimated that in 1983 such overruns, ultimately paid by the U.S. taxpayer, amounted to $318 billion.

Consequences of Misguidance

If a free economy is truly to serve the public interest, the consumer should be able to make an informed and rational selection between competing products. Advertising, as now used, does more to defeat than to serve this vital economic end. Advertisements are primarily concerned with con- veying a favorable image, aided by half truths if necessary, not with providing factual information that will genuinely assist the consumer in making an intelligent choice. They impede the operation of the sort of economy sought by Adam Smith.

To fully appreciate the extent of this problem, it must be viewed in the context of the competence of the American public to cope with advertising obfuscation. An 1975 government study found that one in five American adults is functionally illiterate and unable to deal with many situations common in today's society. The study included tests of the ability to choose rationally between the sort of competing products that are the subjects of most advertising. The problem is compounded by the growing complexity and diversity of the products available. Many Americans, we must conclude, are easy victims.

Another public cost of advertising is its very major role in engendering waste, particularly those forms of waste associated with manufactured products that we discussed in the preceding section. Advertising is the handmaiden of consumerism, which in turn is the child of mass production.

Besides the waste induced, advertising also encourages excessive buying on credit, a wrong discussed in the next section.

Cigarette advertising is the epitome of another damaging consequence of an appreciable part of the advertising spectrum—an ill effect on health. It is well established that cigarette smoking is harmful, yet according to a 1984 report, the tobacco industry was spending at the rate of more than 1 billion dollars per year to promote it. And despite counter campaigns now in effect, the advertising works. Tragically, it works most effectively with youngsters. To cite just one statistic, a 1980 study showed that in the prior decade the number of twelve-year-old girls who smoked cigarettes increased more than tenfold. An increase of lung cancer in women has been an inevitable consequence of this increased use of cigarettes. The American Cancer Society predicts that in 1985 the incidence of lung cancer in women will surpass that of breast cancer for the first time.

Faced with some decline in the sale of cigarettes in recent years as the result of warning education, the industry has come up with a new way to sell tobacco. For some years it has been promoting the oral use of moist snuff, using advertisements that feature star athletes. The users now run into the millions, many of them young boys. The result: a marked rise in oral cancer and other mouth diseases (10).

Advertising of alcoholic beverages aimed at college students and military recruits, many of them below the legal age of purchasing such beverages, is another problem area as pointed out by several dozen consumer organizations in a communication to the Federal Trade Commission.

The health of children and teenagers has also suffered from food advertising. Much of the food promoted is oversweetened or contains more junk than nourishment. A widespread use of inappropriate diets has resulted.

Television commercials can also have a damaging psychological effect on children. This was suggested by Professor Neil Postman of New York University in an interview with the editors of *U.S. News and World Report*. According to Professor Postman's estimate, typical Americans in the first 20 years of their lives see something approaching one million commercials, the equivalent of about 1,000 a week. The message of these commercials, as Postman points out, is that the problems that beset us— whether lack of self-confidence, boredom, or money problems—can be quickly and painlessly solved if only we will avail ourselves of the various products or services being advertised. Pain killers or ointments or break-fast foods or autos take on magical attributes not usually confirmed by the facts of life. Professor Postman concludes: "One has simply got to wonder what the effects are on a young adult who has seen a million of these little vignettes. One has to ask, 'What is being taught?' "

A problem in television commercials not mentioned by Postman but emphasized by other concerned observers is the growing exploitation of sex aimed at younger and younger children. One example cited is the use of sex appeal to sell jeans for 11-year-olds. One observer believes that advertising of this sort has contributed to a lowering of the sexual morality of teenagers (2).

It is interesting to note that although Quebec has a law prohibiting television ads aimed at children under 13, the U.S. Congress has refused to accept such restrictions in this country.

One of the most insidious consequences of advertising is its generation of cynicism and distrust. A society cannot be bombarded year after year through its mail and mass media with intentionally misleading material without losing some of the mutual trust that is so essential for the suc-cessful functioning of a free society. Also, it seems to me, advertising as we practice it may well lead some people to accept deception as respectable.

One aspect of this problem is well illustrated in a variety of training materials developed for teachers and school children to help the children cope with advertising. In *Deception Detection: An Educator's Guide to the Art of Insight,* Jeffrey Schrank writes, "To leave consumer education in the hands of advertisers is nearly as wise as having monkeys guard a banana plantation." He finds that most high school students and many teachers are convinced of their own immunity to advertising, and yet are influenced by it because of the skillful use of indirection that works through impressions or, in extreme cases, subliminal seduction (12).

A publication of the National Council of Teachers of English, the *Quarterly Review of Doublespeak,* is designed in part to arm children against advertisements. One of its approaches is to teach students how to analyze the pictures and implied messages of advertisements as well as the superficially apparent message. Many teachers use the material enthusi-astically.

A study carried out at the University of California at Los Angeles showed that a few brief training sessions can make elementary-school children more discerning in their reaction to television commercials. The researchers, however, note the magnitude of the mismatch between the skill of advertisers and the judgment of small children and are unsure as to the long-term effects of the training.

Besides the mistrust engendered, another damaging byproduct of misguidance in the marketplace is the increasing penetration into our political campaigns of techniques basically similar to those developed by Madison Avenue to promote the sale of commercial products. The promotion process as applied to candidates is particularly insidious because it requires, in addition to direct media advertising, substantial manipulation of the media themselves. The result is a debasement of the democratic process that, if allowed to continue, can undermine the foundations of our free society.

A possible and, if real, frightening consequence of marketplace misguidance is the creeping spread of the disregard for truth. The introduction of Madison Avenue methods into politics just referred to is one of the more convincing examples. Of course American campaigns have never been noted for their scrupulous truthfulness. I doubt, however, if we would find, before the 1980 Reagan campaign, a statement similar to that attributed by the *San Francisco Examiner* to a Republican National Committee candidates' guidebook" "Truth is what people believe—it has nothing to do with fact." I cannot document other examples of the spread, but it is a threat that we should not forget.

An indirect consequence of advertising is that it subjects the news media to an appreciable degree of control by the advertisers. This effect is greatest in the case of television. To obtain the maximum number of viewers for commercials, television programs are carefully tailored to insure a mass audience. The result is a lowering of standards, for example, by the inclusion of many scenes of violence. There also can be direct intervention in the news by advertisers. Thus, as noted by Schrank, Coca-Cola switched a million dollars of its advertising after NBC broadcast a report on unhealthy living accommodations provided for migrant workers by Coca-Cola's Minute Maid subsidiary (12). Even *Time* magazine stooped so low as to run an advertisement in a tobacco trade publication, noting that "Women, singles, 25–49 year-olds, and high school grads" are a hot market and that *Time* was a good way to reach them (4).

The littering of many of our highways with billboards and huge signs is one of the more obviously unpleasant consequences of advertising. Many citizens, including the wife of President Lyndon Johnson, have worked to eliminate this blight, and a Highway Beautification Act passed in 1965 was supposed to accomplish this. The billboard lobby, however, has succeeded through amendments and evasion in largely negating the law, and in 1985 and several prior years more signs have gone up than have come down.

Advertising, where properly used, serves legitimate economic functions, but in its present form it also causes economic and social damage. Since advertising costs are borne by the consumer, advertising not serving the public interest is an unnecessary drain on the consumer's pocketbook. But advertising's indirect effects—its perversion of the Adam Smith economy, its promotion of wastefulness and debt, its penetration of the political process, its corrosion of the trust in people and society so necessary for the successful and beneficent functioning of a free economy and a free society—pose the greatest threats.

D. Debt Encouraged

The nations of the world have allowed debt to become a horrendous monster. We in the United States are among the worst offenders and certainly the most inexcusable one. President Reagan and the Congress are putting politics ahead of national need in attempts to do something about our huge and steadily growing federal debt. A big and continuing drop in the value of the dollar in international markets is an expression of the distrust that this has inspired around the world. Business debts are growing. The takeover of corporations by raiders, a form of high-stakes gambling, lines the pockets of the gamesters but burdens the acquired corporation with high-risk debt. Our bankers, retailers, and advertising agencies, following in the footsteps of their most aggressive members, have made self-indulgence, minimum saving, and buying on credit our way of life. Respectability is a fancy but often unaffordable auto and a load of credit cards in the billfold. Personal bankruptcies and the humiliation that goes with them are on the rise. Our biggest bankers have enriched themselves, and for the time being at least, their banks, by huge but probably uncollectable loans to Third World countries. We have built a house of cards that any unfriendly wind could bring tumbling down.

Borrowing has its appropriate uses; the problems arise when it is used excessively or inappropriately. I am concerned here with the unwarranted uses and especially those unwarranted uses inspired by greed and self-seeking. Farmers who have gone deep into debt because of rising costs and falling food prices deserve our sympathy, not our condemnation. We certainly can forgive a businessman in financial trouble because of an error in judgment even though the damage may extend to others. Unfortunately, not all expansion of debt has these justifications. Lending can be profitable business. The would-be lender may both encourage borrowing and stretch the limits of responsibility to get the wherewithal to lend. Ethical considerations may be overlooked. An ethical failure is particularly clear in a practice common in both business and banking—the encouragement of consumer debt to increase sales and profits. This encouragement of debt will be our particular concern in this section. Because, however, there are other questionable practices that can increase debt, and because inappropriatc debts interact with one another in ways that can multiply the risk, we shall examine debts in general.

Economists typically shun any ethical judgments when examining debts, whether appropriate or inappropriate. In the context of this volume, however, I believe that an ethical approach is warranted despite the risk of treading unfamiliar ground.

INTERNATIONAL DEBT

I examine first some figures on international debt.

Many countries have borrowed extensively from foreign creditors, but the debts that are now causing widespread concern are mostly those of the so-called developing or Third World countries. The World Bank estimates the total 1986 external debt of these countries at $1,035 billion. The debt has been growing rapidly—it was $650 billion in 1980. Some of the biggest debtors and the figures for their 1986 debt are Brazil, $108 billion; Mexico, $102 billion; Argentina, $53 billion; Venezuela, $34 billion; the Phillipines, $28 billion; Nigeria, $22 billion; Chile, $21 billion; and Peru, $15 billion. Merely the interest on these loans runs into the billions for the biggest debtors. Seventeen percent of this money has been advanced by big banks in the United States. Small United States banks, banks in other countries, and international financial institutions such as the International Monetary Fund have all contributed in varying degrees (12,30,53).

While current concerns with regard to international debt center on some of the developing countries, these are not the only nations with debt problems. Israel and a number of Eastern Block countries have borrowed extensively in the West. Poland, which owed $25 billion in 1983, has had difficulty making payments (28) and the United States in 1985 became, for the first time in many years, a net international debtor.

The international loans of big banks involve them in more ways than one in the arcane world of international banking. Not only is the money loaned internationally; much of it also is raised internationally through multinational banking networks. "Eurocurrency" and "offshore offices" become part of the picture. In some small countries (e.g., Luxumbourg, the Bahamas, Singapore) banks or their branches can operate under rules far more relaxed than those prevailing in the United States (21). Their operations can be perfectly legal, but an officer of New York's huge Citibank described some of the offshore methods used by his bank to bypass national regulations as "rinky dink deals" (21). The obvious purpose of these transactions is to get money at low cost, which can then be loaned abroad at high interest rates while at the same time (and not entirely coincidentally) generating wealth for the bankers. As one article put it, "International lending is banking as its most glamorous and often its most lucrative" (46).

Do the hundreds of billions of dollars of debt of the developing countries in any way threaten the health of the world's economy?

While more and more competent observers have been saying that these debts just can't be repayed, people holding official positions in the financial world, at least in their public statements, have claimed that the problems can be solved. Early in 1987, however, even the big United States banks acknowledged the problem. Citicorp, a holding company for Citibank, led the parade on May 19, setting aside $3 billion against possible losses on its loans to Brazil. Other big banks soon followed suit, with the result that their earnings dropped sharply (9,44).

This step implies, however, that the problems are basically manageable. Is this really the case? There are, I believe, valid reasons for believing, as a number of observers are now saying, that the debts will be essentially a total loss. Their repudiation, in turn, may well create other problems.

The only way the debtor nations can get money for repayment is to change their unfavorable trade balance into a favorable trade balance. It is relevant to note, before we analyze the chances of this happening, that the United States is in the same boat. We also have an unfavorable trade balance, and thanks to Reagan's policies, we also are a major debtor nation. Even if the undeveloped nations do generate a surplus of exportable goods, we can't afford to buy this surplus and they certainly can't afford to increase their buying of ours. Any increased outflow will have to be purchased by Japan, South Korea, Germany, and other countries that are now successful net exporters and don't seem anxious to change this situation.

But what is the chance that the presently undeveloped nations can generate an exportable surplus? An impartial examination shows, I believe, that it is virtually nil. All we need to do is examine what was done with the huge sums loaned them. Some of this, undoubtedly, was well used, but much of it also went into grandiose but unproductive projects. Much, also, has gone into armaments, perhaps as much as one-tenth of all loans. Argentina is developing a fleet of attack submarines at a cost of half a billion dollars (6). Large sums going to Brazil went into the construction of steel mills whose output contributed to the world steel glut and the loss of jobs by many American Steel workers. By all odds the greatest problem has been the flow of borrowed money into the hands of corrupt officials who in turn, because they have no confidence in the economy of their own countries, have sent it abroad, often to the United States, where much of it comes from in the first place. A 1986 poll found 8 out of 10 Mexicans blaming the country's economic troubles on the siphoning off of large sums by corrupt or inept officials. It is reported that a Mexico City newspaper, in April 1986, printed the names of 575 Mexican citizens who had each transferred a million or more dollars to foreign banks. According to a Federal Reserve Board report, at least one-third of the $252 billion increase in the debts of Argentina, Brazil, Chile, Mexico, and Venezuela over the years 1974–1982 fled those countries. A report by the Morgan Guarantee Trust Co. suggests an even greater capital flight (2, 7, 15, 23, 33, 36, 52).

Finally, despite all the inpouring of cash, the lot of the average man has not improved. There was some real growth in per capita income in the major Latin American countries in the 1970s, but this was reversed in the 1980s. Also, inflation ran wild in the 1980s, reaching 851% in Argentina in 1985 (1). The rich have got richer but the poor and most of the middle class are, at the least, no better off (2,8,18,45).

The enrichment of the rich by Third World loans has not been confined to the rich of the Third World. The big lending banks have also come out on top. They have been major recipients of billions of flight capital. It has been reported that they hold $30 billion more in Mexican deposits than Mexico owes them, and they have made money both ways (23).

What will be the outcome of Third World indebtedness? Miguel Wionczek, one of the Mexican participants in an international conference on Third World debt, in a summary statement based on studies by international organizations including the United Nations, the World Bank, and the International Monetary Fund, has written, "Sooner or later, but rather sooner, the world economy will fall apart." This outcome might have been foreseen as a likely possibility in the early days of the debt build-up. This is not the first time that extensive loans have been made to Third World countries. In the case of Latin American borrowing, which goes back into the last century, there have been some periods when debt was repaid, but "the debt experience of the 1920s and 1930s was one of pervasive default" (18, 42).

DEBT WITHIN THE UNITED STATES

We turn now to debt within the United States. The principal sources of data are the Board of Governors of the Federal Reserve System and the Bureau of Economic Analysis of the Department of Commerce, but useful summaries can be found in publications from other government sources. Our major objectives in this section are to examine the growth of the debt burden and to compare the growths of its different constituents. We use a figure (Figure 3.1) for this purpose. For a comparison of the constituents, we divide debt into three major groups: government debt, business and farm debt, and consumer or personal debt. Government debt, in turn, can be subdivided into federal debt and state and local debt. Business debt subdivides into the debts of incorporated and unincorporated businesses. Nonresidential, nonfarm mortgages are also usually treated as a separate category, but we do not include this. Personal debt can be subdivided into home mortgages, installment debt, noninstallment debt, and credit card debt.

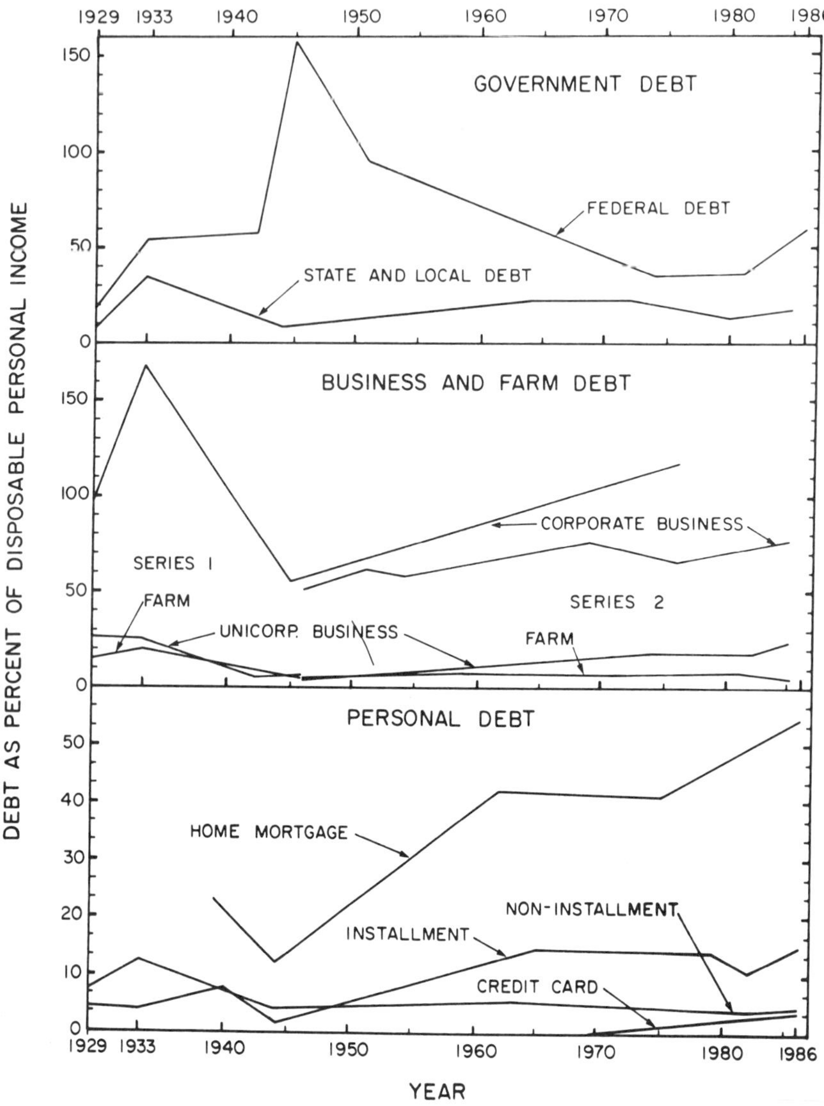

FIGURE 3.1. The magnitude of nine different categories of debt, expressed as a percent of disposable personal income, for the years 1929–86. Only major turning points are indicated. See pages 124 and 125 for sources.

The problem of comparing these components of debt is complianced by a diversity in the published figures. In the case of business and farm debt, I have found no one uninterrupted series that extends from 1929 to 1986. Our figure for these categories of debt is therefore broken into two series, labeled 1 and 2. The discrepancy between the two series is slight for unincorporated business and farm debt, considerable for corporate business debt.

Our objective in using the data as a means of examining the growth of the debt burden is subject to complications from other sources. The growth of debt as measured in dollars is influenced by two factors that are extraneous to the burden debt places on the individual. These factors are inflation and population growth. To correct for these, we shall use as our statistic the debt in each category measured as a percent of Disposable Personal Income or DPI. The total DPI of United States citizens is a statistic published annually by the Department of Commerce. Since DPI increases both with inflation and population growth, using debt as a percent of DPI should cancel out the effect these disturbing factors have on any increase in debt.

Disposable Personal Income can also show a real growth causing a real increase in the money we have to spend, as well as increases owing to inflation and population growth, which do not help the individual. And, in fact, DPI over the years has shown substantial real growth. Whereas, however, we can expect that inflation and population growth will affect DPI and debt in very similar fashions, the nature of the connection between real growth of income and debt is much less obvious. If our income grows, do we use the growth to reduce our debt or do we figure that it will make it safe for us to increase our debt? Our answer to this question is very much a matter of our personal economic philosophies. Our figures should yield clues as to what these philosophies may be.

Figure 3.1 is divided into three sections corresponding to our three major categories of debt. It should be noted that, because personal debts are on the average smaller than either government or business and farm debts, the scale of the vertical axis of the third part of the chart is smaller than the scale in the first two parts. The sources of the data on which Figure 3.1 is based are listed at the end of Chapter 3.

To simplify the figure, only the major turning points in the different debt categories are shown. These turning points generally correspond at least approximately to certain major events of economic significance. The chart starts with 1929, the year of the stock market collapse and the beginning of the Great Depression. Nineteen thirty-three was the bottom of the Depression. The Japanese attack on Pearl Harbor and the U. S. entry into World War II occurred at the end of 1941, and the War ended in 1945. There was an escalation of the Vietnam War in 1964. The Arab oil producing nations declared an oil embargo in 1973, touching off a rise in prices which was repeated in 1979. Ronald Reagan was elected president in 1980 and soon introduced his supply side economic policies, including a major cut in income taxes, with the result that the federal government began to run big deficits. Some of these events had a greater effect on debt than others—World War II was far more significant than the Vietnam War—but they all correspond to some extent with turning points.

A first point to note is that the federal debt generally moves in an opposite direction from the other debt categories. The end of World War II in 1945 marked a major peak in federal debt and 1944 or 1945, marked a major low in other types of debt. Federal debt as a percent of DPI dropped sharply for six years after the war and then more slowly until 1974. This was followed by seven stable years and then, after Reagan took office, to a new period of growth. The deficit for 1986 was $221 billion, or almost twice the total external debt of any Third World country. There may be some reduction in 1987 (13).

Turning now to business and farm debts expressed as a percent of DPI, we note a major drop between the bottom of the Depression in 1933, and a point at or near the end of World War II in 1945. This was followed by a gradual rise. This was uninterrupted in the case of Series 1, but interrupted by two declines in the case of Series 2. In the case of unincorporated business, the long rise continued with only one interruption until 1985, the last year included in the chart. Farm debt, expressed as a percent of DPI, showed a gradual and irregular rise until 1981 and then, as mortgages were foreclosed and loans curtailed, dropped by 27%.

The last section in Figure 3D.1 shows personal or consumer debt. This dropped during World War II and then started a steady climb. There was a tapering off in all categories after 1962–1965, the Vietnam War years, but about this time the banks discovered the credit card and soon consumer borrowing took off again. There was also a big rise in home mortgage debt beginning in 1975.

From 1933 to about 1975, the trends in federal debt, as we have noted, clearly opposed those of nearly all other debt/DPI categories. This opposition, however, does not hold at the early and late ends of our chart. During 1929–1933, the time of the Great Depression, all categories of debt except unincorporated business debt and installment debt, which were essentially stable, were moving up. This shared trend reappears in 1981. There is a hint of the same trend beginning in 1976, but state and local debt during this earlier period are in clear opposition.

Is there a warning in this appearance, in the second year of the Reagan administration, of a trend in our debts which characterized the Great Depression? I shall return to this question in the discussion at the end of this chapter.

The sum of all the categories of debt shown in Figure 3.1 for the last year shown is $7,729.3 billion, or more than 2½ times DPI.

A remarkable aspect of the growth of consumer debt has been its rapidity. During the 26 years from 1945 to 1971, federal debt expressed as a percent of DPI decreased 80%. At the same time, but in a 31 year interval, the corporate figure (Series 1) increased 107%, while the corporate figure (Series 2), in a 23-year interval, increased 51 percent. In only 18 years subsequent to 1944, the home mortgage figure grew 273% and the installment debt figure, in 21 years, grew 733%. Credit card debt grew from a negligible figure in 1966 to $129 billion in 1986. The figure for noninstall-ment debt grew much more slowly after 1944 (57% in 19 years), but this type of debt is much less apt than the other three categories to be used for the purchase of homes or consumer goods.

If the funds from most increases in consumer debt enter the marketplace, they should cause a corresponding increase in business activity. This is confirmed by the figures. The rate at which consumer debt has grown has not been consistent. It shows definite cycles, and the ups and downs correspond quite closely to the ups and downs in the growth of the Gross National Product (31). (These minor fluctuations, a few years in length, are not shown in Figure 3.1.)

Since 1945, consumer borrowing has kept the economy moving. It is thus reasonable to ask: Has consumer debt now reached or closely approached an upper limit, and if so, what does this bode for our economic future?

THE ENCOURAGEMENT OF DEBT

If the growth in consumer debt has outpaced the growth of debt in other categories, there is a very simple reason for it—business has actively and aggressively encouraged this growth (39). Consumer indebtedness is yet another byproduct of the need of mass production industries for a mass market. It keeps the marketplace active.

In considering consumer debt, it is important to keep in mind the distinction between consumer installment debt and consumer mortgages. It may make common sense for a young couple, one or both of them employed with good pay and therefore financially secure, to take out a mortgage to buy a home. This type of borrowing often needs no encouragement. To some degree this applies also to the purchase of an auto, though here there is more room for injudicious extravagance and also for encouragement. It is in the smaller purchases of such items as furniture, refrigerators, clothing, and indeed even such trivia as cigarettes, chewing gum, and lipstick that the consumer has the greatest opportunity to choose between paying cash and buying on credit or not buying at all. It is here that the temptation to use credit buying is greatest.

Over the years, banks and businesses have multiplied and refined the devices for financing installment debt. There are now three major instruments for enabling the consumer to buy on credit: banks, finance companies, and credit cards.

Before 1910, banks typically made personal loans only if secured by collateral. In that year, Arthur J. Morris of a Norfolk, VA bank developed what came to be known as the Morris Plan, which made unsecured loans available to the small borrower. The increased risk and administrative costs were compensated for by correspondingly high interest rates. Within

10 years the plan was in use in banks in 37 states (32). Over the years, banks gradually expanded the availability of credit for consumers. One major area of expansion was the provision of installment loans for the purchase of automobiles, though the concept of such loans was originated by finance companies.

The prototype of the finance company is the General Motors Acceptance Corporation. This was set up in 1919 as a wholly owned subsidiary of General Motors (25). It served several purposes. Most obviously, it provided installment credit for the purchasers of General Motors cars. The interest rates were high, so it also provided a handsome investment for GM's surplus cash. And the use of this route for employing the cash avoided the complications that might have accompanied other and in many ways more socially responsible routes. Cutting prices, for example, could have damaged competition to the point where an antitrust suit might have resulted (34).

The pattern established by General Motors was soon widely copied. Not only did other big companies establish finance companies to spur their sales, but also small firms emerged whose sole function was supplying credit for purchases, offering second mortgages, or even providing unsecured loans in competition with banks. Many of the smaller firms took high risks and charged correspondingly high interest rates. A 1976 survey by the Federal Reserve System reported 3,376 finance companies, ranging from giants with outstanding loans of $100 million or more to small firms with loans under $100,000. Increasingly, the finance companies at this time were moving into the area of loans to businesses, but in 1975 this constituted less than half their total volume (29). Among the large corporations with financing subsidiaries are Ford Motors, General Electric, Sears Roebuck, Alcoa, and Deere and Company.

The third major source of credit is the credit card. The credit card craze was started in 1949–50 by the Diners Club. American Express Company issued its own credit card in 1958 and was soon followed by two large banks, Chase Manhattan and Bank of America. Other banks followed suit, but at first cautiously. In 1966 the rush started (33). In 1967, according to Federal Reserve figures, 390 banks were offering cards and $828 million in credit was outstanding. Ten years later the figures had grown to 2,317 banks and $14,607 million in credit (16).

Issuers of credit cards profited not only in the form of high interest charged on credit repaid in installments, but also from payments required of stores, restaurants, or travel agencies in return for the right to use the cards as a means of extending credit.

The first two mechanisms for financing growth in consumer debt were well in place before the boom that culminated in 1929. They were unimpaired by World War II, and the environment for a period of growth was enhanced in 1945 by the remarkably low level to which the Great Depression and the War had reduced both installment and mortgage debt. Some growth, especially in mortgage debt, doubtless would have occurred without any encouragement, but slow growth was insufficient to meet industry's demand for the sale of its mass-produced goods. Sales had to be encouraged, and the greater the resistance of the consumer to any increase in indebtedness, the greater the encouragement applied.

In October 1964, an article entitled "The Importance of Being in Debt," in *Time* stated, ". . . many businessmen are busy encouraging their customers to plunge more deeply into debt, and producing new and delightful ways in which they can do it. At the Emporium, San Francisco's largest department store, sales clerks have standing orders to encourage each customer who presents cash—which seems to lower one's status in many big stores—to open a charge account. To show how painless borrowing can be, a Los Angeles finance company runs a TV commercial of a man speaking into a pay telephone: 'I wanted to ask, could I borrow . . .' At that point, money pours out of the phone, filling the booth.

"Even staid banks, which used to leave most consumer credit to others, are bombarding customers with new easy-loan plans. In the competition for auto loans, which account for nearly half of all installment debt, banks have pulled ahead of auto-finance companies by offering lower interest rates. Still the competition grows" (50).

Whenever sales lagged, the pressure to buy on installments increased. At a meeting of the American Bankers' Association in the late 1950s, a period of recession, a speaker who headed a major advertising agency stated, "If we are to break the present economic log jam, you installment-credit bankers and we in advertising must do it by working together" (40). And the bankers did advertise, enticing customers considering automobile financing by offering gifts such as a free license plate or a tool kit with each loan of more than $2,500. Down payments were reduced and the duration of loans extended (32). Sylvia Porter, in a 1979 column, reported car loans of up to five years. An advertisement of a Georgia bank announced, "Instant money—cash loans within 20 seconds."

Automobile salesmen were taught to make maximum use of financing and were encouraged to do so by commissions on the order of 3% on a 2-year note. They could usually arrange a deal through either the manufacturer's finance company or a bank (34). I remember well, when I was inquiring about the purchase of a Chevrolet in the early 1960s, how the salesman suddenly cooled off when I said I would pay cash. He didn't like to lose that commission. The auto industry also encouraged debt by promoting a variety of optional gadgetry and the purchase of excessive insurance (34).

Until very recently, the ultimate device for expanding customer debt has been the credit card. It is attractive to the bank or other institution issuing it for two reasons: It makes buying on credit a simple matter of routine, and it returns a higher rate of interest than ordinary loans. Of course the holder has the privilege of paying for a purchase before any interest accrues, but needless to say the issuers do nothing to encourage this. In fact, the whole stress is on the ease with which the credit can be obtained.

Estimates placed the number of card holders in the United States at about 30 million in 1972, 112 million in 1978, and 124 million in 1980. (These estimates are from separate sources and are not necessarily fully comparable.) Often individual holders had more than one card; in fact, in 1979 there were at least four times as many cards as there were individual holders. In 1984, 600 million cards were in use. An interesting reflection of this use of multiple cards was an advertisement of a ladies billfold with room for eight different varieties. One Indianapolis bank in 1967 claimed to have issued some 300,000 all-purpose charge cards. Cards were mailed out in an indiscriminate fashion. Often they were sent to individuals who had not requested them. A method used by Visa, one of the two major cards, was to get lists of merchants' charge customers and offer them credit cards with the names of those merchants on the label. Mayer, in 1974, quoted Eric Younger, marketing director for the Interbank Card Association (which later became MasterCard) as saying that in Chicago "babies were getting cards when they were born; cards were being sent to dead people. Dogs got cards. People would follow the postman to the house to steal cards from the box—you could tell they were cards because they practically printed it on the envelope" (33).

As the competition between the different cards increased, new forms of bait were devised to lure the consumer. In the early 1980s, several cards appeared that made it easy for the user to substantially extend his or her line of credit. American Express, Visa, and MasterCard all put out "gold cards" offering higher credit and convenience services. CitiCorp and an electronics retailer, Radio Shack, teamed up to produce a CitiLine card that enables the holder to buy thousands of dollars worth of equipment at Radio Shack with no cash payment. Ralph Nader has denounced the card. "This," he says, "is corporate gouging. The rate is excessively high, and there are going to be people hanging on three and four years on repayments."

When I started work on this section, I assumed that the credit card was the ultimate devise to induce consumers to buy on credit. I was wrong; I underestimated the ingenuity of the banking moguls. In 1986 they began pushing a new device for making installment purchases, the home equity loan. This uses a second mortgage on a home as security for a line of credit on which the home owners can draw to buy goods and services over and above what they might finance by the older methods. Real annual interest rates are estimated at 12.43%. The go-go bankers who are offering these loans promote them through big ads with lines such as:

> ". . . the instant phone loan . . . the fastest, most convenient way to get a loan for almost any good purpose—a new car, a boat, a recreation vehicle, a vacation, you name it!"

> "I can hear the neighbors. 'You're putting in a pool? How marvelous!' . . .
> Oh, they'll kid me, but they'll be jealous."
> "We'll help you live your dreams."

The unincumbered equity in residential real estate has been estimated at about $2 trillion. A bank offering Home Equity Loans found that 45% of its customers planned to use the new line of credit to take on new debt rather than to consolidate their existing debt. Economists, it is said, are revising their view that consumers just can't go beyond their 1986 debt limit, the servicing of which absorbs 18% of their disposable income. Bankers have found a way to put more dynamite in the debt bomb (11, 43).

How Our Growing Burden of Debts Has Been Financed

The money to finance our growing debts has come from various sources and sometimes by complex routes. I shall try to give some conception of the way in which new lending sources have been developed and old ones expanded without, however, going into the economic complexities. I shall mention some causal relationships, but the reader should recognize that this is an area where even economists often disagree. I think what I say is valid, but it probably is an oversimplification.

Our financial institutions are well regarded both here and abroad. Bankers are among our most respected citizens, and officers of most local banks, I am sure, have the welfare of their communities at heart. The Federal Reserve Board, moreover, has proved to be a generally effective regulatory agency. Nevertheless, the urge to make a profit combined, probably, with pressure resulting from devices adopted by the more aggressive banks have led to some possibly dangerous practices. And the Federal Reserve has at times yielded to pressures from both banks and the government (17, 38). So far the damage has been slight, but the future may hold greater risks.

We have already noted how the banks in this country that are heavily involved in international lending were able, through their offshore offices, to raise money in ways not subject to the regulations under which most of our banks operate. Our concern now is our own debts. We start with the debts of the Federal Government which, as we have seen, have been growing rapidly since President Reagan took office.

Our recent huge federal deficits have been largely financed by an inflow of money from abroad. This inflow has been due to a variety of causes, but a chain of events going back some years has been largely responsible. Excessive wage increases in the 1970s, rising food prices, and the oil shortages of 1973 and 1979 helped bring on a period of inflation. When prices are rising, lenders expect higher interest rates. Our growing debt and the resulting demand for money put further pressure on interest rates and the Federal Reserve Board encouraged the rise to cool the economy and thereby bring inflation under control. Since interest rates were higher in this country than abroad, the United States became an attractive place for foreign investment. This led to a demand for dollars and a rise in the price of dollars on the international market as compared with other currencies. This in turn encouraged purchases by Americans abroad and discouraged purchases by foreigners in this country.

Our balance of trade became very unfavorable. In 1986 our trade deficit reached $170 billion. The dollars for these heavy foreign purchases in turn had to be found somewhere. They, like the money for financing our deficits, came largely from a relative increase in foreign investments in the United States. During the years 1981–1985, foreigners invested an estimated $100 billion a year in this country. And during the same period, American firms and bankers reduced dollar holdings abroad by about the same amount. This inflow of foreign funds has been the primary source of financing of our growing national debt (4, 14, 20, 26, 32, 35, 51).

A final result of this chain of events is that, sometime during 1985—the exact timing is uncertain—the United States became a debtor nation. The last time our debits exceeded our credits was in 1914 (5, 20).

The growth in consumer debt has been financed primarily in two ways: the formation of finance corporations and an expansion of credit by banks, building and loan associations, and other financial intermediaries. Banks have also been a major source of financing of new business and farm debt, but big corporations have often resorted to other, less conventional methods of borrowing.

I have already mentioned General Motors Acceptance Corporation and the many other finance companies that have been set up by big corporations to provide their customers with credit. These finance companies have to start out with substantial funds and a considerable proportion of this money (I have no precise figures) has to come from the profits of the corporations that create them. This use of profits is a wonderful way to bring in more profits, but we have to ask if it is the most socially responsible way to use them. If a corporation has a $100 million surplus, it could pay it out as dividends or wage increases—setting up a finance subsidiary is by no means the only option. If the $100 million were paid out, consumers would have that additional money to spend and would have less need to borrow. Dividends going to the well-to-do might not be used to reduce borrowing needs, but certainly wage increases would substantially reduce these needs. Thus finance corporations create a need for credit as well as satisfying this need.

Banks have increased their lending capacities by a gradual change from more to less conservative lending policies. Besides using their funds as a source of loans, banks have traditionally invested some of them in government and other securities. Loans bring the greater profits, and banks therefore are under some pressure to use their funds in this way. And indeed, over the years banks have made a substantial switch from bonds to loans. Thus between 1948 and 1976, while investments in securities increased 3.4 times, loans increased 12.6 times (17). Or, to cite a slightly different comparison, between 1954 and 1984, while investments in government securities increased 2.9 times, loans plus investments in non-government securities increased 20 times, reaching in 1984 a total of $1.6 trillion (10).

As Figure 3.1 shows, the climb in business, farm, and personal debt began in 1944 or 1945. The figures cited above show that the shift of bank funds from government securities to loans began not later than 1948 and has continued into the 1980s. The implication is clear that the shift in bank investments was a source of the steady growth of debt.

Three other policy changes of a somewhat more technical nature have contributed to the growth in debt. Banks are required to keep a certain percent of their deposits as reserves. Over the years, the Federal Reserve Board, which has the authority to set reserve requirements, has gradually allowed this required figure to decline. This trend began in about 1945, and has been a major source of expanded lending capacity. The banks themselves normally keep some funds in reserve in addition to the required reserves to give themselves a safe working margin. It is possible to cut this margin and the banks, again with some help from the Federal Reserve, have elected to do this. A third factor contributing to the increase in lending was a change made by the Federal Reserve in 1960 in the nature of the reserve requirements (17).

An economic change not under control of the Federal Reserve or the banks that has helped in the loan expansion has been a decrease in the use of currency. In 1945, currency made up 20% of so-called Money Supply II; by 1976 this had dropped to 13%. Loans and deposits rose correspondingly. It is interesting to note that a similar trend preceded the 1929 economic collapse (17).

These factors all help explain the increase in lending by banks over the past 40 years and the corresponding increase in business, farm, and personal debt.

We have noted the general high regard in which American banks are held. Nevertheless, problems have developed. The concern expressed by Leonard Silk, economist and columnist for *The New York Times,* is not untypical. "Banks," he writes, "have rushed to swell their loans and deposits, demonstrating great ingenuity in evading the efforts of the monetary authorities to curb too rapid an expansion" (47). And Hyman Minsky, another economist, noting "the major shift . . . from bank portfolios dominated by government debt to portfolios dominated by business debt," comments: "The growth of business indebtedness to banks relative to gross business profits, and sharp increases in interest rates in the context of highly indebted businesses, tend to increase the instability of the bank financing process" (37).

The banks have had other problems. The falling prices for oil and farm products have led to defaults on many loans to farmers and oil-related industries. Savings and loan institutions are also in trouble. *The Wall Street Journal* has commented on the high proportions of assets of some go-go Sun Belt thrifts committed to dubious construction loans (27). The result of these problems has been a growing number of failures of banks and savings and loans.

In assessing the risks imposed by our monstrous debts, we need to review them as a whole, not separately. There are potential problems in our international debts, our federal debts, and our business, farm, and personal debts, but it is the accumulated total and the potential interactions between the separate parts that makes the whole problem so serious.

SOCIAL AND ECONOMIC PROBLEMS CREATED BY CONSUMER DEBT

In his *The Theory of Moral Sentiments,* Adam Smith asks, "How many people ruin themselves by laying out money on trinkets of frivolous utility?" Clearly, he was aware of the fallibility of human preferences and hence, presumably, their susceptibility to exploitation. We also find him writing: "What is agreeable to our moral faculties, is fit, and right, and proper to be done; the contrary wrong, unfit, and improper" (48). I cannot say to what extent the encouragement of debt in the American mind is improper by this standard, but I think the evidence shows that it has a variety of hurtful consequences and therefore is properly judged as unethical.

We have seen that the encouragement of consumer borrowing has caused consumer debt since 1945 to increase at a faster rate than disposable personal income. It has also caused many consumers to borrow beyond their reasonable needs and sometimes beyond the point where repayment is still possible. This totally unwarrantable growth of consumer debt has had a variety of social and economic consequences, most of them undesirable. A few figures will help put the subject in perspective.

The total of the personal debt in 1986 included in Figure 3.1 was $2,399 billion. This amounts to $9,888 per capita, or 80% of per capita disposable personal income. Merely the interest on a sum this large adds up to big money. The interest rate on mortgages, one of the cheapest forms of personal borrowing, was 11.3% in August 1985. The interest rate on credit card debt, according to a Federal Reserve Board survey, was almost 19%. With interest rates in this range, the average family of four may be paying well over $3,000 per year in interest (19).

Let us now consider some of the consequences of this debt. The extension of credit generates administrative and sometimes various indirect costs that are passed on to the consumer, increasing the price of the goods and services that he or she requires. Generally, the higher the return on credit for the banker, the greater the costs to be passed on. Credit card loans are particularly costly to administer and correspondingly costly to the customer. Theft and fraudulent use of credit cards, which are common, generate their own costs. It was estimated in 1983 that banks would lose $200 million in credit card fraud alone. Stores and their suppliers can also lose through purchases that are never paid for. According to a study by the Federal Reserve Board, credit card use in 1983 added about $6 billion to the total cost of goods and services. The rise in these costs, along with the rise in debt, is a significant element in inflation.

Money spent on premature or unessential purchases because of the easy availability of loans might otherwise be saved. Savings provides funds for industrial growth. The United States has one of the lowest savings rates of any of the industrial nations. The Japanese save more than three times as much as we do.

A third effect of the growth of personal debt is a magnification of inequalities of wealth. Debts cost money and decrease wealth, savings earn money and increase wealth. There can be partial exceptions; the well-to-do can use the purchase of mortgaged property to get tax breaks and as hedges against inflation. It is most often the poor who incur debt they should not and cannot afford and who pay the penalty.

A *1977 Consumer Credit Survey* prepared by the Board of Governors of the Federal Reserve System has some revealing statistics. The survey showed that moderately well-to-do families with incomes in the $20,000–$25,000 range were most apt to have debts, but their debts were generally of manageable proportions. Fewer than 5% of them had 20 or more percent of their income obligated for debt payments, a level of obligation regarded as dangerous. A much lower proportion of low income families had debts, but a much larger proportion of these debts were in the danger range. Only about 20% of the poorest families, those with incomes under $3,000, had debts, but one-fourth of them had debts amounting to 40% of their income (22). The upper end of the income scale was not covered in detail, but a study by a Georgia State University professor indicates that a considerable proportion of the wealthy are heavy borrowers because they tend to have their funds tied up in search of financial return.

These data show where debts hurt most, but they give no clues as to what part of personal debt can be regarded as unnecessary and amendable to conversion to savings. From what we have seen earlier in this section concerning debt encouragement, however, it is safe to infer that a good deal of it is. In the absence of mass persuasion to spend, many low to moderate income families could have deferred spending, saved more, and in the long run been better off.

Another consequence of injudicious debt is its effect on families that incur it. It has been a cause of thousands of unhappy, broken homes. There are many intangibles here, but there are clear indications of the problem and its magnitude. These can be seen in the many families resorting to several hundred Consumer Credit Counseling Services scattered across the country; in the considerable proportion of families whose debts are such that they start to juggle bills, paying off one this month, delaying another until next month; in the calls of debt collectors, the garnishment of wages, and the repossession of cars or their simple removal, without notice, from a parking lot; and in the ultimate humiliation, the foreclosure of a mortgage or the declaration of personal bankruptcy (41).

Statistics give some hint of the magnitude and growth of the problem. According to the *Consumer Credit Survey* already cited, about one family in 14 in the United States in 1977 had debts in the danger range where the juggling of bills and calls of the debt collector are likely. Other figures indicate the growth of the problem. Over the period of 1946–80, when personal debts were multiplying, personal bankruptcies increased from 10,000 –314,905. Part of this increase, but only a part, was due to a change in the bankruptcy law, making the declaration of bankruptcy a more favorable option for the debtor (22,49).

Another effect of easy credit has been described by Gamblers Anonymous, a group concerned with helping compulsive bettors. According to this organization, easy credit has caused a dramatic increase in the proportion of addicted women. In the words of one woman the group had helped, "They were very good to me. They extended me a lot of credit—too much. That's what really ruined me financially."

We cannot reduce to statistics the misery produced by the encouragement of debt, but I think few people would question its significance.

This next effect of debt encouragement is unproven, but certainly highly probable. If people are led to regard being in debt as a normal way of life, they are likely to be unconcerned about government debt. Polls have shown that if families can pay their installments and other bills on a regular monthly basis, they feel there is nothing to worry about (3). If people then carry this attitude to the national scene, it can dispose them to be unconcerned about growth in federal debt. And the growth of federal debt, especially since President Reagan took office, is the greatest single threat to our economic future.

A final effect of debt encouragement is its predisposition to a boom that has to be followed by a bust. Just as an individual becomes dependent on a drug, so the economy becomes dependent on the continual injection of buying power that results from growing consumer debt. If either type of stimulus is cut off, withdrawal symptoms ensue. As early as 1958, John Galbraith wrote in his *The Affluent Society,* "As we expand debt in the process of want creation, we come necessarily to depend on this expansion. An interruption in the increase in debt means an actual reduction in the demand for goods" (24). In 1960, the Federal Reserve Board expressed concern that too many consumers were going too deeply into debt (40). Up to the time that I write this (August 1987); however, while there have been recessions, there has been no real crisis, and many economists and bankers are skeptical of any dire predictions. Yet we have to ask, is it reasonable to assume that personal debt can grow indefinitely faster than income? It would be pointless to try to guess here the form that any reaction will take, but we would be fooling ourselves if we assumed that it will not be serious.

The growth of debt at a faster rate than income raises an interesting question: Why was this growth seemingly necessary for the economy to grow? The phenomenon is old enough that economists have been asking this question for more than one hundred years. An answer, as Galbraith notes, was offered early in the nineteenth century by Sisondi, a Swiss historian and economist. It was his thesis that the working public simply were not paid enough to purchase all the goods and services they produced. Either an artificial stimulus had to be provided or some goods and services went unsold. Most recent economists, according to Galbraith, favor this same hypothesis (24).

This leaves us with another question: Why is there an imbalance between income and the cost of available goods and services? Some possible clues will be offered in the next section.

E. Unneeded Wealth Misused

INTRODUCTION

My income this year will be approximately $460,000. I have been able to keep my taxes down to less than $100,000. My family and I live simply; our expenditures will run about $60,000. That leaves me with $300,000 to invest or spend.

I have a wide range of options to choose from. In the past, I have invested a good deal in tax-exempt bonds, but in recent years there have seemed to be better opportunities. When prices were rising sharply, I purchased some real estate as a hedge against inflation. My wife also got into the act; she had fun stocking up with antiques and art. The real estate also served as a tax shelter and helped lower my taxes. Until the tax reform act of 1986, I, like many of my friends, put a good deal of money into shelters. When the new tax law went into effect, I shifted a lot of money into stocks. I figured other people also would be doing this, so I was prepared. With the help of my broker, I had picked out some stocks whose market price seemed reasonable in terms of their actual value. I made my purchases right at the beginning of 1987, beating the sharp rise in prices that followed. I bought for the long haul, but sensed an air of public uncertainty and did a little selling shortly before the October bust. Since by 1988, prices had recovered to the January 1987 levels, and since my stocks performed a little better than the average, I haven't done badly.

Public uncertainty is still evident, however, and I plan to have promising real estate lined up in case I decide to sell.

One purchase my wife and I have talked of is a yacht suitable for extended cruises. We already have a lovely sailboat, but some of our friends have yachts and seem to have a lot of fun with them. Such a purchase would of course add substantially to our annual expenses.

A very different but appealing use is investment in new, high technology companies. I know, for example, of a young man who has interesting ideas for the use of genetic engineering to improve crop plants. Of course I might get no return for 5 or 10 years, and the risks are high.

Finally, there is the area of philanthropy. A small local college of which I am a trustee could very well use $300,000 in uncommitted funds. I expect to give something to them anyway, but I am not sure that this is the time for a major donation.

Very few of us will ever have to make the sort of decisions that I have ascribed—I admit with very imperfect knowledge—to our imaginary man of means. But there are a substantial number of citizens with his or greater wealth, and many more with lesser incomes but still sizable sums to spend, invest, or give away. Included in this group are some persons whose wealth, in whole or in part, is derived illegally. There are also corporations with sums, vast in their totality, in excess of their immediate needs and hence open to wide discretion in their use. The decisions these people and corporations make can have a profound affect on the economy and the general welfare.

I have titled this section *Unneeded Wealth Misused*. The wealth we are concerned with is income rather than principal. But what is *unneeded income?* A precise definition is neither possible nor necessary; it is sufficient to say that it is income over and above normal living expenses. Since the well-to-do spend much more on themselves than do people of moderate means, normal living expenses are themselves a variable. Most Americans, I suspect, accept these differences in personal consumption, at least up to a point, quite willingly. Leadership deserves reward. Perhaps we can say that personal expenses are justified insofar as they are approximately representative of the importance of our jobs and are sufficiently tempered as to avoid any suggestion of excessive self-indulgence or any attempt to say "see how successful I am." Personal expenses may also appropriately include a reasonable level of savings for future personal and family needs. Any income above this level I classify as unneeded.

Economists use the expression *discretionary income* in much the same sense that I have used unneeded income, and I shall occasionally use this term as a synonym. Economists, however, do not let any ethical judgment as to what is acceptable spending on self and family enter into their definition, so I am somewhat altering this use of discretionary. Economists, also, do not extend the term to corporate income, but I shall do that. Here again, a sharp definition is difficult, but I shall regard any income beyond that necessary for a corporation to hold its own against both foreign and American competition as unneeded or discretionary.

Economists, besides accepting uncritically any personal spending, seldom attempt any kind of ethical judgment concerning discretionary spending. What people do with their money, so long as they stay within the law, is viewed as strictly their own business. The sums involved, however, are enormous, and I believe that it can be shown—and here will be shown—that there are inappropriate but commonly practiced uses that are both socially and economically damaging. Ethical judgments, therefore, are very much in order.

Some Data Concerning Income in the United States

There are numerous sources of information concerning wealth and income in the United States. Two sources are broad in their coverage. The Census Bureau conducts an annual field survey in which one of the subjects covered is income. A second source is the Internal Revenue Service, which issues an annual bulletin showing the number of tax returns reporting an adjusted gross income in each of a range of income groups. All other sources are confined to a limited spectrum of incomes, usually at the upper end of the income scale.

All these sources are subject to significant deficiencies. Important categories of income can be legally omitted from income tax returns. Social Security payments and income from tax-exempt bonds are examples. Philip Stern tells of a widow with an income of $5 million, derived from tax-exempt bonds left her by her husband, who was required to file no return at all (60). There are also billions in income that are illegally unreported. Some of this comes from interest payments and dividends; far larger sums come from the *underground economy,* a catchall name for sources of income that never get into the federal statistics. Some of this underground income comes from fully legal sources such as odd jobs or waiting table. Much of it goes to people of limited means. Substantial sums, however, go to lawyers and other prosperous individuals who hide receipts by bartering their services. In some cases legal income is hidden by illegal means such as the use by a storekeeper of two sets of books. Another very major part of the underground economy is income from crime. Much of this is white-collar crime. Experts estimate that white-collar criminals take in $200 billion annually, far in excess of the 11-billion-dollar cost of violent crimes (26).

By its very nature, the magnitude of the underground economy is difficult to gauge and estimates vary widely. The criminal component, especially, is carefully obscured. Guttmann, one student of the subject, placed the total in 1976 at $176 billion and, in a later report, forecast figures of $450–$470 billion for 1982 (27). These figures equal approximately 10%

TABLE 3.1. Some figures on the distribution of income in the United States in 1983; the table shows the percent of incomes falling within each of four income classes.

	Under $15,000	$15,000– $25,000	$25,000– $50,000	Over $50,000
IRS[a]	35.8%	34.8%	23.8%	5.6%
Census[b]	34.7%	22.2%	31.7%	11.5%

[a] These figures from the Internal Revenue Service are derived from individual income tax returns. The total number of returns filed in 1983 was 96,293,634 (70).

[b] These figures taken from the Census Bureau are derived from the annual field survey and are based on individual households. The number of households was estimated to be 85,290,000 (59).

of the gross national product for 1976 and 15% for 1982. Estimates as to the amount of money lost to the U.S. treasury from all illegally unreported income vary substantially, but ran as high as $100 billion for 1982.

If the treasury is losing sums of this sort in taxes because of unreported income, it follows necessarily that published data on income underrepresent the true figures. It is likely also that the underrepresentation is greatest for the higher incomes. In the case of Internal Revenue Service figures, this underrepresentation is in addition to that arising from legally unreported income.

Table 3.1 gives some data on income distribution for 1983. Two sets of figures are given, one derived from Internal Revenue Service (IRS) data taken from tax returns, the other from the Census Bureau. The two sets are not strictly comparable (see footnote for figures), but are sufficiently alike in design for our purposes. They manifest one notable difference. In the middle income bracket, $15,000–$25,000, the IRS data show 57% more entries than the Census data do. In the two top brackets the situation is reversed; the Census data show 33% more entries in the $25,000–$50,000 group and 105% more entries in the over $50,000 group. The IRS figures thus substantially underrepresent the concentration of wealth at the top. This is exactly what we would expect from the deficiencies in IRS data that we have already noted (59).

The year 1973, when the Arabs clamped on their oil embargo, was a turning point in income growth. In the 1960s and early 70s, real family income grew rapidly. Thereafter, while there were marked fluctuations, average family income, measured in 1984 dollars, showed a slight decline. The decline, however, was not evenly distributed. The two highest income groups showed gains, the three lowest income groups showed losses that increased progressively as income went down (Table 3.2).

The change for the worse in the economic status of the lower income groups is further emphasized if we take their debts into consideration. Not counting mortgages, the average household owed, in 1985, $6,166 to banks, retailers and other creditors, compared with $4,695 in 1975, and $4,070 in 1965. The figures are in constant dollars (11).

TABLE 3.2. Percent change in income in constant (1984) dollars, 1973 to 1984, or families with children; families divided into 5 equal income groups.[a]

Lowest one-fifth	Next-to-Lowest one-fifth	Middle one-fifth	Next-to-Highest one-fifth	Highest one-fifth
−26%	−12.4%	−5.0%	+2.8%	+7.9%

[a] Data from study by Danziger and Gottschalk (43).

The United States, once a model middle class society, is losing that status as wealth inequality grows. What has caused this shift?

A major factor has been the growing inability of U.S. industry to compete in the world market. Millions of semiskilled workers are feeling the effects of layoffs, plant closings, and wage concessions. Newly created jobs tend to pay less than the jobs they replace (53). The growing cost of Social Security has been particularly hard on the middle class, and the Reagan tax cuts were so distributed as to be of little help. Many middle income families have further complicated this situation by striving for a life style that they are increasingly less able to afford (11).

Another important factor in our growing wealth inequality has been the greed of our business leadership. Top executives of major U.S. corporations get huge salaries and a variety of perquisites while at the same time the discrepancy between their incomes and that of workers and middle level executives has been increasing. A survey by *U.S. News and World Report* (71) of 202 of the largest companies revealed a median pay at the top of $780,769 at the end of 1984, an increase of 15.6% from 1983. Fifty-five of the executives were receiving more than $1,000,000. Perquisites ranging from club memberships to huge life insurance policies added substantially, and additional income undoubtedly came from ownership of company stock. General Motors in 1987 allocated $169 million for executive bonuses although profits were down and GM workers, unlike those at Ford, where profits were up, got no bonuses. Japanese and most European executives get far less. Pay at the top contrasts even more sharply with that of middle level executives, whose earnings generally were of the order of $20,000–$50,000 a year (1978 figures) and were increasing only slowly (61, 65, 71, 73). Undoubtedly a job at the top requires a demonstrated skill in maximizing profits and demands long hours and often work under stress, but it is also true that it gives considerable freedom in setting ones own salary and that this freedom is widely exploited.

Celebrities in professional sports and TV are another group with both big salaries and big increases, and successful doctors, lawyers, and lobbyists are well up there too. And so too, we may be sure, are top figures in crime.

At the other extreme are the millions of U.S. citizens living below the poverty line. The Census Bureau defined this line, for 1983, as an income of $10,178 for a family of four, and placed the number of poor at 35.3 million. The number has been growing. The Bureau qualified these figures by noting that most poor families receive financial aid in the form of food stamps, school lunches, etc. (63). On the other hand, as several senators noted in debating a bill known as the Access to Health Care Act of 1986, an estimated 37 million Americans, most of them in the lower income brackets, were without health insurance (17).

Besides the studies of income, there have been a number of studies of wealth, most of them prepared by unofficial sources. Perhaps the best known of these is an account, prepared annually by *Forbes* magazine, of the 400 people presumed to be the richest Americans. In its 1985 report, Sam Walton, founder of Wal-Mart Stores, was placed at the top of the list with an estimated wealth of $2.8 billion. A number of individuals at the bottom of the list had fortunes in the $150 million range. The total wealth of the 400 was estimated at $134 billion (24).

The Census Bureau in 1986 published for the first time an estimate of wealth in America. The 12% of Americans in the top bracket were found to own 38% of all household wealth, and to control 63% of all stocks and 53% of all bonds and money market funds. At the other extreme, 11% of all white families and 30.5% of all black families reported no assets or a negative net worth.

The concentration of wealth at the top, as well as the concentration of income, appears to be increasing. This was the conclusion of a study prepared by the staff of the Democratic members of the Joint Economic Committee of Congress and released 24 July, 1987, by the Committee Chairman, Rep. David Obey, D-Wisconsin. The study was based on surveys by the Federal Reserve System. The major finding was that the top 420,000 American families, comprising 0.5% of all families, saw their wealth grow from slightly more than 25% of the total in 1963 to 35% in 1983, an increase of 38% in 20 years. It should be noted that the Census Bureau figures do not indicate an equal degree of concentration at the top. Data on wealth are, I suspect, both inherently less reliable and more likely to be underreported than data on income. This could account for the considerable discrepancy in the two estimates.

THE EFFECT OF TAXES

Federal, state, and local taxes took 35% of national income in 1982. Federal taxes alone took 24%. Taxes thus substantially reduce the amount of income available for discretionary uses. The tax burden, moreover, falls unequally on different income classes, and thereby affects the distribution and amount of discretionary income. An examination of the effect of taxes on income is therefore relevant to an understanding of discretionary income and the uses to which it is put.

The Sixteenth Amendment, adopted in 1913, gave Congress the power to tax incomes. Its intent was to make possible a progressive tax structure. A federal income tax law was promptly enacted and subsequently, 40 states have established income taxes. Inheritance tax laws also have been passed, both nationally and by many states, and are intended to be progressive.

While these statutes, in their sum, have produced a shift of the tax burden to the upper end of the income scale, hundred of revisions in the original law plus a variety of other taxes at all administrative levels have largely thwarted the original intent of the income and inheritance tax laws. Sales taxes, for example, are strongly regressive. Studies of the distribution of the tax burden, state and local taxes included, were prepared by Pechman and his collaborators in 1972 and again in 1984. According to the more recent study, the 10% of families with the lowest incomes—the very poor—paid 42% of their income in taxes. For the higher tax brackets, the percent of income paid ranged from 23% up to 29% The total tax burden was thus slightly progressive at the top but regressive overall. It was assumed in this study that the total burden of the corporate income tax was borne by the corporations and their stockholders and not passed on to the customers. Pechman and Okner in the earlier study note that information on this point is hard to come by. If some of the tax was in fact passed on to consumers, which may well have been the case, this would shift more of the burden onto the lower income groups (38,39).

While taxes thus do nothing to reduce the tremendous income gap, the poor are helped by a variety of public assistance programs designed specifically for this purpose. The net effect of taxes plus these programs is progressive (38).

The many revisions in the income tax law have caused fluctuations in the tax loopholes available to the rich, but they have never closed them. President Reagan's economic and tax policies have worsened the inequality in income distribution (13, 21, 35, 36). The Tax Reform Act of 1986 presumably will close some of the worst loopholes. We may no longer find one in every 1,000 people with incomes of over $200,000 paying no tax as a 1987 Treasury Department study found to be the case (32,54), but overall fairness will be decreased. This is the conclusion of a study by the nonpartisan Congressional Budget Office requested by Senator George Mitchell of Maine and released in November 1987. The study concludes that the poorest one-tenth of Americans will pay 20% more of their earnings in federal taxes in 1988 than they did in 1977 and the richest will pay almost 20% less.

Inheritance taxes are no more effective than income taxes in reducing wealth at the top. Loopholes in our inheritance tax laws have been noted by competent authorities, and the potential for tax avoidance that these loopholes provide is attested to by the number of families in this country with huge inherited fortunes. Of the 400 families on the *Forbes* list of the wealthiest Americans, 181 had wealth that was mostly or entirely inherited. The *Forbes* article also lists 78 additional families with wealth of $250,000,000 or more but which, because the inherited fortune went to several family members, did not place in the top 400 (9,23,24).

What do these figures tell us about the amount of discretionary income in this country? Income inequality probably increases the amount of income available for discretionary uses. If income were shifted from the upper to the lower brackets, most of it would probably be devoted to consumption without a corresponding reduction in consumption at the top. As it is, the amount of income of the well-to-do that is not used for consumption and which thus, by definition, is discretionary, is certainly enormous. I will not attempt to place a figure on it, but it must run into the billions. Big corporations also have substantial unneeded income as I have defined this expression. How this income is used can have a profound affect on the national economy and the national welfare. We now examine how responsible, in actual fact, these uses are.

LEGITIMATE USES OF UNNEEDED INCOME

Edwin Land, the founder of Polaroid Corporation and the developer of the Land instamatic camera, is a living example of wealth well used. His wealth was both made and spent creatively. *Forbes,* in its 1984 listing of great fortunes, includes a section on nine Americans who might have made the list of the 400 wealthiest but did not because of the munificence of their donations. Land, one of the nine listed, at the time of writing held 550 patents. Not only was he creative, he was also generous. His gifts include $100 million or more to a foundation, and $50 million to universities, hospitals, and other institutions. He is still a wealthy man, but considers wealth incidental to his scientific pursuits (8).

Joseph Kennedy II, son of Robert Kennedy and nephew of President John Kennedy, provides another example of the responsible use of wealth. Kennedy founded and, for a number of years ran, the $400 million Citizens Energy Corporation. The purpose of the corporation was to bring cheap home heating oil to the poor of New England. Kennedy traveled the world, shopping for crude oil at the best possible price. He would then locate a refinery to process the oil, and sell at a profit the gasoline and all other products except the heating oil. The profit thus generated financed the operation of the Corporation. The heating oil was shipped to Massachusetts and distributed with the aid of that state's fuel assistance project. A 1984 shipment of oil was sold to needy consumers at 55 cents a gallon compared to the going rate of $1.10–$1.15. Kennedy himself drew a salary of $60,000 (44).

The careers of Edwin Land and Joseph Kennedy II illustrate two of the three legitimate uses of unneeded income—philanthropy and creative investment. The third, I suggest, is paying taxes ungrudgingly and in full. Joe Kennedy's Citizens Energy Corporation was a rather special case of creative investment. It served a social need but also provided good publicity for a later entry into politics. I would include as appropriate, investment in any company providing a useful product or service and at the very least staying within the law in such matters as toxic waste disposal. The appropriate uses of unneeded corporate income are similar but not identical and will be considered separately.

PHILANTHROPY

In 1985, individual U.S. citizens gave approximately $66 billion to organizations dedicated to good works. These organizations received another $9.6 billion from foundations and bequests, for a total of $75.6 billion originating from individual charity. Corporations, which have increasingly participated in philanthropy, gave another $4.4 billion, bringing the total to $80 billion. Individuals also contributed through volunteer services. According to a study by Independent Sector, Inc., 89 million Americans participated, giving $110 billion worth of their time. Universities, research laboratories, and other similar organizations received not only private donations, but also aid from government sources such as the National Science Foundation (41).

The $75.6 billion in donations derived in 1985 from individual sources amounted to 2.67% of disposable personal income and to $513 per person 25 or older.

How are donations distributed among different income groups? An article in the December 1986 *Town and Country* provides some information on this point (45). According to a study which they cite, individuals who live on less than $10,000 a year give away twice as much, proportionate to their income, as those making $50,000–$100,000. Most of this giving is to religious organizations. The relative generosity of the poor is not a new phenomena. An analysis of 1970 income tax returns with itemized deductions showed that gifts, as a percent of disposable income, peaked at the high and low ends of the income distribution. Individuals in the $30,000–$50,000 income group gave less than half as much as those in the lowest bracket. Only people with incomes of over $1 million gave more than the very poor (66). *Town and Country* lists 186 people who have given $5 million or more. The rich give not only directly but also through bequests and the creation of foundations which, in their sum, have done an enormous amount of good.

The evidence suggests that donations, perhaps particularly by the wealthy, are motivated at least in part by the tax breaks they generate. Both our inheritance and our income tax laws have been written to encourage philanthropy. These laws have unquestionably encouraged the creation of foundations. A major role of income taxes in lifetime giving is indicated by a 35% drop in donations by the 16,300 top income receiving Americans subsequent to 1981. It was in that year that Reagan's tax cuts reduced the amount of income that could be figured into deductions (22). An account of John D. Rockefeller III's philanthropy provides a rather extreme example of generosity combined with tax avoidance. His total personal philanthropy was estimated in 1981 at $93 million. At the same time, testimony that he gave in 1969 before a Congressional Committee revealed that between 1961 and the date of the hearing he paid no income taxes (46). Insofar as donations are offset by tax reductions, the net effect is to pass the cost of the donations on to taxpayers in general. This is not necessarily a bad thing, but it does emphasize the very calculating way in which the wealthy use their discretionary income.

Turning to investment, the purchase of stocks and bonds in progressive and socially responsible companies is certainly an altogether suitable use of unneeded income, especially for persons of moderate means. Providing venture capital for small, start-up, high technology companies is a particularly creative area for investment, but because of the high risk and long delay before there are any returns, it is most appropriate for the well-to-do. Venture investing has had its ups and downs, but generally seems to be an expanding field. Aided by several hundred firms concerned with venture capital investing, by venture capital clubs, and by at least one publication specializing in this area, *Venture Capital Journal,* investors raised $4.5 billion in venture capital in 1986 alone (25, 28, 48, 62, 68). An area of investment open to people of substantial means is starting and financing companies of their own. In their most creative form, these companies are run by their initiators. Rockresorts, a group of exclusive resorts in choice locations conceived and financed by Lawrence Rockefeller is an example, but atypical in that while Rockfeller maintains control, he does not actively participate in management. Also the concept, while profitable, is hardly creative. We cite another, more creative example, Centocor, when we discuss research and development.

The well-to-do, as we have noted and shall note again, have often found ways to greatly reduce their taxes. Most of this tax avoidance, though by no means all of it, is perfectly legal. It nevertheless means that other taxpayers, many of them with lesser incomes, have to pick up the tab. United States officialdom undoubtedly wastes or makes injudicious use of substantial amounts of tax money, but the education, research, national parks, highways, legal systems, social security, national defense forces, and the like financed by our taxes are indispensable services. It probably is expecting too much of human nature to assume that the wealthy can be persuaded to avoid legal tax loopholes, but at least we might hope that they will strongly support reforms that close these loopholes. Paying a fair share of taxes is one of the very best uses of unneeded income.

We turn now to legitimate uses of unneeded corporate income. While philanthropy is increasingly practiced by corporations, it is and certainly will remain a very minor use of spare funds. Investment is and should remain the major concern. The question then becomes: What are the creative uses of available corporate funds? My very inadequate answer to this broad question is a list of five objectives worthy of being furthered with available funds.

1. *Development of good worker-management relations and a safe and congenial workplace.* First and foremost, workers should be adequately paid. Profits may well be shared or paid out as bonuses, and especially so if bonuses are going to management. Child-care facilities could be a useful expenditure. The attitude of management, expressed in such ways as the sharing of a common lunch room, can be extremely important and of itself costs nothing, though a concerned management is likely to recognize needed expenditures.

2. *Unfailing high quality in the products produced.* This is one of the areas where the commitment of management is a major factor, but spending for such things as appropriate research and careful design can contribute. Japanese products excel in quality but, curiously, the Japanese learned many of their methods from an American, Dr. W. E. Deming, who had developed quality assurance skills in this country. Beginning in 1950, he was repeatedly invited to Japan to provide instruction. His work has been honored there by creation of a Deming Prize. His basic thesis: "Quality is made in the boardroom" It cannot be left to subordinates or "made on the factory floor" (52).

3. *Long-term company health.* It has often been noted that American companies tend to be run by managers with a financial or sales background rather thn a background in production. This has been accompanied by an emphasis on quick profits rather than long-term health and growth. There is no single factor more important in long-term company welfare than a dedication to well-designed and adequately financed research and development (R&D) (2). This aspect of the long-range view is so important that it deserves examination in some detail.

Two American industries illustrate the importance of R&D for long-term economic health. The Bell Telephone Company, from the day of its founding by Alexander Graham Bell, was dedicated to research. Bell Telephone Laboratories was widely recognized as one of the finest if not the finest industrial research laboratory in the world. It not only did research, but saw it through the developmental stage. Its success was a major factor in Bell's essential monopoly of telephone communication, a monopoly which the U.S. government has recently terminated, with dubious benefits to the telephone user.

United States Steel and the other large American steel companies stand at the other pole. Starting out with state-of-the-art technology, they soon dominated the American market and developed substantial sales abroad.

But complacency set in. The handsomely paid corporate officers did not recognize the inevitability of change. A mere pittance was directed to R&D, spare cash going instead into often quite unrelated acquisitions. When Europe and Japanese steel companies, sometimes drawing on American scientific findings, developed improved methods, their American counterparts were slow to adapt them. Meantime, global steel capacity was expanding, leading in due course to an overcapacity in steel. Inevitably, American steel, with its backward technology and high wages, became unable to compete and lost most of the home market to foreign competition. Seven steel companies have gone out of business and the United Steelworkers Union has seen half of its 1.4 million membership disappear. In the meantime, however, some start-up steel "minimills," drawing on the latest developments and concentrating on specialized products, successfully penetrated the steel market, adding fresh evidence for the power of R&D (7,47,51).

J. Fred Bucy, a physicist who served as president and chief executive officer of Texas Instruments from 1984–85, has stated: "R&D fuels technological innovation, which in turn drives productivity and growth. And it is those countries with higher rates of growth of R&D expenditure that have experienced stronger productivity gains and more rapid GNP growth" (14). Is the United States competitive in this regard?

The R&D expenditures of American industry grew from 0.6% of gross national product (GNP) in 1953 to almost 1.4% in 1986. There was a significant spurt beginning in 1979 (18). Despite the growth in our R&D effort, Japan outspent us during the years 1970–82 by 33%. The Japanese government does more to encourage R&D than our government does (12). The United States awards a smaller proportion of first university degrees in the fields of science and engineering than any of the other industrial nations do. Yet other countries are exploiting our educational system. More than half the engineering Ph.D.'s awarded in this country have been going to foreign nationals. The Japanese also have made use of American universities at the research level. A growing number of university research projects are funded by Japanese corporations. As in the case of similar arrangements with American corporations, the funding corporation has first rights to the findings. American researchers do not make these arrangements from any lack of loyalty, but from necessity (5,37).

R&D is no cure-all; attempts to use it can fail for various reasons. Projects must be well conceived and well executed. Since most of the basic research is done in universities and research institutions and most of the development in industrial laboratories, one of the problems is the availability of appropriate channels for the flow of the basic findings to industry. Interesting plans have been developed in this country for the development of centers where academic and industrial personnel can cooperate and where engineers can be trained for careers in research rather than applied engineering (50). At present, however, Japan appears to excel in achieving the university-industry flow.

R&D needs to be more intensive in some industries than in others. The oil industry spent only 0.4% of its income on R&D (1979 figures) (1). From the long-range point of view, I think much more might have been spent, but certainly this industry has less need for such spending than some others. The potential of R&D for young, high-tech firms is well illustrated by the case of Centocor. This company, founded in 1979, develops mono-clonal antibodies for diagnostic and therapeutic purposes. Among its products are antisera for diagnosing several forms of cancer and the heart damage caused by heart attacks, and antisera for the treatment of pan-creatic cancer and blood clot-related cardiovascular disease.

To speed the growth and transfer of the basic knowledge essential for the development of these techniques, working relationships have been established with universities and research laboratories both in this coun-try and abroad. These associations are fluid and generally informal. About 20 were in operation in 1985. So essential are they for Centocor's plan of operations that founder and board chairman Michael Wall has devoted full time to their negotiation and sustenance. He can speak with finality for the company. He can also use the contacts established to accelerate the transformation of information into end products.

In 1984, Centocor devoted $7,635,000 of a $12,141,000 budget—more than one half—to research and development. Essentially, that proportion has continued through 1986. About one-third of the R&D budget goes to Centocor's partners. Centocor may provide the stipend for a graduate student or to finance a major program. It may also contribute needed antisera. When discoveries with potential for medical application result, scientists at Centocor work out the needed technical details. This work at Centocor consumes the remaining two-thirds of the R&D budget.

Besides its cooperative arrangements for research, Centocor formed marketing arrangements with companies that had established distribution networks in the United States, Europe, and Japan. Japan was its biggest customer in 1984, buying about 50% of its total product.

Chairman Wall put up some of the initial capital, but it took substantial additional venture capital to get the company established, and it wasn't until 1984, five years after the founding, that Centocor turned a profit. But shares that sold for as little as 5 cents in 1979 brought $21–$45 on the Over the Counter Exchange in 1986 (15).

Centocor and many other new, high-technology companies like it clearly have an eye on the future. But the record of many older American companies has been pretty sad. As Malcolm Baldrige, Secretary of Com-merce in the Reagan administration until his tragic death in a horse accident, put it, "Our major industries gave little thought to long-range strategies. Management rested on its laurels" (7).

4. *Confining growth to areas where the management has or can achieve personal familiarity.* We shall examine the problem of conglomerates, which combine a diversity of industries under one corporate roof, when we discuss inappropriate uses of wealth. Suffice it to say here that internal growth or, to a limited degree, growth by acquisition of related companies, is the one largest legitimate use of available funds. It should be noted that internal growth is a natural consequence of expenditure on R&D.

5. *Protecting the environment.* Industry produces a great variety of toxic products and these have all too often been improperly disposed of or, in the case of gases, allowed to escape into the atmosphere. Water supplies have been contaminated, carcinogens left where people can be exposed to them, fresh-water fish killed, and forests threatened by the acid rain caused by auto exhausts and smoke from industrial smokestacks. Billions will have to be spent to clean up contaminated dumps, yet we still leave the problem of proper disposal inadequately resolved. We need both legislation to establish guidelines and a willingness by industry to shoulder at least a major share of the costs.

WEALTH MISUSED

We turn now to unjustifiable uses of discretionary income. Several of these uses are common to both individuals and corporations, so we cannot sharply separate the uses according to the source of wealth. In general, however, we move from personal to corporate sources.

A dubious use that is both conspicuous and contentious is the pursuit of luxury, especially luxury flaunted as a badge of wealth and privilege. No other use is more likely to raise the ire of the poor or to be staunchly defended by those who indulge in it. The inflation of the late 1970s and early 1980s may have encouraged such use, but it has not ceased now that inflation has slowed. The concentration of wealth in an era of growing scarcity, according to some sociologists, has had the perverted effect of making extravagance more appealing (20).

Examples are legion. Sales have boomed in jewelry priced from $1,000–$40,000 and above. Dior furs costing $60,000–$100,000 are popular, as are men's belts priced up to $30,000. The building of luxury homes is a thriving business. The most expensive and snobbish hotels have been the most successful. Dealers in Rolls-Royces and other cars in the $100,000 and up price range have done better than dealers handling economy models. While there was some slowing in the sales of Rolls-Royces in 1982, the general trend has been upward. At an Atlantic City show of customized limousines in 1984, $1.25 million worth of vehicles were snapped up by customers (20).

At least some wealthy families do not draw the line in extravagance when the spending is for their children. In Dallas, coming-out parties priced above $10,000 have been common. At the wealthy Grosse Pointe South High School in suburban Detroit, according to a 1979 survey reported by the Associated Press, the average yearly spending by students amounted to $4,388.52. Almost four times as much as was used for alcohol and drugs as for lunch. A wealthy San Francisco manufacturer of jeans bought his 16-year-old son a $35,000 Porsche Targa as a reward for making good grades in high school (3).

An instructive indication of the spending propensities of the well-do-do has been provided by the tax cuts introduced by the Reagan administration in 1981. These tax cuts benefited primarily the wealthy, and they were justified on the ground that the rich tend to be big savers and that any extra income they received would provide the capital needed for economic growth. This would give a long-term lift to the ailing economy. What actually happened was a rise in spending and a drop in savings. As Barry P. Bosworth, a Brookings Institute senior fellow, speaking in 1982, put it: "It's peculiar, against this backdrop of the recession, about how taxes were going to increase saving. It has gone the whole other way. It is a consumer-led recovery, not an investment-led recovery" (29). The increased corporate income from the tax cuts has gone, not into modernization and internal growth, but into bigger dividends, boosts in cash reserves, and the acquisition of smaller companies (4).

While considerable discretionary income goes into splurging, larger sums are used for investments made in the hope of financial return. The well-to-do, as we have seen, make many wise and appropriate investments, but the search for profits has led also to the diversion of billions of dollars to a variety of economically harmful uses. The temptation to make inappropriate investments varies with the economic climate and generally increases in periods of economic instability.

We now examine some of the specific areas of nonconstructive investment. During the 1970s and early 1980s, the wealthy put huge sums into presumed hedges against inflation. They bought art, antiques, real estate, jewels, and gold, and other minerals—anything whose price they hoped might rise with inflation itself. The most famous case was the purchase by the Hunt family of Texas oil billionaires of perhaps 60 million ounces of silver. Prices were pushed sky-high, but the Hunts failed to get the corner on the silver market that, according to common assumption, they hoped to achieve. (When charges against them were brought to trial in 1987, their lawyers denied this intent on their part.) In any case, they lost millions when they were forced to sell. The last of their silver hoard was not unloaded until 1985. All told, billions of dollars annually went into these inflation-hedges with very little benefit and some harm to the economy as a whole. In 1982, with inflation greatly reduced, this money began to flood back into the stock market (19).

Hoarding is an unproductive use to which wealth is often put in times of economic uncertainty. In the years from 1978–82, which included an oil crisis and two recessions, there were reports that both individuals and corporations were holding unusual amounts of money as cash or in other easily available forms. As business recovered, much of this money moved back into the economy. Much wealth from criminal sources is hoarded in foreign banks. We can only guess at the amount, but it undoubtedly is large.

For many years, a major inappropriate use of personal income has been investment in tax shelters. The tax reform law passed by Congress in 1986 presumably will make most of these unavailable. Because they are now a thing of the past, I treat them briefly. I draw heavily on Meyer's *Running for Shelter* (34).

Tax shelters are devices by which individuals and to some extent corporations evade taxes. They are made possible by provisions (loopholes) written into our tax laws, usually with good intent, but which have been perverted by the wealthy and their high-powered lawyers into devices for evading taxation. The loopholes and the means of exploiting them are many and diverse, but the secret of many tax shelters is that they produce artificial losses that can be used to claim a tax deduction. A small investment in a shelter can produce a very big return. The shelter can be mortgaged property suitable for rental, a farm, or a share in an oil well, all of which can be so managed as to permit the entry of a financial loss on an income tax return.

Because of the complexity of the shelters devised by professional tax dodgers, syndicated tax shelters were developed and became big business. In 1983, $13 billion in tax shelter equity was offered for public sale. An even larger sum was probably handled privately by professionals. All told, the well-to-do may have put almost as much money into shelters as into new issues of corporate stock.

Federal tax revenues are lost not only by way of tax shelters but also through so-called *tax expenditures* permitted under our laws. The tax exemption permitted on state and local bonds is the largest source of tax expenditures, but exemptions allowed businesses for such things as entertainment expenses cause further tax losses. The value to society of these particular tax loopholes is debatable, but at least they do not require the shenanigans involved in shelter-seeking. There is not the diversion of money into such unconstructive uses as buying farms and putting them under corporate management so that a paper loss can be claimed on the money invested, or the use of *leverage,* the borrowing of two or three dollars for every dollar put up as cash.

There is no firm evidence as to the use the well-to-do made of their spare cash after January 1, 1987 when a tax law closing many tax loopholes went into effect, but much of it probably went into the stock market. The bulls have dominated the market since August 12, 1982, but there was a particularly sharp rise in January of 1987 which in all probability reflected an inflow of cash withdrawn from tax shelters. Despite a precipitous reversal in October of '87 there has been no repeat of 1929. There is still a good deal of bullish sentiment, and many wealthy individuals, foreigners as well as Americans, probably have billions of spare cash in stocks. Speculative buying has pushed many stocks beyond their true worth, a dangerous situation. The point of importance here is that while investment in stocks and bonds that reflects real growth is constructive, speculative investment is not only unproductive but dangerous.

And what were people of modest means doing with their money while many of the well-to-do were gambling in the stock market? They were, on the average, cutting into their savings to continue their buying binge. This kept the economy going, but at the expense of a further mortgaging of our future.

We turn now to corporate growth by acquisition, a use of discretionary income that involves primarily, though not exclusively, corporate spending. Acquisition may well be justified when it involves related companies; it goes wrong when it builds conglomerates or results from raids on unwilling corporate victims. Many American corporations have been expanding by this route.

Growth by acquisition has produced some weird corporate mixtures. Senator Thomas Eagleton, who had investigated the subject, is quoted as stating: "According to my information, oil companies either have acquired or attempted to acquire every type of industry from highway construction to insurance to almond farming to real estate development. At one time an oil company even attempted to buy a circus." The inefficiency of the resulting conglomerates has been thoroughly documented by Michael Porter and by Peters and Waterman (40), and I shall not repeat their evidence here. The reason conglomerates don't work is very simple. The management may be able to read the financial reports of their acquisitions and spot areas of high profit, but they cannot possibly adequately master, in a group of unrelated business, the activities of the laboratory and factory floor necessary for product quality and long-term growth. Management emphasis on the quick return is reflected in the miserable record of conglomerates in spending for R&D (1).

What has been the reason for most of the conglomerate formation? Greed. It makes money for the corporate management and other money managers. As corporations grow, and particularly if they grow by acquisition, the size of the management grows. This justifies bigger salaries, more perquisites, and more income from stock holdings for the top management. Investment-banking firms specializing in the financing of mergers and acquisitions also rake in huge sums. First Boston Corporation, one such specialist, took in something on the order of $200 million in fees in 1985, and other similar firms were not far behind. In 1986, some of these firms put several billion dollars of their own cash into acquisitions (55, 56). Commerce Secretary Malcolm Baldridge, before his death, and Richard

Darman, Deputy Treasury Secretary until his resignation in April 1987, have both criticized our business leaders for their emphasis on quick profits rather than long-term business health. And Norman Lear has written: "The societal disease of our time, I am convinced, is America's obsession with short-term success, its fixation with the proverbial bottom-line." Growth by acquisition, he says, produced 14 new billionaires in 1986 (7, 30, 47).

A practice which has become increasingly common in recent years is the take-over by "raiders" of companies whose management has not consented to the union. The raiders may be either individuals or corporations.

How does the typical raider operate? Allan Sloan, writing in *Forbes,* has given us a good description of raiding by individuals (58). The raider first has to acquire 51% of the target company's stock. He can't hope to get this much at the going market price—he has to offer a premium, probably something on the order of 15%. Huge sums may be involved. He undoubtedly is a wealthy man, but even his total fortune might not cover the cost. In any case, why put up a lot of cash yourself when you can borrow most of what you need? Use what is called the *leveraged buyout,* a buyout mostly financed with somebody else's cash. If you offer a high enough rate of interest, there are many potential buyers. Even though the securities which you offer the investor fall into the category sometimes referred to as *junk bonds,* there are people willing to take the risk. Investment firms have organized pools that permit the small investor to get into the act. But individuals are not the only takers. Presumably conservative money managers have become involved. Pension funds, banks, savings and loans, and a few insurance companies have been buyers. The face value of successfully marketed junk bonds has gone from $27 billion in 1980 to $120 billion in 1986 (57). If the takeover attempt succeeds, the raider, now owner of the company, very generously transfers his junk bond debt to his new acquisition (49, 58).

A high proportion of takeover attempts do not succeed, but like hurricanes that spin off tornadoes, their destructive effect is not limited to their own immediate path. To resist the takeover attempt successfully, the takeover target has to resort to drastic action. Many routes to resistance have been used, but they virtually all share one property—they cost money. The intended victim, whether it succumbs or resists successfully, ends up with a greatly increased debt load. Stockholders who sell out to the raiders may gain, and the raider has ways to insure that he is an almost certain winner. Raider T. Boone Pickens received in one year $22.8 million in salary, bonuses, and long-term compensation from his Mesa Petroleum Company, which has served as the medium for his takeovers and takeover attempts. The big question is the ultimate effect on the company and its condition.

To pay the high-rate of interest on its new debt and to make payments on principle, quick income is needed. This has been sought in a variety of ways, which I cannot adequately describe. One source of quick cash is a raid on pension funds, a practice which has put at risk the retirement income of millions of Americans (6, 75). Another very common route is the sale of subsidiaries. Defenders of takeovers claim that most takeover targets have been inefficiently managed, and that raiders, by reorganization, can make them more profitable. Perhaps they are top-heavy at the management level and can perform better and at less cost with a leaner management staff. This claim may be justified in a few cases, but almost certainly long-term benefits are the exception and not the rule. Can an activity that Sloan describes as "Wall Street's equivalent of the philosophers' stone, that mythical substance that turned ordinary metal into gold," and as "a game that everybody seems to win," be anything in the long run but another swindle? (58, 72, 74).

It is noteworthy that since takeover targets end up with an increased debt, corporate raiding is not only an inappropriate use of wealth, but also a practice that encourages debt. It is wrong because it encourages debt. The rise in corporate debt that began in 1976 (Figure 3.1) and has recently accelerated is due in substantial degree to the contemporaneous increase in takeover attempts.

A spin-off from the corporate raids has been the use of inside information about these raids to make a killing on Wall Street. The use of such inside information is illegal. The revelation, that raider Ivan Boesky and merger specialist Dennis Levine had made millions by such trading in stocks caused a near panic in the stock market (67).

Two final dubious uses of wealth are the hiring of lobbyists and the donation of money to political campaigns to influence legislation.

Lobbyists undoubtedly have a major influence on legislation. In 1984 there were over 7,000 registered lobbyists in Washington, and perhaps an equal number were unregistered. The top ones had salaries in the $400,000 range. Insofar as their efforts are confined to purveying to members of Congress the viewpoints of business groups, labor unions, and citizens organizations concerning legislation, they are fulfilling a legitimate role. When their logic is backed by donations to campaigns, this role becomes dubious.

Getting elected is expensive. In the 1986 campaign, the average House winner spent about $300,000. Congressmen complain that during elections much of their time has to be devoted to money-raising. The major source of money for campaigns is Political Action Committees or PACS. There were 4,100 of these in 1986, each representative of some pressure group. Their total donations in 1986 were about $140 million. Their largest donations go to congressmen whose committee membership gives them particular power to influence legislation of major concern to the donor. The donations presumably are made with two intents: to elect congressmen with views favorable to the donor, and to influence the votes of congressmen in office—essentially a buying of votes. While the first of these objectives has some legitimacy, both objectives reflect a perversion of the democratic process. There have been some attempts at reform, but so far little in the way of results (10, 16, 33, 69).

Consequences of Wealth Misused

Inappropriate uses of unneeded or discretionary income by corporations and wealthy individuals cause serious economic, political, and social damage. The harmful economic effects are diverse; we shall examine primarily the effect on debts, dividing this into three parts: the effect on U.S. national debt, the effect on business debt, and the effect on personal debt.

A major misuse of corporate income, as we have noted, has been investment for quick return and the aggrandizement of management rather than investment for the long haul. This, in turn, has been the major cause of the poor competitive showing of our products in the international market, and this the cause of our unfavorable trade balance and our return in 1985, after many years as a creditor, to the status of an international debtor. By the end of 1987, our external debt is expected to reach $400 billion, more than that of any other nation. This building up of debt has required an inflow of foreign money, and this has been occurring in the form of foreign investment in the United States. The net effect of this has been somewhat reduced by U.S. investments abroad. In 1984, foreigners purchased $192 billion in U.S. treasury securities and $96 billion of corporate stocks. They also spent $160 billion in the actual acquisition or development of businesses (64).

The next step in the chain is the effect of these foreign investments on U.S. internal debt. The foreign purchase of U.S. treasury securities has helped to finance the growth in our internal national debt. The side effects of this are uncertain, but it is reasonable to assume that, since it has helped to provide a ready market for our bonds, it has reduced the pressure on Congress and the President to balance our budget. The investment in stocks certainly contributed to the recent bull market. If foreign nationals should decide for any reason on a major withdrawal of these liquid funds, which is certainly not impossible, this could cause a crash in the stock market in which millions of Americans would be the losers. The October 19, 1987 drop of 502 points in the stock market and subsequent lesser drops, even though they were followed by a recovery, is a frightening illustration of what might happen. Moreover a withdrawal of foreign investors from the market for government securities could force a major increase in interest rates. The average American has been doing quite well, but at the expense of a mortgage on our future.

The creation of conglomerates has been harmful in two ways. It has contributed to the poor quality of our products, and it has produced, when the result of take-overs, a major growth in corporate debt, much of it high risk debt. Again, we find our future put at risk.

The uses to which personal discretionary income is put have an important but indirect relationship to personal debt. Wealth misuse reduces the extent to which money accomplishes its one really ultimate function—the purchase of the goods and services that the economy generates. The money may change hands a number of times, but from the point of view of economic growth, it is chasing shadows. I shall return to this and its role in debt creation in the conclusion to this chapter.

The wages and salaries of people of moderate means are spent quite promptly for the necessities of life. The Friday pay check usually has all been used for groceries, gasoline, utilities, and other needs by the time the next check arrives. Some is deducted for Social Security and perhaps a pension, and people receiving middle-range salaries may save and invest some in one way or another. Even this money, however, probably finds its way back into the economy quite soon. As a bank deposit, it may be one of the resources that enables the bank to make loans for installment purchases.

In contrast with these economically constructive uses of wages and a considerable proportion of salaries, much unneeded wealth, both personal and corporate, goes to unproductive uses. I know of no study that has attempted to trace the course of money spent on inflation hedges, tax dodges, speculation in bull stock markets, or corporate takeovers, but both common sense and expert opinion indicate that money so used must, for days or weeks, be engaged simply in recycling old goods or in purely paper transactions.

Typical inflation hedges are the purchase of gold, real estate, or antiques. Money spent for gold may buy old gold or may go to South Africa or Russia where the mines are. If the money is spent for antiques or real estate (excluding new construction), it is recycling existing goods. The recipient of the money may or may not use it to buy goods and services, but in any case, there is a time lag.

With respect to tax shelters, I can do no better than to quote Meyer:

> "Tax shelters may be the single economic issue where most liberal and conservative economists find themselves in complete agreement. By granting investments in certain sectors of the economy special tax status, the government has tampered with the way the free market allocates precious investment capital. The investment dollars of individuals are diverted into unproductive tax shelters and away from products and services that better serve the public needs and wants expressed through the market" (34).

Meyer also quotes a congressman and a financial consultant as referring to shelters as "paper shuffles" and deals in existing assets. Speculation in stocks, as contrasted with long-term investment in stocks, is certainly a form of paper chasing. Corporate takeovers and also many mergers, like the purchase of inflation hedges and shelters, move money around, but without profit to the economy.

I have suggested that these misuses of wealth can account for the growth over the years of personal debt. For this relationship to exist, misuses must have grown correspondingly. I am not aware of any specific figures on this point, but I think that from all we know about these misuses, this is a plausible assumption. I postulate, then, that the wrong of wealth misused, at least in many of its forms, and the wrong of debt encouraged, constitute a pair, one being the cause or a major cause of the other.

The use of wealth most clearly damaging to our political system has been the donation of large sums to congressmen by way of Political Action Committees. An inevitable result has been legislation that grants or protects special privileges. It is, I suggest, also a reasonable assumption that, by adding to the many other pressures to which congressmen are subject, it delays decision making and reduces the effectiveness of the legislative process.

A final consequence of wealth misused is harm at the personal level. This harm is partly financial but perhaps more properly classified as social. Takeovers and mergers can cause a loss of jobs, many of them useful and necessary. Pension funds may be drained to pay the cost of takeovers. When small, local banks are merged with large city banks, local businessmen may find it harder to borrow. Acquisition of a TV station may change its local character. In some extreme cases, a whole community may suffer.

Farmers have also been victims. Farms are one of the favored tax shelters, and syndicates have been set up to purchase them for the benefit of wealthy individuals seeking tax avoidance. An Iowa farmer has described the result. "Right next to me, five families used to farm 870 acres. Now it's been consolidated by some outside syndicate. Now there's nobody living there. You have no neighbors to work with." Meyer writes that a growing number of experts regard these purchases of farms by outsiders in search of tax breaks as the most dangerous long-term enemy of the family farm (34).

The potential for widespread personal and social damage is illustrated by the case of the Pacific Lumber (PL) Company, described on the MacNeil/Lehrer News Hour program on Public Television. The PL Company was a family owned corporation founded in 1905 and dedicated to the long-term viewpoint. Instead of clear cutting on its 200,000 acres of redwood forest in northern California—the method adopted by most companies—it cut selectively so that its supply of timber would never be depleted. Scotia, where the company's headquarters and lumber mill are located, is a town of 1,200 citizens, many of whom are second- and third-generation PL workers devoted to the town and its founders. The company was cash rich and debt free. This, plus the wonderful stand of lumber, made it a tempting takeover target, and because it had gone public in 1975, making its stock available for purchase, Charles Hurwitz, a Texas millionaire, with Wall Street assistance, moved in for the kill. The PL board was unable to fight him off; Hurwitz became the new owner and, following the usual practice of raiders, passed the huge debt acquired in the purchase on to the company. To clear up the debt, only one future course is possible—clear cutting, which will reduce 200,000 acres of fine redwood forest to barren land and deprive Scotia and its citizens of an assured future, leaving them instead perhaps 20 years during which work will be available, but with little of the satisfaction and pride in their jobs that was their lot under the old management (31).

An indirect effect of unneeded income misuse is damage to the mutual trust so necessary for the successful functioning of a free society.

In their sum, the forms of unneeded wealth misuse that we have described have significantly hurt both individuals and the economy. The greatest threat, however, is to the future. And this threat is very great indeed.

F. Security Sought Through Power or Privilege

Security. Everybody wants it and almost everybody, nowadays, expects the government to play a major role in providing it. Yet despite the generally felt need, the sense of insecurity is widespread and growing.

What do we mean by security? How much security can we reasonably expect? What are the keys to its more general attainment? What has gone wrong that we have fallen short of our reasonable expectations? An attempt to answer these questions will require, as in other sections of this chapter, an exploration of some important moral issues facing us today.

Webster's Dictionary defines *security* as freedom from danger and freedom from fear or anxiety. There is no sharp line between the desire for security, as thus defined, and the desire for comforts and prerequisites that go beyond the reasonable bounds of security. This perhaps has been one of our problems.

High on most people's list of threats against which they seek security are want in old age, ill health, unemployment, crime, and aggression by foreign powers. Less likely to be articulated but probably quite important in many people's thinking are declines in living standards and declines in status. As an offshoot of the wish for these forms of security, business managers may search for financial security for the firms they manage.

Threats against security may be inherent in nature, as in the case of old age, or they may originate in social and economic circumstances or in the errors and failings of the person threatened. We need then to ask if social insurance can protect in this whole range of cases. If a mismanaged corporation is threatened with bankruptcy, can it reasonably expect a federal bail-out? Should an unmarried woman with children she cannot support expect a secure living from the government? Or is overprotection in areas such as these counterproductive in the long run?

The availability of security in any society stems from many sources, both individual and governmental. Two ingredients, I suggest, can be singled out as paramount: mutual trust and, except for the incapacitated, hard work. Trust is the lubricant that makes our social institutions work without breakdown. In its absence, the most artful shelter in the long run fails to shield. Work provides the physical and economic means to security and, if fairly shared, without shirking, it enhances trust.

There is one necessary exception to the reliance on trust. Security from the hardened criminal or sociopath can come only through protection by appropriate agencies of law enforcement.

We seek security in many ways. Chosen routes vary with time and place, with social group, and with the individual. In Colonial America, people relied for security, especially in old age, on family and community. Trust was almost a necessary component under these circumstances. As the country became urbanized and the economy more complex, people turned increasingly to the government. But people also often sought security through any special leverage they possessed or could procure. Sources of leverage are many, but most of them can be grouped under the two headings, power and privilege. These routes to security tend to be self-centered. They do not depend on trust and indeed they can erode it. They often do provide security in the short run and often can be used to provide added layers of protection, but their long run effect is the creation of dangerous social and economic instability.

Power and privilege are not independent variables—they tend to be associated in various ways. Despite the overlaps, it will be convenient to consider them separately.

Security Sought Through Power

The ultimate example of security sought through power is the enormous commitment of virtually all nations to armed force. Nations do not trust each other; the result has been a steady growth in their capacity for mutual destruction. Armaments have at some periods in history given some nations a sense of security, but certainly in this last quarter of the twentieth century the balance of terror produced by vast nuclear arsenals has done the exact opposite.

It is the great tragedy of our times that an ultimate solution is not in sight. To achieve disarmament, we must first achieve a genuinely *mutual* trust. Until all parties concerned are willing to make the concessions necessary to build a true basis for trust—and these concessions will have to be broad-ranging and substantial—any progress beyond limited arms reductions will remain a futile dream.

This is not to say that international cooperation is not possible in some areas, but there must be a willingness to make the *mutual* concessions necessary to create at least a minimum of mutual trust. If full trust could be attained, the resulting sense of security would immeasurably exceed that attainable through power.

All other examples of the search for security through power are overshadowed by the international resort to arms, but there is at least one other example in the United States that has assumed alarming proportions. Many Americans, as a presumed protection against crime, are turning to handguns. An example of the prevalence of this practice comes from Nashville, TN. When the local sheriff's department offered a one-night handgun safety course for a $10 fee, 3,000 women signed up.

The U.S. Treasury Department estimated in 1980 that 140 million rifles, shotguns, and handguns were owned by private citizens. A good case can be made for the ownership, with appropriate restrictions, of rifles and shotguns by hunters. It is handguns, because they are easily concealed, that are the problem. Their wide ownership is certainly a major cause, either through intention or accident, of an alarming number of deaths in this country. Comparative figures for 1979 showed eight deaths caused by handguns in Great Britain, 21 in Sweden, 34 in Switzerland, 42 in West Germany, 52 in Canada, 58 in Israel, and 10,728 in the United States. Polls consistently show that Americans want some form of handgun control, but the National Rifle Association has so far been able to block nearly all attempts at regulation.

SECURITY SOUGHT THROUGH PRIVILEGE

Privilege, as I shall use the term here, refers to power of a nonphysical nature. It is derived from belonging to a particular group that has maneuvered itself into a favored position whereby its members have advantages not generally available. The group may be a labor organization, the members of a profession, an ethnic group, or a large corporation. Any group that has acquired a position of privilege can use this position to enhance, at least for the short run, the security of its members.

Unions have sought security for their members through contracts that require the employment of unnecessary workers and prevent or delay the introduction of labor-saving machinery. Railroad unions years ago won contracts that required the hiring of unnecessary breakmen and mandated the presence in the locomotive of firemen long after steam locomotives had been abandoned for diesels. Similar practices have developed in urban transportation. A 1979 investigations by the *Boston Globe* found that the contract of the local Carmen's Union, through a clause prohibiting the changing of established practices, made it impossible to eliminate duplicate overtime or to shorten the time spent on routine repair jobs. Newspaper printers' unions, both in this country and in England, have often successfully resisted the introduction of labor-saving printing methods. Some of the most glaring, though also perhaps some of the most understandable, examples of featherbedding are found in the uncertain world of the theatre. In one case a man doing a solo act was required to hire 18 unneeded musicians and accessory personnel.

Another feature of contracts won by powerful unions in search of security for their workers has been wage increases in excess of increases in productivity. This has been an important contributing factor in inflation (3).

It is illuminating to note that labor relations in this country have generally been based on confrontation rather than on cooperation. This has encouraged the search for security through privilege rather than through trust. Fortunately, there now seems to be a trend towards labor-management relations based on accommodation and helpfulness. Management has contributed by adopting more considerate, open, and cooperative management-worker relationships. There are indications of new attitudes on the part of labor as well. This is particularly true of workers who have been laid off and then, with an upturn in business, taken back on the job. Both employers and union officials report that these workers show a more positive attitude, work harder, and are less frequently absent. Both sides are beginning to realize that true security lies in mutual trust and a job well done.

Another area of American life where security has been sought through privilege is in race relations. For years, the white community sought security through the creation of barriers between itself and blacks and other minority groups. During the second half of this century, there has been a substantial amelioration of the situation; some people would say that in some areas there has been over-correction. Trust has increased, but difficult problems remain.

Similar situations exist in many countries. The situations deserve mention because they illustrate the basic nature of one aspect of the problem. In many countries, two groups—one relatively well educated, prosperous, and unprolific, the other less educated, less prosperous, and more prolific—live contiguously. The Protestants and Catholics in Northern Ireland, the Walloons and Flemmings in Belgium, the Christians and Moslems in Lebanon, and the While Russians and Moslems in the U.S.S.R. are examples. I think it is safe to say that wherever this combination exists, friction has resulted. There is perhaps no more difficult context in which to establish mutual trust, but that does not remove our obligation to generate conditions that will permit it.

This subject could take us into the problem of population growth, but this is outside the scope of this chapter. Suffice it to say that there can be no long-term security unless population growth is controlled, and that if mutual trust is to be sustained, all parties have to contribute their fair share to the control process.

Our legal system was designed to provide security through justice. Over the years it has served this purpose with notable success, and in general it still does. Like all good things, however, it is susceptible to abuse, and in recent years the abuses have become substantial. Both the citizen as litigant and the lawyer who carries out the litigation have used their respective privileges in a search for security unjustified in degree or kind.

The worst abuses have involved civil law. In 1977, 12 million suits were filed in state courts, which handle most such cases. The cases now come in so fast that it has not been possible to complete more up-to-date figures. In federal courts, a 1982 count showed 200,000 cases filed in a year. The increasing resort to law is reflected in the number of lawyers. In 1982, members of this profession numbered 610,000. Law has been the fastest growing profession, with an increase of 83% in a 10-year period. We have five times as many lawyers per-capita as Germany, 10 times as many as France, and 20 times as many as Japan.

The motives that induce Americans to file so many lawsuits are not easily analyzed. Many factors may be involved, but a major one is certainly the erosion of mutual trust. Such units as families, churches, and neighborhood groups have weakened. People more often feel estranged from and distrustful of their doctors, their employers, their school and college leaders, and the members of their families. In these circumstances, the opportunities that our current legal system offer to obtain substantial financial recompense for actual or imagined damages seems to many people an adequate reason for a suit. The privilege of justice, thus exploited, provides security of a sort but at heavy social and financial costs.

The harmful social effects of too many lawsuits include prolonged delays in the disposal of cases, an increased *insecurity* among the members of groups subject to suit, and the adoption of defensive rather than progressive, forward-looking, and economically sound practices.

The dollar costs of lawsuits are tremendous. In successful suits against corporations, the damages assumed may run into many millions of dollars. Malpractice awards against doctors and hospitals averaged $450,000 in 1981. Substantial awards have been assessed against cities, universities, and lawyers themselves. And the awards have been steadily increasing. A Rand Corporation study of civil suits in one county in Illinois showed that the amount of awards, even after correction for inflation, has more than doubled in a 20-year period. Along with the rising cost of the awards has gone a corresponding rise in the cost of insurance against awards.

The fees of lawyers constitute the other major cost. The hourly charge of experienced trial lawyers is often in the $100-an-hour range and may be several times that. Southwest Airlines spent more than $1 million defending itself against sex-discrimination charges by men who claimed they were denied jobs as flight attendants and ticket agents. Costs in this range are not uncommon. The Asbestos Compensation Coalition, a group of companies facing suits by workers and other persons with asbestos-related diseases, has said that its members pay out an average of $150,000 in costs for each $28,000 that is paid to a successful claimant. The litigant, even if he wins the suit, also can be a loser. In one case a lawyer, after leading his clients to expect a $10,000 award, settled for $1,000 and charged a fee of $2,000. This is an extreme case, but insurance data indicate that lawyers usually get a larger share of the jury award than the victim does.

In financial terms, lawyers are certainly the major beneficiaries of our legal system as it exists today. This gives them a stake in its perpetuation and, indeed, in its elaboration, and there is evidence that they have acted accordingly. They have used the privilege of their position, not only as lawyers, but also as legislators and judges, to enhance their security to the point of opulence.

An association operating under the name Help Abolish Legal Tyranny (HALT) has been a prime mover in pointing out abuses of the legal system by the legal profession. I have drawn much information from its publications, but it is not alone in its judgments. Lawyers themselves are expressing concern. At a conference sponsored jointly by the American Bar Association Section of Litigation, the Harvard Law School, and the National Institute for Dispute Resolution, the existence of a problem was clearly acknowledged. As James Vorenburg, Dean of Harvard Law School, stated in his keynote address, many Americans believe "the system of justice serves its own functionaries and officials better than those it should be most concerned with: the clients." Former President Carter, former Attorney General Griffin B. Bell, Harvard President Derek C. Bok, and Chief Justice Warren Burger, among others, have also expressed concern.

Bok has described the American legal system as "grossly inequitable and inefficient" and has called for fundamental changes in how lawyers practice and how they are trained (1). Burger has stated the problem bluntly: "We may be on our way to a society overrun by hordes of lawyers hungry as locusts." And again, in speaking before the American Bar Association in February 1984: "The entire legal profession—lawyers, judges, law teachers—have become so mesmerized with the stimulation of the courtroom contest that we tend to forget that we ought to be healers of conflict . . . Trials by the adversarial contest must in time go the way of the ancient trial by battle and blood" (2). The potential for reform is attested to by the success of Maine's Court Mediation Service. Begun as an experimental program in 1977 with private funding, the Mediation Service was formally adopted into the state judicial system in 1981 and has been widely used. In fiscal year 1983, the service handled 980 cases, of which 163 were divorce and child custody matters. More than three-quarters of the nondomestic cases and all but 29 of the domestic cases were resolved without the need for a trial (4).

The complexities of the legal system that give added business to lawyers are many. Attempts at simplification have been consistently opposed by the legal profession. We can cite only a few examples.

Probating a will can cost up to 100 times more in the United States than in England. It can take 17 times as long. The process can be simplified, and Wisconsin is one state that has done so, reducing the cost of probating a small estate from several thousand dollars to $45. A 1982 attempt to reform probate law in Maryland, though passed by a large margin in the House of Delegates, was defeated in the Senate. Lawyer-legislators led the opposition. The annual take by lawyers for probating estates amounts to about $2 billion. It is an obvious conclusion that profits motivate the opposition to reform.

The Maryland legislature in 1982 also defeated an attempted simplification of procedures for uncontested divorces. The law that was passed was actually a reversal of a Court of Appeals ruling that would have accomplished simplification.

Legal complexities in settling claims arising from auto accidents are a source of large legal fees. No-fault auto insurance is a proven method of reducing these costs, yet change has been slow because of consistent opposition by lawyers. The HALT publication *Americans for Legal Reform* notes with enthusiasm, however, that in 1982 the District of Columbia adopted a no-fault plan that HALT had strongly supported.

One of the reforms advocated by Chief Justice Warren Burger, so far without result, concerns those civil procedures involving prolonged, pretrial interrogations. As Burger remarked, we tend to try such cases twice. Both parties in the proceedings pay accordingly and the lawyers benefit (2).

While the legal profession has its share of crooks, lawyers in their personal ethics probably are not much different from the public at large. Other professionals have also continually sought fatter fees. If lawyers have used the privileges of their position to expand beyond reason their opportunities and earnings, it is because their privileges are unique. Their roles as judges and as legislators give them virtual control of the legal system. It is not surprising that they have used to their own advantage the powers thereby conferred.

Consequences of Security Sought Through Power or Privilege

The consequences of the search for security through power and privilege rather than through accommodation and honest labor range from annoyance to waste and from rising costs to loss of life. Security sought through power is the source of the loss of life, with the many deaths caused by handguns in the United States its most inexcusable manifestation. Security sought through privilege leads to wasteful practices and hence adds to the cost of goods and services. As the practices have increased in recent decades, as has been the case with litigation in the United States, they have caused rising costs and hence have become contributors to inflation. Union contracts providing wage increases that outpace the growth of productivity also produce inflation. Besides its economic effects, the search for security through privilege can be a source of tribulation to its victims. Thousands of people have been caused unnecessary annoyance or suffering when caught in avoidable legal delays or litigation.

G. Dishonesty Condoned

Doubtless most Americans today, if asked, would render lip service to honesty. In actuality, many of them both practice and condone dishonesty. Over the years, we have lost the restraints imposed by a Puritan ethic and by the personal relationships of the small community. Now many of our relationships are with big government and big business or, if we are in business, with people who are and will remain strangers. It is easy to convince ourselves that we hurt no one if we make these the target of theft and cheating.

Dishonesty exists at all levels of our society. An appreciable number of people in highly responsible positions in government, business, and the professions—people who certainly should know better—practice it and accept it in their associates.

While the impersonal character of this dishonesty helps those who perpetrate it to convince themselves that they do no harm, this, of course, is not true. Short-changed individuals suffer personal loss. The financial losses to government and business reappear in higher taxes and higher prices. We all have to pay these, plus added indirect costs, but it is the honest who are hurt most. Perhaps more important than the monetary loss is the damage to the social fabric. Dishonesty is the great eroder of the trust in people and institutions so essential for the functioning of a free economy and a free society. The total damage is enormous.

The current tolerance of dishonesty is perhaps most evident in the widespread practice and infrequent condemnation of cheating of the government. Many small instances of cheating add up to billions of dollars in losses.

In my previous comments on unneeded wealth misused, I noted the prevalence of incomplete reporting of income on income tax returns. Much of this is tied to the underground economy. Where the underground income is derived from moonlighting at odd jobs or waiting table, the failure to report may be almost as much inertia over the chore of record keeping as any desire to avoid the tax, usually small in any case for the odd job worker. In these situations, the tax law itself may be at fault. But there is also underground income that is specifically planned to avoid tax. The keeping of two sets of books and business done by barter are examples. Collusion is often involved, reflecting the degree to which tax cheating has gained general acceptance.

Cheating the government takes a variety of other forms. People with substantial savings collect welfare payments supposedly only for the genuinely poor. One woman with $89,000 in the bank was partly supported by federal food stamps. Government employees cheat the government. In Ohio, a woman who headed a federally supported but county-run anti-poverty agency received $44,847 to buy blankets for the poor. She and two associates bought 3,000 substandard blankets for $15,000 and pocketed the difference. There is substantial fraud within the armed services. Military reservists received and kept some $774 million in payments for exercises they never attended. The government is robbed by businesses that serve it. A dairyman making regular deliveries of butter to an Army post arrived each time with a partly full truck that he then proceeded to fill from the storeroom holding the previous delivery. Only then was the load presented for checking to commissary officials. Cost to the government: $80,000.

Big businesses as well as small cheat the government. Admiral Hyman Rickover, in testimony before the Joint Economic Committee of Congress, spoke scathingly of the extent to which corporations fail to meet obligations specified in their contracts with the federal government:

". . .many corporate executives," he said, ". . .seek to evade moral and legal liabilities for the companies they own and control by insulating themselves from the details.

"It used to be that a businessman's honor depended on his living up to his contract—a deal was a deal.

"Now, honoring contracts is becoming more a matter of convenience. Corporations are increasingly turning to high-priced law firms which, by legal maneuvering, obfuscation, and delay can effectively void almost any contract . . . Under these circumstances, government contracts with some large companies are binding only to the extent the company wants to be bound" (8)

Unpaid loans to students and veterans are another source of loss. A Department of Education lawyer, twice promoted and with a salary of $34,000, had made no payments on a student loan. Another federal employee with a similar loan and a similar record of nonpayment had bought a $20,000 luxury car. All told, about 25% of all loans, or a total of about $1 billion, has gone unpaid.

In some cases states have passed laws, touted by their advocates as a good source of state income, that have had the unintended effect of encouraging law-avoidance. New York State, in 1970, legalized a system of offtrack betting. The result, over the next 4 years, was an estimated 62% increase in the dollar volume of illegal gambling on the races (2).

No one really knows how much the government loses from fraud, but George Egan, an associate director of the General Accounting Office, mentioned several billion dollars as a minimum figure. This would not include unpaid loans and probably not evasion of contracts. One billion dollars is said to be taken illegally each year from the food stamp program alone. Fifty billion has been suggested as a possible top figure for the annual cost of fraud plus waste.

Businesses suffer from losses caused by dishonesty quite comparable to those that afflict the government. Two major sources are shoplifting by customers of retail outlets and theft by employees. These practices are so common that we have to assume that they are often viewed with tolerance by fellow customers or fellow employees.

The annual cost of shoplifting was placed in a 1977 study by the American Management Association at $2 billion. The number of individuals involved runs into the millions and includes children and adults of all ages. About 55% are female (9). Shoplifting is particularly prevalent during the Christmas shopping months, November and December. Lawrence Conner of Shoplifters Anonymous, an organization that tries to help convicted individuals, estimated that during these months in 1981, three million shoplifters would be caught. There must be many more who go uncaught.

The tolerance with which shoplifting is viewed is illustrated by an experiment carried out by the marketing class of a suburban Chicago high school with the knowledge and consent of a local store manager. For several hours, three students stuffed their pockets with a variety of merchandise within plain view of dozens of customers. They got a dirty look from one woman shopper but, in the words of the teacher, other shoppers ". . . either walked away or looked away." No one reported the presumed thefts to the store manager.

The same study that placed the cost of shoplifting at $2 billion annually estimated the cost of employee pilferage at $5–$10 billion. These sources of loss represent only a part of the total of business crime. Other crimes, such as embezzlement, arson for profit, and burglary, are usually carried out by hardened criminals and probably are not generally condoned. There are, however, some crimes against businesses practiced by persons of so many backgrounds and with such frequency as to suggest a distorted view of the wrong involved. The victims of one such form of crime are the electric power companies and the crime is tampering with meters so that they do not register all the electricity used. Losses to the companies were estimated in 1978 at $3–$4 billion annually. Insurance companies are the victims of another type of graft: falsely reported automobile thefts. The devices used vary, but one common method involves establishing an apparently authenticated theft by having the car stripped of salable parts and then left on the street where the police will find it. The owner reports the car as stolen and has the parts replaced by the collusive auto shop, with which he then shares the theft insurance. According to the National Automobile Theft Bureau, 10–15% of the 1.1 million auto thefts reported annually are fraudulent, and the loss to insurance companies is about $400 million. These costs are of course passed on to the public.

This practice of false auto theft reporting has been growing, a tragic testimonial to the tolerance with which it is viewed. In the words of the Independent Insurance Agents Association of New York, a "byproduct of our ailing economy is the apparent trend for decent, law-abiding citizens to feel perfectly justified in engaging in criminal acts designed to swindle insurance companies." The insurance companies themselves have not helped. Rather than working to eradicate the fraud they have taken the easy route of passing the costs on to their customers (4). By 1980, according to federal and state officials, organized crime had moved in as an active participant.

Total losses to business from all sources of crime combined are estimated by the American Management Association to amount to 15 percent of the retail price of all U.S. merchandise. The crimes are believed to account for one out of five business failures (9).

Another source of loss to business is loafing on the job. This does not come under the heading of crime, but it is fair to classify it as a form of dishonesty since, in effect, it cheats the employer. It is an area where accounts may be prejudiced. It is also probable that, in many instances, it is due to poor handling or inept labor policies by management rather than to unconscionable laziness. Nevertheless, there is evidence of a real lapse in any sort of work ethic.

Robert Half, an employment specialist and head of a firm that recruits trained workers and executives, interviewed 400 personnel directors and managers throughout the United States about worker attitudes. Common among complaints concerning work habits were the prevalence of arriving late for work, of leaving early, of habitually long lunch hours, of claiming illness and taking unjustified sick days, of excessive personal phone calls, and of slowing down the pace of activity to create overtime situations. Robert Half estimates the cost to business at more than twice that due to actual crime.

There could be an element of prejudice in this account. Nevertheless, the reality of the problem is confirmed by interviews concerning younger workers carried out by members of the staff of *U.S. News and World Report*. Youngsters themselves were included in the interviews. As in the case of adults, work attitudes varied greatly, but shirking on the job was not uncommon. According to the manager of a New York City employment agency, easy jobs were preferred and menial jobs were avoided like the plague. Lateness and absenteeism were common complaints. A University of California survey of working high school students in Orange County, California, found that 32 percent had phoned in sick when they were not actually ill.

We could go on and cite endless examples of corruption. Labor unions as well as business and the average citizen are involved. An area where we might hope for a clean record, the members of the professions, is also tainted. A 1981 pilot study in two states by the Department of Health and Human Services found that doctors had presented $757,000 in improper Medicaid claims. Another government study found that 5,716 doctors, some of them clearly affluent, were seriously delinquent in payments on government loans they had received to help with their medical education. Studies of fraudulent auto accident claims, estimated to cost insurance companies $2 billion a year, led to the conviction of more than 350 doctors and lawyers. These are tiny samples of the total problem. Even some clergymen have been guilty of proven wrongdoing.

Inevitably, the soiling hand of corruption reaches to some extent into the higher levels of government. The power of office undoubtedly subjects government officials to more than average temptation to accept illegal favors. The temptation may be particularly great in the case of some state legislators who receive totally inadequate salaries. The tragedy is the prevalence and public tolerance of wrongdoing.

A study of building contracts in Massachusetts in the 1960s and early 1970s turned up massive corruption. The study commission, headed by former Amherst College President John Ward, found that contracts for the design and construction of public buildings were regularly obtained by bribery of public officials, including members of the legislature. The Chairman of the Senate Ways and Means Committee was charged with extorting $39,000 from an architectural firm for a design contract. Many of the buildings produced under these contracts were grossly defective. In the words of the Commission, "the state was for sale." The record varies greatly from state to state, but Massachusetts is (or was—the record may have improved) by no means alone in the presence of deep-rooted corruption.

The United States Congress has many outstanding members, but it also has more than its share of corruption. According to the *Congressional Quarterly* report of 9 February 1980, 42 members of Congress were criminally indicted in the preceeding 40 years. Twenty-eight of them were indicted between 1970 and 1980 (1). An editorial writer calculated that the fraction of congressmen involved in wrongdoing throughout approximately this 1970–80 decade is one in 84. Some degree of tolerance of wrongdoing by elected officials is indicated by the considerable number of cases in which officials known to be dishonest or whose honesty is in doubt have won either primary or final elections.

Any substantial punishment of congressmen guilty of wrongdoing has usually come through the courts, not from Congress itself. The case against Congress as a judge of its own conduct has been well stated by Leon Jaworski, prosecutor in the Watergate trials. "Congress," he said, "has proven itself to be a poor investigator and even less qualified to sit as a tribunal considering appropriate punishment."

Ethical problems at high levels of government are not confined to Congress—they occur also in the administrative branch. According to a 1984 report, 33 political appointees of the Reagan Administration resigned or were dismissed in the face of allegations of political, criminal, or financial wrongdoing.

Dishonesty has been reported in the upper levels of major corporations as well as in government. A select team of accountants and finance-law experts of the Securities and Exchange Commission (SEC) has examined the accounting practices of a number of businesses. The suspicion has been that financial statements available to the public have been distorted to aid in the sale of stock. The investigation has led to allegations by the SEC that a number of corporations have indeed been "cooking the books."

The extent to which dishonesty is condoned in this country stems from many sources. The depressed economy, the unequal distribution of wealth, a widespread disillusionment with business leaders, the light punishments imposed for white-collar crime, and the artificial stimulation of wants by advertising are all contributory causes. More fundamentally, we have substituted complex theories of social justice for firm rules of what is right and what is wrong. The result, according to University of California sociologist Gilbert Geis, is that "people seem to be able to rationalize anything. Those who steal don't see themselves as thieves— they're oppressed people or adventurers who are trying to outwit" an unfair system (9).

These attitudes are reflected in, and to a considerable degree are cultivated in, our schools and our homes. Schools of course vary greatly. Many suburbs and small towns have excellent schools, and many communities are striving to raise standards of both scholarship and discipline. In some of our city schools, however, both theft and violence are all too common. According to a study of the mid-1970s by Dr. Walter B. Miller of the Harvard Law School Center for Criminal Justice, schools in at least six cities have been terrorized by gun-toting teen-age gangs. Miller found gangs extracting fees of a quarter to a dollar from students for the "privilege" of passing through hallways or using school facilities. New York City police investigators reported that gangs in the City's schools in the 1970s were modeling their organization and operations after the pattern of organized crime (3). Detroit schools in 1982 reported locker break-ins, restroom rapes, arson, and malicious destruction and theft of school property. Students, some of them brandishing switchblade knives, were found milling around in unsupervised hallways to act as a cover for drug dealing and locker break-ins. More than 800 people, most of them students, were arrested in or near Detroit schools in 1981 on charges ranging from loitering to murder. According to an official of the Senate Subcommittee on Juvenile Delinquency, loss of and physical damage to school property in 1978 cost more than $600 million for prevention plus repair and replacement.

One of the most disturbing problems is the frequency, in some areas, of physical attacks by students on other students and on teachers. A study by the Massachusetts Teachers Association showed that one of every six teachers in Massachusetts, at one time or another, had been physically assaulted. Dealing with the violence of school bullies is a problem for which many teachers are by temperament perhaps particularly ill suited. It is a major cause of teachers leaving their jobs (5).

What has brought about this level of violence in our schools? The causes are complex; they include, for example, violence in movies and on television (6), but high among them is certainly a failure to make discipline and sound education top priorities in the schools. Teachers often feel that they operate under inadequate and inconsistent policies in regard to pupil conduct, or that they are insufficiently supported by school administrators and school boards. Teachers were also quoted, in the Massachusetts study already cited, as feeling that they are unfairly criticized by taxpayers and politicians.

Attempts to deal with the discipline problem in schools have achieved substantial success. Most of them rely on two basic methods—greater involvement of the parents and dismissal or transfer of problem students. Los Angeles, according to a school official, effectively reduced vandalism by instituting the rule that parents of the students involved be required to make restitution. Some Miami schools succeeded by using the carrot rather than the stick. With federal support, a group of parents was paid to ride on school buses with the students, to check on repeatedly tardy students, and to help with other sources of trouble. Some teachers' groups have won contracts forcing school boards to either transfer or permanently dismiss students who attack teachers. The suggestion has also been made that school systems include alternative schools to provide special education for frustrated and resentful students. This proposal was featured in a 1977 report, based on a three-year study, prepared by the Senate subcommittee on Juvenile Delinquency chaired by Senator Bayh (3)

Quite as disturbing as school violence, which is not confined to but tends to be the worst in poor city areas, is the lack of moral sense, prevalent at least since the 1970s, even in schools serving the affluent. Some relevant statistics were contained in a 1977 report prepared by researchers in Illinois. Of all youths questioned statewide, 13% admitted taking part in a robbery, 40% admitted keeping stolen goods, and 50% admitted shoplifting. Very similar figures on shoplifting were revealed in an earlier report from a Delaware high school. A researcher involved in the

Illinois study found "a near vacuum of morality" among youngsters in one of Chicago's wealthiest suburbs.

Educators attending a 1987 symposium on the U.S. Constitution took the occasion to emphasize the need "to teach students about moral responsibility." Robert Coles, one of the speakers, emphasized the need by recalling a conversation with a 10-year-old boy. Responsibility, the boy said, "means you should know how to take care of yourself and be on time." Coles went on to say, "At this moment, moral squalor is everywhere, in politics, religion and business, and we certainly do need to take responsibility to change that. We need moral integrity."

While there seems to be a widespread lack of moral understanding among students, many of them, like their parents and teachers, do recognize the need for more discipline. This was the finding of a 1981–82 survey of 12,000 junior and senior high school students by the national school periodical, *Read*. The magazine introduced the request for students' opinions with an article that discussed some of the possible causes of vandelism and lack of discipline. The magazine editor, Dr. Terry Borton, wrote of the survey: "What surprises us about the discipline survey is both the large response—twice the normal—and the overwhelming support for school discipline." Eighty-seven percent of the participating students said they would appreciate "a lot of" or "some" discipline, and 80% said violence could be curbed with discipline. A typical comment came from a boy in Grand Blanc, MI: "The teachers are afraid of the students, so they don't discipline. The kids know it and think they can get away with anything—merely by threatening the teacher." Various other sources of trouble in the schools, including parental failure, were mentioned by the students, but the lack of discipline was the overwhelming concern (7).

The lack among many of today's students of firmly inculcated ethical standards did not come up in this survey. This may have been due to the absence in the introductory article of any cues suggesting the importance of such standards, but it also could have resulted from the relative uniformity of student mores and their subjective nature. Discipline is something external, and its presence or absence is easily sensed by the student. Ethical standards are internal, and people can be unaware of the possibility of, and even more of the desirability of, alternative moral attitudes. As we noted in Chapter 2, moral teaching in public schools has been steadily declining throughout this century. A reversal of this trend may be a key to eliminating or at least lessening the wrong of dishonesty condoned.

H. Major Crime Unpunished

In 1967, a Commission on Law Enforcement and the Administration of Justice, appointed by President Lyndon Johnson, reported: "Organized crime is not merely a few preying on a few. In a very real sense, it is dedicated to subverting not only American institutions, but the very decency and integrity that are the most cherished attributes of a free society.

"The extraordinary thing about organized crime is that America has tolerated it for so long" (5).

Twenty years later, this national blight is still present, and despite a recent setback, seems likely to retain its powers for corruption and evil. While the prevalence of crime in any of its many forms is a cause for concern, the persistence of organized crime is a wrong of unique dimensions. It is this particular failure of our system of justice that we shall consider here.

THE MAFIA: HISTORY AND ORGANIZATION

There are two chief organized crime groups operating in the United States, the Mafia and South American drug smugglers. The Mafia has been here since the 1890s. Its members are spread across the country, have penetrated labor unions and many businesses, control various frauds and rackets, and have bought officials who thenceforth protect rather than expose and prosecute. The South American drug rings, operating from Bolivia and Columbia, are concerned primarily with smuggling cocaine and marijuana into this country. The Mafia also is in the narcotic business, especially the heroin trade and drug sales (11). Despite the magnitude of the threat that the South American drug interests represent, they are less of a challenge to our basic institutions than the Mafia. It is on the Mafia, therefore, that we shall concentrate.

Crime loves the dark, and the Mafia has done its best to operate unseen or, when seen, to appear respectable. Facts have been hard to come by, and there have been responsible people who have gone so far as to question the existence of a well-structured and closely-knit organization based on family groups. Mafia killings make the headlines but tell us little about who perpetrated them.

Firm information comes from three principal sources. Mafia members, usually to escape or lighten a threatened conviction, may be persuaded to talk. State or local police or members of the Federal Bureau of Investigation (FBI), using telephone taps or electronic bugs, may listen in and record revealing conversations. Sometimes this involves actual penetration of the Mafia by law agents posing as criminals. A third uncommon but important source is the occasional citizen who has been contacted by or has become involved with the Mafia and who takes the risk of revealing the contact to and subsequently cooperating with the FBI. In all these cases, great personal risk is likely to be present and in all of them electronic recordings are likely to provide the key evidence. By piecing together separate facts derived over many years in these various ways, a rather complete picture of the Mafia has been constructed.

Numerous books have been written about the Mafia. I have drawn principally on two: Donald Cressey's *Theft of the Nation: The Structure and Operation of Organized Crime in America* and Jonathan Kwitny's *The Mafia in the Marketplace*. I have also drawn on many newspaper and magazine accounts of separate incidents or developments.

The Mafia in the United States is a lineal offshoot of an ancient, still powerful, and remarkably similar criminal body in Sicily. It is held together by family ties, by the lure of the great wealth derived from its illegal activities, and by the probability of quick death for trespass or disloyalty. A few Italians from mainland crime families, known as the Camorra and based in Naples, have been taken into the inner structure. These core units have been expanded by the addition of many non-Italians, but the power is centered in family groups of Sicilian or Neopolitan origin.

Some of the earliest reports of the Mafia in the United States came from New Orleans, where the murder of the chief of police in 1890 was attributed to this group (6). The major influx, however, occurred later. Among the many Italians that came to the United States early in this century were numbers of Sicilians. These immigrants, for the most part, settled in eastern cities. Soon there were reports of murders—men shot in the streets—and references to "The Black Hand." The great majority of Italians, though totally without involvement in the crimes, failed to come forward as witnesses. Terror and a lack of faith in the law served to protect the perpetrators.

The adoption of the Prohibition Amendment to the Constitution in 1920 offered a gold mine of opportunity to organized crime. Bootlegging gangs of various origins—Irish, Jewish, Polish, German, Italian, and Sicilian— moved in to mine the treasure. Gang wars soon developed. A bloody contest between Sicilian and Italian gangs in 1930–31 led to a confederation of these two, and thereafter the Sicilians-Italians dominated the field. The wealth they had accumulated by the time prohibition was repealed in December 1933 gave them an entrenched power that they still successfully maintain.

During the later years of Prohibition, the Sicilian Mafia in the United States was reinforced by an immigration, often illegal, of Mafia members who fled Sicily and southern Italy when Benito Mussolini ordered a crackdown. Italy's gain was our loss; many of these immigrants seem to have assumed positions of leadership in the American Mafia. The organization thus emerged from the Prohibition era strengthened with both new money and new blood.

While most accounts agree on the importance of family in the structure of the Mafia, there are some differences as to detail. It seems reasonably clear that most organization members are of Sicilian descent, though some mainland Italians may be involved; that marriages are often and perhaps usually within the group; and that family connections play a role in determining status within the hierarchy. The requirement for membership, according to a Mafia hitman from Los Angeles who had become an FBI informant, is relationship, or sometimes friendship, with an existing member. Also, according to his statement, "You have to be Italian." Asked "How do you get out?" he replied, "There's no way out, sir. You come in alive and you go out dead."

From its beginnings in east coast cities, the Mafia has spread across the country. In 1980, Justice Department lawyers were assigned to anti-Mafia work in 26 cities spreading from Boston to Miami and Los Angeles to San Francisco. Attorney General William French Smith, in testimony before the Senate Judiciary Committee, referred to 25 Mafia families, each usually operating in a single metropolitan area.

The number of members in Mafia families is one of the less certain statistics. One estimate places the number at 2,000, another at 5,000 (8). Patterns of family organization doubtless vary, but it seems clear that they tend to be hierarchical and despotic. Each family is headed by a boss or don whose authority is maintained by his control of promotions, of funds, and of the hitmen who execute the killings he alone can order. In at least some instances, there appears to be an organized hierarchy below the boss, with the higher positions going generally to relatives of the boss (5). According to the hitman turned FBI informant already cited, the Mafiosi at the bottom of the hierarchy, the "soldiers," are new mob members who "do all the dirty work." Above them are five *capos* or captains, each supervising ten soldiers. There is also an underboss who gives orders in the absence of the boss, a *consiglieri* or counselor who acts as advisor to the family and, according to Cressey, at least one family member who specializes in the corruption of public officials.

It is interesting to note that the hierarchical features of this structure are often found in other absolute systems. Hitler initiated a somewhat similar hierarchy and so did the Incas in Peru. Another similarity between the Mafia and other despotisms is the relative frequency of succession by murder.

While control of organized crime is centered in the 25 or so Mafia families, its practitioners include other sizable ethnic groups. This larger organization, with an estimated membership of 20,000 includes, among others, Jews, Irishmen, WASPs, and Eastern Europeans. To distinguish this larger organization from the core structure—the Mafia—it is usually referred to as the "mob" or the "syndicate." Non-Sicilian syndicate members, if they can blaze the way to greater sources of illicit wealth, may rise to positions of substantial power. Arnold Rothstein, Bugsy Siegel, Meyer Lansky, and Lepke Buchalter are persons in this category, well known to the police and the press but seldom troubled by the law. Despite the power some non-Sicilian syndicate members have achieved, the ultimate power of life and death has remained in the hands of the Mafia bosses (7).

According to a 1983 report, four "clubs," derivatives of the Hell's Angels, Outlaws, Banditos, and Pagans—youth motorcycle gangs of an earlier decade—have recently become allied to the Mafia. The clubs are said to have nearly 4,000 members and many additional "associates." The club chiefs, though apparently subordinate to the Mafia dons, are major crime figures in their own right.

Gangs in some of our inner city schools are probably turning out more future mobsters. School gangs are nothing new, but an officer in a New York City police intelligence unit, cited by Senator Birch Bayh, reported in 1977 that gangs were modeling themselves on the Mafia, and that their operations had become correspondingly more sophisticated and effective. It would be hard to find a more devastating tribute to the pervasive influence of the Mafia.

Inevitably, there have been jurisdictional disputes between different Mafia families. Sometimes these have led to bloody gang wars (11). In an attempt to curb these, some leading families have set up a shadowy but apparently well-substantiated national commission. According to testimony given in 1980 before the Senate permanent investigations subcommittee by James Nelson, unit chief in the FBI's organized crime section, the membership of the commission consists of the bosses of the five New York City families, plus one each from Chicago, Philadelphia, Detroit, and Buffalo.

The Mafia in Sicily

The ancestral, or Sicilian, Mafia is far older than its American derivative but is no less alive today. Recent examples of its malevolence and power have led to new efforts at control. Mafia murders have increased; in 1982 the police recorded 230 in Palermo, the Sicilian capital, and its environs. Most of these killed were Mafia members, but also included were judges, local politicians, policemen, and journalists concerned with exposing Mafia activities. All Italy was outraged when news came from Palermo of the murders of General Dalla Chiesa, appointed by the government to lead the anti-Mafia drive, and his 32-year-old wife. In December 1981 Pope John Paul II threw the weight of the Catholic Church into the crusade. Speaking to 21 Sicilian bishops assembled in the Vatican, he said:

> "There exist unfortunately certain aberrant phenomena, dating back centuries. These are the so-called Mafia mentality and structure which create, at certain levels, and manifested in various ways, misdeeds which are harmful for the good name of Sicily and its people. Such deviating mentalities claim to be outside the law and to be able to violate it with impunity."

He urged the bishops to fight against the fear-inspired silence of the Sicilian people, which shields offenders from the police and encourages murder and violence.

The Camorra, or Neapolitan crime organization, is also very active. Since the destruction of large areas of Naples in the earthquake of 1980, it has greatly expanded its power and influence in the city. It is estimated that it has 5,000 members and takes in over $2 billion a year from drugs, smuggling, arms deals, and prostitution. The city's murder rate has tripled.

The Sicilian Mafia, and in general also the Camorra, is remarkably like its American counterpart in both organization and methods. It is based on autocratic family groups. It does not hesitate to murder when murder serves its purpose. It uses, with callous calculation, the power thus achieved to extort a substantial tribute from the law-abiding population. And despite the nearly universal recognition of the evil it represents, it remains above the law.

As noted by Pope John Paul II in his statement to the Sicilian bishops, the Sicilian Mafia is centuries old. Its origins are obscured both by time and by its rule of silence. In any attempt to analyze the character and traditions of its members, we have only a few clues to guide us.

Sicily, because of its exposed Mediterranean location, has been invaded and conquered many times in its long history. Cressey lists nine foreign powers that have ruled it (5). Settlements by the ancient Greeks, beginning in about 735 B.C., played a particularly important role in its development. For a while, under Greek influence, the people enjoyed some degree of freedom and the arts flourished. But through much of its history, Sicily has been governed by native tyrants of foreign despots. In the second century B.C., under Roman rule, most of the population was reduced to slave gangs that tilled the estates of rich natives and Roman settlers. The harsh conditions led to two major slave wars. In the second century A.D. there was another revolt of slaves and bandits. Subsequent to the period of Roman domination, Sicily was conquered and for considerable periods ruled by Saracens and Normans. There also were lesser invasions (7).

Perhaps partly in response to foreign oppression, the extended family has come to play an important part in Sicilian life. Intermarriage in small peasant villages probably has also played a role (5). The distribution of wealth has been uneven and most families are very poor (7).

Out of this amalgam of tyranny, foreign oppression, slavery, poverty, and strong family groups emerged one family that was set apart by its successful accumulation of wealth through extortion and terror. Though outside the government and the ruling class, this Mafia family became virtually a clandestine government that may have provided some service in the form of protection against bandits, but that also displayed the worst aspects of despotism (5).

CRIMES WITH BIG RETURN, LITTLE RISK, AND MURDER IN THE BACKGROUND

The focus of Mafia activities is the acquisition of wealth. Over the years, the American Mafia has successfully followed many routes to this goal. Four of these, which trace back to the early years of this century, are extortion, loansharking, illegal gambling, and pornography.

Lesser criminals are the easiest victims of extortion—they dare not go to the police for help. Over the years, Mafia families have collected millions in tribute from bookies, gamblers, prostitutes, pornographers, bank robbery rings, and other illegal groups. Cooperation may bring help in case of police intervention, but the penalties for noncooperation are brutal and swift. Legal businesses can also be victimized. Night club owners have been a common target. They either turn over 10% of their income to a Mafia collector or their persons or property suffer attack (5).

Persons desperate for money because of gambling losses, a business crisis, or other, similar reasons are likely targets for Mafia loansharking. Interest rates can run up to 50% per year or more. Security for the loan is the realization on the part of the debtor that nonpayment can bring anything from a broken hand to death.

Our propensity for gambling means treasure for the Mafia. According to Attorney General Smith and FBI Director Webster, gambling ranks along with narcotics as the top source of illicit profits. Mafia families not only extort huge sums from illegal gambling activities, which they almost totally control, but also derive illegal income from legally operated casinos. According to FBI information reported in 1980, members of a Kansas City Mafia family headed by Nicholas Civella maintained a secret interest in five well-known Las Vegas casinos and, through devious means, skimmed thousands of dollars from some of their own operations as a means of evading federal and local taxes. The opening of casinos in Atlantic City in 1978 led to a Mafia struggle for control. Angelo Bruno, who had dominated the numbers racket in Atlantic City for almost a quarter century, was slain, reportedly on the order of East Coast dons. Nineteen other killings in Philadelphia and New Jersey families, according to underworld testimony obtained by the FBI, were also inspired by the struggle. It is reported that the casinos are providing easy credit for gangsters.

Pornography is mentioned in a number of accounts as one of the oldest and largest sources of Mafia wealth. According to a Jack Anderson column, it ranks third after gambling and drugs as an income maker. The methods of operation are typical. Rather than engaging directly in the sale of pornography, Mafia families collect tribute from the small operators. The toll is said to run into the billions.

Though not among the older, first four areas of Mafia crime, drug dealings were noted by Cressey in 1969 as a Mafia activity and by 1980, it has assumed a major place (5). Heroin produced by Sicilian Mafia families is smuggled into the United States and sold by American families. The value of the trade has been estimated at $6 billion annually and the profits at between $500 million and $600 million. In October 1984, U.S. Justice Department officials, acting under a new extradition treaty with Italy, ordered the arrest of 28 Sicilian-born American residents involved in the international heroin trade and wanted in connection with Italy's crackdown on the Mafia.

According to reports published in the 1970s, the hijacking of trucks carrying valuable freight was a source of substantial income for some New York, New England, and Midwest Mafia families. Trucks were held up in broad daylight, often as many as 100 a day in the New York metropolitan area. The annual value of stolen goods was estimated at $800 million. A friend of mine in Maine who had built up a business processing poultry for the Boston and New York markets remarked on the substantial losses he took on New York shipments.

Two increasingly important areas of Mafia involvement are the theft of cars, which are then cut up to provide parts, and the illegal disposal of toxic industrial waste. Experts in the problem of auto theft place the number of stolen cars in the one million range and the profits per car at about $5,000. The role of the Mafia in the dumping of toxic waste is less well documented, but several reports point to a connection. New Jersey's deputy attorney general in charge of toxic waste enforcement said, in testifying before a House subcommittee, "I believe that the waste industry is probably one of the most violent industries in the nation today. There have been murders, threats, arson. Organized crime plays a major role."

An activity of the New York Genovese crime family is said to be contract killing and the acquisition of arms for this purpose. The Kansas City family headed by Nicholas Civella is said by the FBI to be engaged in bombing and arson (8).

Three areas of Mafia activity have a particularly malign influence on American society: their penetration of labor unions, their penetration of business, and their corruption of public officials.

THE PENETRATION OF LABOR UNIONS

The Teamsters Union, with about two million members, is the most powerful union to have been infiltrated by the Mafia. The early history of the infiltration is obscure, but it probably goes back to the early years of the century and started with one or two locals, whose heads could be threatened or bribed into cooperation. In due course, corruption touched the union leadership. As of 1983, three of the union's international presidents had been convicted of federal crimes. A son of a fourth president was convicted in 1979 of accepting $100,000 in kickbacks when he was an officer of a Detroit local.

There is a long history in this union of beatings, bombings, and murders. In January 1983, underworld financier Allen Dorfman, who had been associated with the Teamsters for more than 30 years, was the victim of a gangland-style killing. In the 1960s, a former Kansas City Teamster official, Floyd Haynes, was shot to death in a bowling alley parking lot shortly after talking to FBI agents. His testimony implicated Roy Williams, then president of Kansas City Local 41 and subsequently Teamsters Union president. According to Haynes's account, members of the Nick Civella family demanded that Williams enroll his local in a fraudulent medical plan and backed the demand with the threat to "kill his children, then his wife, then him." Williams, already too involved with the Mafia to resist, succumbed to the threats and let the union treasury take the loss.

Of all the sources of illegal gain from union control by the Mafia, the most important have been the multibillion dollar union pension and welfare funds. While the details have been carefully obscured, there appear to have been huge, unpaid loans to mob-related businesses and Las Vegas casinos. Corrupt union trust fund service providers have collected excessive fes. There also has been outright embezzlement. All told, the Mafia has stolen many millions of dollars.

Another major union with a long history of Mafia penetration is the International Longshoreman's Association. The movement of enormous volumes of valuable freight through the docks of a few waterfront cities has provided a vulnerable point for extortion and theft. A shipping executive, Joseph Teitelbaum, told a Senate subcommittee in 1981 how he first resisted making payoffs to union officials, yielded to buy labor peace and to prevent sabotage, then profited from kickbacks, and finally turned FBI informant. He also told of direct theft of valuable cargo moving through the docks (6). Millions of dollars have been extracted by these methods from waterfront companies (9).

Various other reports point to the Amalgamated Meat Cutters' Union and to construction unions working on publicly financed projects as additional victims of Mafia penetration (6).

THE PENETRATION OF LEGITIMATE BUSINESS

Like other people with accumulated wealth, Mafiosi seek profitable investments. Increasingly, they have turned to legitimate business. The story of one honest man who accepted the risks of a double life as Mafia collaborator and FBI informant tells much about their methods (2).

Louis Peters grew up in Maine, moved to California as a young man, and worked his way up to the ownership of a prosperous Cadillac dealership. In 1977 he was approached by a local contractor whom he knew who said he had a buyer for the dealership. The interested party, Peters was told, "has all kinds of money," so Peters could name his own price. He did—$2 million—a good deal more than he thought the business was worth. His acquaintance came back in a few days and said the offer was accepted.

A meeting with the purchaser, Joe Bonanno, and his son was arranged. The impressions gained during this meeting, the price, and hints that Bonanno wanted to buy other dealerships in the area aroused Peter's suspicions. He went to the local police who advised him to contact the FBI. At the FBI office, he was shown pictures, among which he recognized the purchaser of his dealership. It was then that he learned that Bonanno was in organized crime. He later came to know Bonanno as a native Sicilian and a Mafia don.

At the request of the FBI, Peters agreed to work as an undercover agent. He had recently been divorced and was living alone, hence was free of family obligations that might otherwise have made the assumption of this role more difficult. For two years he was accepted as a member of the Syndicate. He gained the confidence of the Bonanno family and rose in the Syndicate hierarchy. Throughout the period, the FBI was listening in. Agents had rented an apartment for him and filled the floor above with recording and surveillance equipment. It was a difficult role to play. Peters said that he sometimes felt like two people. "It took a lot of my time and a lot of my effort to gather evidence. It takes all of your mental capacity to concentrate on what's going on and what's happening."

For about two years all went well, but eventually the organization became aware of Peters' dual role. Overnight he was transformed from friend of the boss to intended victim of a Chicago hitman. Fortunately, Peters and the FBI learned of the transformation. With the aid of 6 months in hiding, a bullet-proof vest, and eternal vigilance, he survived the threat. His ex-wife and daughters were also guarded.

During the 2 years that he worked with the California Mafia, Peters learned a great deal about the plans and purposes of its members. The prime reason for acquiring his auto dealership was to use it as a means for laundering dirty money. Criminal transactions are generally on a cash basis, but at times the best place for even criminal wealth may be a bank or conventional investment. Opening a bank account or buying stock with cash, however, may involve risk. Banks and brokerages are required by the Internal Revenue Service to report any cash transactions in excess of $10,000, and in general they probably do, although there are reports of lapses. Criminals with large sums of cash at their disposal, therefore, often look for means to hide their source.

It was for this reason that Peters could be valuable to Bonanno. Once he was accepted as a safe accomplice, he had frequent night visits at his apartment from mob members who launched into discussions of ways illegal wealth could be made to appear as if derived from legitimate business deals (11). It then could be deposited in banks or otherwise used without leaving records that would suggest its criminal source. Peters also helped the Bonanno family in another matter, destroying some business records dangerous to Bonanno Jr., who at the time was under investigation.

In addition to the methods used by the Mafia, Peters also learned something of the magnitude of Mafia operations. The Bonanno family planned to spend $30 million to $40 million in the Lodi-Stockton area, the quiet, out-of-the-way county where he lived. It would become the headquarters for their activities in dope pushing, prostitution, racketeering, and extortion. Other auto dealerships would be purchased.

In a phone interview with a reporter from the *Bangor Daily News,* Peters summed up his experience. "I wouldn't trust the organization against anything. They're a bunch of animals. They have no feelings about you, me, or anyone else. And they are just as powerful today as they have ever been."

Joe Bonanno was convicted in 1980 of conspiring to obstruct justice. He was found to have tampered with a special grand jury investigating his son's business dealings. He was 76 years old and retired. Never before had he been convicted of a felony. Louis Peters could feel that his efforts were not entirely in vain but, as he said, the mob remained just as powerful as ever.

The acquisition of legitimate businesses to launder dirty money is only one of many forms of Mafia involvement. We can scarcely do more here than outline their activities.

Illegal profits can be made from a business through fraudulent bankruptcy. A company with a good reputation dealing in salable goods is acquired or established by a front man. A large inventory is then built up with borrowed money, the goods sold through a "fence," and bankruptcy declared. The Mafia has the profits and the creditors are left holding the bag.

The Mafia has penetrated banking, often through the leverage provided by a mob-dominated union local. An ambitious young banker, in return for large union deposits at favorable interest rates and sometimes even personal favors, indicates a willingness to make loans to new "customers" the union official says he will recommend. The loans are made and, of course, go unpaid. Big sums are involved. In a period of a few years subsequent to 1973, according to a federal investigation report summarized by Kwitny, such Mafia raids caused at least $10 million in losses and the failure of at least four New Jersey banks. Other banks both in and outside of New Jersey were reportedly in trouble. The Mafia has also mulcted the stock market through the agency of stockbrokers whose personal gambling debts had made them victims of Syndicate loansharks (8, 11).

Mafia acquisition and corruption of meat processing and distributing firms in the late 1970s has led to profits through the sale at normal prices of low grade or contaminated meat. One such case involved Merkel, a meat distributing firm that was in financial difficulty in part because of depredations by a Mafia-controlled local of the Amalgamated Meat Cutters' Union. This provided an opening for the Mafia, and the firm was purchased by mobster Nat Lokietz. Merkel soon started distributing ground beef to which 25% horsemeat or spoiled beef treated with formaldehyde had been added. Purchasers included the New York City school system, state hospitals and prisons, the Army, the Air Force, restaurants, and hotels. Low grade meat was also stamped and sold as being of a higher grade. These activities all required the corruption of public officials, especially meat inspectors, and of purchasing agents (5, 8).

Other Mafia business operations requiring the collaboration of officialdom are garbage collection, school busing, and public construction. Profits come through excessive charges and, in the case of construction, inferior products (5, 11).

THE CORRUPTION OF PUBLIC OFFICIALS

All crime hurts society, but the power for corruption unique to organized crime represents an evil far in excess of other forms of lawlessness except perhaps some crimes in the white-collar category. Mafia families have a long record of gaining, through intimidation and bribery, the compliance of the public officials who should be tracking them down. This collaboration of officialdom with the criminal represents a corrosion of the system of law and order on which our whole society is based.

Collaboration is not easily proved, since every effort is made to hide it. Nevertheless, largely on the basis of recorded conversations, many cases have been established.

Information was slow in surfacing. In the first part of this century, according to Kwitny, a loose political coalition managed to suppress evidence even of the existence of the Mafia. This was the era when some city political machines were Mafia dominated. The cover-up was sponsored by congressmen who either belonged to these machines or were dependent on their votes, and by others who were sincerely but overzealously concerned with the threat to civil rights created by wiretaps and listening devices (8).

In the 1960s, substantial evidence of Mafia influence in the courts and legislatures was published. A 1967 *Saturday Evening Post* article, based on government wiretaps, cited a considerable list of New York State officials who were in the service of the Patricia family. Included were a high-ranking state official, a police chief, two licensing officials, a high-ranking court administrator, and several powerful state senators and representatives.

Legislators at the federal level have also been implicated. Kwitny cites FBI bugs and taps linking Representative Gallagher of Bayonne, NJ, with a local Mafioso. And Victor Riesel, a reporter whose dogged pursuit of organized crime led to the loss of his eyesight when sulfuric acid was thrown in his face, told in 1967 of the ties of Senator Ed Long of Missouri with Mafia-connected union officials. One of the senator's friends was Roy Williams, subsequently one of the three presidents of the Teamsters Union to be convicted. Long had been a consistent opponent of wiretapping by federal law enforcement agencies (9).

Corroboration of successful manipulation at the government level was provided by a study carried out by the American Bar Association. In the words of the association's report, "The largest single factor in the breakdown of law enforcement dealing with organized crime is the corruption and connivance of many public officials" (5).

WHY THE MAFIA STILL THRIVES

The war between the law and the Mafia has been going on for more than 90 years. So far, except for a recent setback, the Mafia has been the winner. If we have failed to wipe out this blight on our national integrity and honor, it has not been for lack of investigations and promises. There has been one investigation after another. Each time the public's hopes have been raised; each time the accomplishments have been minimal.

President Hoover ordered an investigation in 1931. Senator McClellan initiated one in 1957. Our quotation at the beginning of this chapter is from the 1967 report of a commission appointed by President Lyndon Johnson. A Commission on Organized Crime appointed by President Reagan in 1983 wound up an extensive investigation in January of 1986. Major attempts at prosecution were carried out by John Dewey in New York City in 1935–37 and by Attorney General Robert Kennedy in the early 1960s. Both Dewey and Kennedy achieved some success; Kennedy might have done real damage to the Mafia were it not for his assassination in 1964 after four years in office (5, 7, 11). The most successful attempt to date to break the Mafia was launched by Rudolph Giuliani in 1983 following his appointment as U.S. Attorney for the Southern District of New York. This investigation is still going on. I shall return to it later, but first I will consider some of the reasons for the earlier failures.

The first reason for the survival of the Mafia is fear. If there is any one thing at which the organization excels, it is murder. Mob killings run into the thousands—almost 1,100 in Chicago alone over a period of six decades according to the Chicago Crime Commission (11). There were six recorded murders in a single Teamsters Union local in Philadelphia between 1967 and 1976. One of those murdered was the leader of an attempted reform. The perpetrators of the killings are seldom caught.

Another and very major reason for the failure of attempts to break the Mafia is its enormous wealth. According to a 1986 estimate, if the Mafia were a public company its revenues would top General Motors' on the Fortune 500 list of major U.S. corporations. Wealth on this scale is very effective in buying protection. As William French Smith, Attorney General in the early years of the Reagan administration, said in testimony before the Senate Judiciary Committee, "The dollar amounts are so great that bribery threatens the very foundations of law and law enforcement." Money hires high-powered lawyers who can befuddle juries and who know all the stalling tactics that can drag out cases and delay convictions year after year. It makes bail almost meaningless. South American drug smugglers frequently put up a $1 million bond and then head for home. Skipping bail may be a little less attractive for American-based Mafiosi, but it does occur.

A third source of Mafia invulnerability, a direct byproduct of its wealth, is the ability of the dons to buy the semblance of respectability. Mafia chiefs, when it serves their purpose, give generously to appropriate good causes. A Magliocco family don listed by the Justice Department as an organized crime figure made large gifts to Brandeis University and was rewarded by being named in 1977 as Brandeis "Man of the Year" (8). Carlos Marcello, member of a New Orleans Mafia family, who has sunk profits from extortion and other rackets into businesses that range from real estate to music companies, successfully built a new image as a philanthropist, a churchgoer, and a strong family man. Forty solid citizens vouched for his good character when he had a scrape with the law. That was in the 1960s (11). In the early 1980s, at age 1972, he was finally convicted.

A fourth reason for the survival of the Mafia, somewhat related to our third reason, is, or at least has been, disbeliefs as to its reality. Representative of this attitude is an article by Professor Dwight C. Smith, Jr., "Mafia: The Prototypical Alien Conspiracy," published in the *Annals of the American Academy of Political and Social Science* in 1976. In the judgment of Professor Smith, the Mafia is solely the creation of a conspiracy phobia common to many Americans (10). Donald Cressey, when serving in the 1960s as an organized crime consultant to a federal commission, found skepticism widespread among his friends. Even J. Edgar Hoover, for many years the capable but sometimes controversial head of the FBI, was for a long time among the skeptics or apparent skeptics. It was not until Robert Kennedy became attorney general that he revised his position. Cressy speculates that information from wiretaps and new powers given to the FBI by Congress played a role in his conversion (5).

The image of the Mafia as solid citizens can serve as a source of protection. U.S. District Court Judge Terry Hatter, Jr., in sentencing five Southern California Mafia bosses convicted of racketeering and extortion, praised the mobsters as being devoted family men who could not break a "bond with their crime family." He imposed sentences ranging from 2–4 years in prison plus fines, although sentences of 20 or more years would have been within the law. Actual imprisonment was delayed by a lengthy process of appeals. Such light sentences are not a rarity. A 1985 study by the General Accounting Office showed that sentences given Mafiosi and drug traffickers are usually less than the allowable maximum and often substantially less (1).

Where skepticism does not protect the Mafia, indifference often does. Known crime records have not significantly hampered the careers of a number of Teamsters Union officials. Millions of Americans make use of the availability of illegal gambling, pornography, prostitution, and loan-sharking under the control of the Mafia. An unidentified federal investigator has suggested that "Because so many people will knowingly associate with organized crime, it will never be eradicated." This, I think, is an exaggeration: With the partial exception of narcotics, these widely sought illegal goods and services are provided by small time criminals; the Mafia enters the picture primarily as an extorter of profits. Nevertheless, it certainly is true that patronage creates indifference.

Another deterrent to the successful pursuit of the Mafia is a widely felt concern that the methods employed infringe on civil rights. The use of wiretaps and bugs, the storage of personal data in computers, the availability to the FBI of income tax returns, and the employment of stings, which some people view as inciting guilt, have all come under fire (4, 8). Undoubtedly, the use of these methods imposes some risks, even with well-devised controls, but do these outweigh the enormous economic costs and pervasively corrupting influence of a powerful Mafia? An examination of the facts, it seems to me, can lead to only one conclusion—the existence of major crime unpunished is a wrong so great that we can no longer afford to tolerate it.

A final cause of the Mafia's invulnerability has been a considerable degree of indifference at the top. Which brings us to the one attack on the Mafia which verges on a real success story (12). A record of dedicated and brilliant prosecution of criminals has marked the career of Rudolph Giuliani, whose appointment in 1983 as U.S. Attorney for the Southern District of New York gave him an opportunity to take on the Mafia. Besides his devotion to the job, he had a number of things working in his favor. The legal background has become more favorable. Giuliani has mastered and made maximum use of the Racketeer-Influenced and Corrupt Organizations Act, passed in 1970, which authorizes prison sentences of up to 20 years and maximum $25,000 fines if a "pattern of racketeering can be demonstrated." Perhaps in part because of the complexity of this law, previous prosecutors have made little use of it. Other recent laws now permit the Internal Revenue Service to share information, which can be used in various ways by the prosecution. Reagan has also removed limits on court-ordered wiretaps and the number of these has doubled, a very important step forward.

State and federal officials are cooperating more effectively than in the past, and the Reagan administration has increased cooperation with Italy to detect areas of collaboration between the American and Italian Mafias in the heroin trade.

The Reagan administration has also, in general, provided the large sums necessary for successful pursuit of the Mafia. The number of FBI personnel assigned to the investigation of organized crime (drug smuggling included) has jumped 19% since January 1981. According to a 1987 report, however, there are plans to cut back on a very expensive but important program. Underworld figures turned informant are a major source of the information necessary for the conviction of the top crime figures. A major inducement for Mafia figures who face imprisonment to turn informer is a witness-protection program which provides new identities, houses, fake biographies, and other forms of help from the FBI. Giuliani has made effective use of this program. The program costs $19 million a year, and it has been stated that, to save money, government help will be extended to each informant for a limited period only. This could substantially reduce the number of informers.

A final development which has strengthened Giuliani's hand has been a steady improvement in listening devices and other technical resources. A giant computer installed during the Carter administration provides virtually instant access to all known data concerning mob figures. More important are improvements in the "bugs" that are the single most vital source of the sort of information necessary for the conviction of members of the mob. Their potential is indicated by the many conversations made part of the record by way of a listening device installed in the Jaguar owned by the chauffeur of a Mafia don. Bugs and wiretaps have enabled agents to listen in on nearly a million conversations a year.

Additional crucial information comes from undercover agents who gain the mob's confidence. Creating a suitable role for an agent usually requires costly accessories such as yachts or fake businesses. This is both an expensive and a risky approach.

Making maximum use of these methods, Giuliani obtained convictions of 231 members of the Mafia in the 18 months before April 1986. Some top crime figures were included. Not all prosecutions have been successful. John Gotti, reputed ruler of the nation's largest crime family, and six other alleged members of this family, were acquitted by a Brooklyn jury in March of 1987, much to the consternation of those who had followed the case. But there have also been additional convictions.

Giuliani's successes have led to talk about the end of the Mafia, but few experts would go that far. There are still hundreds of members of Mafia families at large, with billions of dollars at their disposal despite some heavy fines and seizures of assets. As long as this combination exists, the threat of the Mafia, with all of its power for crime and corruption, cannot be dismissed.

I. Conclusions

COMPLEXITY OF ECONOMIC CAUSATION

As a biologist, I have come to accept the extraordinary complexity inherent in human nature. I shall document the biological background of this complexity in Chapter 4 when I examine the human genetic and neurophysiological mechanisms. And besides the complexity inherent in our neurophysiology, the brain has the capacity, through the peripheral nerves and sense organs, to react with and be modified by the environment, thereby compounding the potential for complexity. Our behavioral potential, moreover, is not only diverse, but characterized by individual differences.

Like many other aspects of life, human nature, besides its diversity, manifests uniformity. Our shared need for companionship, sex, food, water, clean air, clothing, shelter, and a means to get about imposes shared behavior patterns, though again with some individual variation.

If the determinants of individual behavior are complex, the determinants of social behavior should be additionally complex by another order of magnitude. The potential for causal interactions are magnified as we go up the structural scale. Observation confirms this. But again we may expect uniformities. An airline passenger flying over Los Angeles during the morning and evening rush hours can see, in the streams of traffic flowing along the highways, an almost ant-like patterning of behavior.

This patterned complexity is reflected in our economic behavior. Just as our tastes and needs show both uniformity and diversity, so also does our production and exchange of the goods and services that satisfy them. Some of the most significant uniformities are imposed by the tools that we have designed to facilitate and organize our economic activities. We use money as a medium of exchange. We have banks and other financial organizations to serve as financial intermediaries between savers and borrowers. We have stock exchanges. We have regulatory authorities created by law, such as the Federal Reserve Board and the Securities and Exchange Commission. We have governments endowed with the power to tax. It is these man-made agencies that give our economic behavior much of the uniformity that to some degree justifies the claim of economists that theirs is an exact science.

Despite their accomplishments and the prominent role they have achieved in government and the marketplace, economists during the last decade have come in for substantial criticism, some of it from their own colleagues. We read in a 1985 article that President Reagan is ignoring the advice of economists because their predictions so often have turned out wrong. Also, economist Lester Thurow has commented on the current low standing of his brethren and British economist Sir Alec Cairncross stated at a meeting: "Often, economists have little to contribute to policy; their elaborate chains of reasoning are of little relevance to the actual problems and aims of decision makers" (1). In 1974, Tibor Scitovsky, professor of economics at Stanford, took exception to the assumption of economists that people in a variety of economic situations display a considerable rationality in their choices. He lists three illusions—three circumstances in which people act contrary to economic assumptions (12). More recently (1982), James Dean, professor of economics at Simon Fraser University and Columbia University School of Business, has pointed to the fallability of economists in an article, "Why Economists Disagree." Noting that "Increasingly over the last decade, economists have formulated theory in terms of *expectations*," he goes on to state that "millions of individuals' expectations play a key role. Yet our ability to predict how these expectations will change is woefully inadequate" (4).

The Moral Component

The determinants of economic behavior are complex, and among them is the degree to which the members of any society adhere to the standards set by moral law. We shall examine ethics and moral law in detail in Chapters 6 and 7. Suffice it to say here that the ethicality of human behavior, which, by definition, is measured by its social consequences, is necessarily of significance to the functioning of the economy. Ethics *is* relevant to economics. And as we shall note, religious leaders and moral philosophers over the years have formulated principles that serve as guides to ethical behavior. Of all these principles, the Golden Rule, "All things whatsoever ye would that men should do to you, do ye even so to them," is perhaps the most universal in its applicability. These moral laws are quite as relevant today as they ever were. Used as a test of behavior in the economic domain, they reveal many instances of conduct that is unethical, though easily condoned because much of the harm falls on strangers. We can quite properly take such acts into consideration in our search for the causes of our economic problems.

Amitai Etzioni, Director of the Center for Policy Research at George Washington University concurs in this judgment. He notes, for example, that "the higher the ethical commitments of individuals who transact with each other, and the stronger the shared norms of proper conduct, all other things being equal, the lower the production costs and the higher productivity" (5). William Simon, Secretary of the Treasury during the Ford administration, has also noted the link between ethical standards and the successful functioning of an economy (11).

Neither Etzioni nor Simon is an economist, but Wilbur and Jameson, who are trained in this discipline, write in *An Inquiry into the Poverty of Economics:* "At the most fundamental level, the economic problems of the current day originate in a moral crisis. . . . The moral crisis has occurred because of the erosion of the moral base of society that economists have simply assumed was there, an assumption which is less and less viable" (13).

Our Seven Wrongs

In this chapter, we have named and described seven categories of unethical behavior common in the United States today. Some of the consequences of these behavior patterns have been discussed. The evidence shows, I believe, that these wrongs have contributed to a variety of economic and social problems. Their damaging action may in some cases be quite indirect because their causal role is further back in the causal chain than is that of the strictly economic determinants of our problems. Nevertheless, their role is very real. They are undoubtedly more difficult to alter and manipulate than strictly economic causal factors such as the money supply. They are, however, not altogether beyond our control, and are so fundamental to our well-being that we cannot afford to neglect them.

The hurtful consequences of these wrongs can be social, political, economic, or some combination of these three. Thus all of the wrongs, from waste to our tolerance of organized crime, exact a financial toll. If and insofar, therefore, as we could reduce them, we could be economically better off. We could save funds that could then be spent for such useful ends as the control of air pollution and improved education. Since I have already examined the consequences of the individual wrongs, I shall not repeat this here. I can, however, profitably consider the risk imposed by the economic problems they have helped generate.

An Economy in Trouble

How serious are these economic problems? Do we face the possibility of an economic collapse? Having just noted the difficulties of economic prediction, I am not going to offer presumably final answers to these questions. I do believe, however, that the evidence confirms a very real possibility that we will be faced with a major depression in the not very distant future. I will not even venture to guess the precise timing. It should be noted, however, that the Reagan administration has a variety of tools at its disposal that can be used to delay though not to prevent an economic downturn. These tools, we may be sure, will be put to use.

I have virtually no support from recognized authorities in such a prediction. Economists, particularly if they are in positions of responsibility, though they have often pointed out the seriousness of our problems, have shied away from drawing the ultimate conclusion. Instead, they have played a game which we might call, "If we do this and if we do that and if we do the other, everything will be all right." I agree that a crash can be averted. The problem is that valid solutions require substantial belt-tightening on the part of the American public, and when advertizers for years now have all too successfully exhorted us to practice splurging and self-indulgence, what, in all honesty, is the chance that articles by some economists urging national frugality will suddenly reverse this trend?

By an interesting coincidence, when in October 1987 I reached this particular point in an updating and revision of this volume, the stock market, after five years of rising prices, suddenly took a plunge that, for two days, was worse than anything in the October 1929 collapse that initiated the Great Depression of the 1930s. Economists have given good reasons why the linkage between the stock market and the economy as a whole is weaker now than it was in 1929. There has been a substantial recovery from the low of October 26, but the market drop has scared our politicians into more serious talk of reducing the federal deficit than any we have previously heard. Insofar, however, as any effect on the public is an increased frugality, this will damage rather than help the economy unless the holders of wealth correspondingly increase their spending for truly constructive uses, and all precedents indicate they will do the exact opposite.

What evidence warrants the view that we may be headed for a major depression? The major dangers center in our growing debts and the misuses of wealth are discussed in this chapter. As we noted in our discussion of Figure 3.1, our debts since 1981 have been repeating a pattern last seen during the Great Depression. Our federal, our state and local, our business, and our personal debts are now all moving up together. These debts, huge in their totality and still growing, together with that portion of Third World debt financed by our banks, all hold dangers that could be unloosed by any one of a number of circumstances capable of jolting the economy. Economists have expressed concern that growth in personal debt has reached its limit and that consumer spending will therefore fall, triggering a recession. The average American family now spends one of every five disposable dollars in payments on their debt. So far, consumers, with aid from bankers and merchants, have found ways, such as cutting down on saving, to keep up their spending, but there has to be a limit.

One factor limiting consumer spending, though a slow acting one, is the growing inequality in the distribution of wealth. As a recent newspaper headline, referring to the 1987 Forbes list of the 400 richest Americans, says, "Billionaires busting out all over." The combined wealth of the 400, $220 billion, was enough to wipe out the 1986 U.S. budget deficit. At the same time, incomes at the lower end of the scale have been dropping, thereby cutting the spending potential of the millions whose purchases keep the economy going. In one of the two recent books in which a depression is predicted, *The Great Depression of 1990* by Ravi Batra, it is noted that the crash of 1929 was preceded by a growth in wealth inequality. Batra's formula for a permanently stable economy includes strict limits on personal fortunes (2). Amasa Walker, whose 1866 economics text we cite in the first section of this chapter, also argued that excessive wealth inequalities are harmful (12).

Another fundamental problem that repeats history has been noted by John K. Galbraith in an article, *The 1929 Parallel* (6). This second problem is the growth of huge corporations and conglomerates. The methods used in the 1920s were somewhat different from those used recently, but the objectives were the same. I described this process as it is occurring today in my discussion of the misuse of wealth. This "merger mania," as it has been called, is inspired more by greed than by any well-designed plans for long-term growth. Bigness confers power on the top management—power, for example, to buy political favors—and it usually means fat salaries and splendid perquisites, but any such returns to management come at an economic and social cost (3). Conglomerates are demonstrably inefficient, and takeovers and some other forms of acquisition can leave the corporate giant saddled with unhealthy debt. An increasing proportion of U.S. corporate financing has come from bank loans rather than the sale of stock, the normally preferred method of raising capital. Corporate credit ratings have been dropping (8, 9).

Galbraith points out a third parallel with 1929. Just as Reagan reduced taxes on the wealthy, so also Coolidge and his Secretary of the Treasury Andrew Mellon changed the tax laws to favor the affluent. The theory in both cases was that this would stimulate the economy but, as Galbraith says, "There is every likelihood that a very large part of the enhanced prsonal revenues . . . simply went into the stock market, rather than into real capital formation or even consumer demand."

Wealth inequality and corporate gigantism are fundamental attributes of our system today, attributes with major ethical implications and a potential for causing a diversity of economic, social, and political problems. It is our debts, however, that carry the most immediate threat of economic collapse.

What sort of event could trigger an economic catastrophe in one area of debt that then might spread through other areas of debt like the falling of a line of dominoes? S. Islam has suggested on possibility. Because of our low rate of saving, the poor showing of many of our products in competition with products made abroad, and the unproductive uses to which much of our unneeded income is put, much of our high rate of consumption is financed with foreign funds flowing into this country from abroad. Most of this inflow is used to purchase government bonds, thereby financing the growth in federal debt, but much of this finds its way into the hands of individuals by way of reduced taxes and heavy government spending. Islam points out the danger in this method of maintaining our comfortable life style. ". . . The longer the federal deficit remains high and the United States imports capital to maintain a higher standard of living, the greater the risk that investors will suddenly lose confidence in the U.S. economy." The resulting cessation of the inflow of foreign money could, Islam suggests, create "a macroeconomic crisis" (7). The recent plunge in stock prices in the New York exchange is one of the sort of events that could lead foreign investors to look elsewhere for a place to put their funds.

Michael Moffit, a New York investment advisor, is another commentator who has taken note of our debt bomb and the folly of basing our high living on borrowing from abroad. He also has been led thereby to join the small but growing circle of economists who believe that "a worldwide economic collapse" is likely (9).

If we do have a collapse, we in the United States will be denied, to a considerable extent, a remedy used by President Franklin Roosevelt to pull us out of the 1929–33 depression. Roosevelt used increased federal spending to revive the economy. Figure 3.1 shows the resulting growth in debt. The huge federal debt built by President Reagan to keep the current ailing economy alive and avoid taxing his rich friends will limit the use of spending in any subsequent slump.

PLANNING AHEAD

Major depressions have occurred periodically in the past. That does not prove that they will occur again, but in view of the economic portents just described, we should view one as possible and act accordingly. The regulatory institutions created after the 1929 crash have enabled us, up to now, to forestall any major economic disruptions, but these successes may be only a delaying action that will make the crash, when it does come, worse than it otherwise would have been. An economic collapse in our highly technological, socially and ideologically divided, and comfort-loving society could be disastrous. The one most important step we can take now to avert total disaster is advance planning. Any examination of our problems and of potential remedies will need to be far-ranging but, as we have attempted to show in this chapter, it certainly must include—I might well say *begin with*—our ethical standards (14). These must be clarified, adapted to the needs of our increasingly complex society, and put to work.

One requirement of an ethic that is to work is that it be based on a realistic perception of human nature. The next two chapters are devoted to developing such a perception.

SOURCES USED FOR FIGURE 3.1, PAGE 54.

Disposable Personal Income (DPI) 1929–1986; (Economic Report of the President, 1987, U.S. Govt. Printing Office, Washington, D.C., p. 275).

Federal debt (Outstanding debt held by public) 1929–1933 (Economic Report of the President, 1978, U.S. Govt. Printing Office, Washington, D.C., p. 337).
   1939–1986 (Economic Report of the President, 1987, U.S. Govt. Printing Office, Washington, D.C., p. 331).
State and Local 1929–1973 (Economic Report of the President, 1978, U.S. Govt. Printing Office, Washington, D.C., p. 337).
   1974–1985 (Board of Governors of the Federal Reserve System, Bull. Z.1, Sept. 1986. Flow of Funds Accounts. Financial Assets and Liabilities, year-end, 1962–1987. Washington, D.C., Credit Market Debt, p. 2). There are considerable differences between these two series during the early part of the interval of overlap, but the two are nearly identical for 1973–1976.
Corporate business, Series 1—1929–1974 (Economic Report of the President, 1978, U.S. Govt. Printing Office, Washington, D.C., p.337).
   Series 2—1946–1985 (Board of Governors of the Federal Reserve System, Bull. C.9, Oct. 1986. Balance Shets for the U.S. Economy, 1946, 1985. Washington, D.C., pp. 21–25). The data for Series 2 are listed under the heading "Nonfinancial Corporate Business."
Unincorporated business and farm, Series 1—1929–1946 (Economic Report of the President, 1978, U.S. Govt. Printing Office, Washington, D.C., p. 337). The figures for farm debt include farm mortages and farm production loans.
   Series 2—1946–1985 (Board of Governors of the Federal Reserve System, Bull. C.9, Oct. 1986. Balance Sheets for the U.S. Economy, 1946, 1985. Washington, D.C., pp. 16–20).
Home mortage (nonfarm 1 to 4 family houses) 1939 to 1986 (Economic Report of the President, 1987, U.S. Govt. Printing Office, Washington, D.C., p. 328).
Installment 1929 to 1965 (Economic Report of the President, 1983, U.S. Govt. Printing Office, Washington, D.C., p. 242).
   1966 to 1986 (Economic Report of the President, 1987, U.S. Govt. Printing Office, Washington, D.C., p. 330). Differences between the two sets of figures in the region of overlap are slight.
Noninstallment 1929 to 1965 (Economic Report of the President, 1983, U.S. Govt. Printing Office, Washington, D.C., p. 242).
   1966 to 1986 (Economic Report of the President, 1987, U.S. Govt. Printing Office, Washington, D.C., p. 330).
Credit Card 1968 to 1987 (Economic Report of the President, 1987, U.S. Govt. Printing Office, Washington, D.C., p. 330).

4

The Biological Background of Social Behavior: Genetic Factors

Introduction

It has been said that we are fearfully and wonderfully made. The study of the life sciences has not lessened the validity of this statement; it has documented and enhanced it. Indeed, the more we learn about life in all its many forms and facets, the more wondrous it seems.

In this chapter, I shall try to convey some of the wonder of life as revealed by one of the life sciences—genetics. There are vast sources of relevant material, so a very selective approach is required. I shall draw data almost exclusively from two species of mammals—men because they are our real concern, mice because they are easily studied and yet, as mammals, share so many fundamental structures with man that there can be an extensive information carry-over. We shall center our attention on the enormous complexity and scope of the genetic machinery, the extent of genetic variation between individuals, and the significance of genetics for human behavior and especially human society. We start with the mammalian genome.

The Immensity and Diversity of the Mammalian Genome

The *genome* is an inherited set of instructions for making a new individual. It is passed from the fertilized egg to all the billions of cells that make up the adult, though whether in whole or with some deletions is at present not entirely certain. In some ways it is like a book. The instructions that it conveys can be broken down into separate units that are analogous to letters, syllables, and words. The genomic alphabet contains four "letters," each a chemical compound of a family known as *nucleotides*. Each "syllable" is a group of three letters that codes for a particular one of the 20 *amino acids* from which *proteins* are made. The "word" is a group of syllables that codes for a complete protein. Proteins are large molecules, and the words are correspondingly polysyllabic. Since a *gene* (or more precisely, a *structural gene*) is defined as a group of nucleotides that codes for a protein, our "words" are, by definition, genes. In the hierarchy of molecules characteristic of life, genes and proteins hold first place.

As already noted, the genomic alphabet consists of four letters. It might seem at first glance that this would limit the complexity of the information that can be conveyed, but this definitely is not the case. Computers are not handicapped as information processors, storers, or conveyers by their digital, two-letter alphabet. All that is necessary is an increase in letters per word.

The presence in the genome of units comparable to letters, syllables, and words is about as far as we can press the analogy to a book. In its other aspects, the genome has its own remarkable and often unique properties.

We start with two of the most fundamental properties that must be mentioned, though we cannot even begin to go into the fascinating details.

The genome can reproduce itself, and it does this every time a cell divides. Reproduction is a characteristic of all life, and the gene is the Adam and Eve of the reproductive process. The reproduction is usually perfect, but occasionally a new gene is formed that is a variant. This is how life acquires its diversity. So fundamental are these two aspects of the gene that we can define life as the capacity for mutable self-reproduction.

The second most fundamental property of the gene, and one already mentioned, is its function as a code for proteins. We cannot go into the steps involved in the process of gene translation. Suffice it to say that in the appropriate cells at the appropriate time, each gene produces multiple copies of the specific protein for which it stands.

There is a third less fundamental but important property of the genome that we must mention: In sexually reproducing organisms, each cell contains two smiliar, but usually not identical copies, one derived from each of the two parents. In technical terms, all cells except the germ cells are *diploid*.

The information in a book is printed on the pages of which the book is made. The information is linear in the sense that it is presented in successive lines printed across the page. In the genome, the nucleotides are strung together in a long thread, so that the information is totally linear in arrangement. Collectively, the nucleotides in the thread are known chemically as *deoxyribonuclein acid,* or DNA. We shall refer to the sum total of the DNA in each cell as the *gene string.*

Man-made twine consists of multiple fibers twisted together. As so often

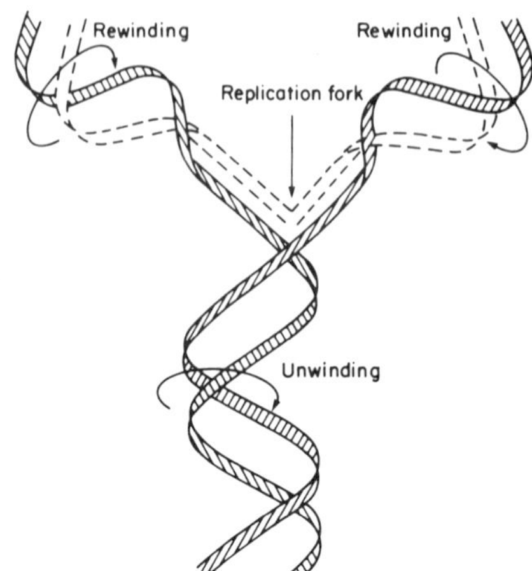

FIGURE 4.1. The DNA double helix in the process of reproducing itself (replication) before cell division. [Redrawn from George Brewer and C.F. Sing, 1983. Genetics, and used with the kind permission of Benjamin/Cummings Publishing Co.]

happens, human invention was anticipated by nature. The gene string consists of two fibers twisted together to form a *double helix*. In the course of cell division, the helix uncoils and each strand immediately thereafter reproduces itself to restore the double state (Figure 4.1).

The gene string, in all except the simplest organism, is subdivided into pieces and packed, together with some specialized proteins, into separate, sausage-shaped bundled called *chromosomes*. The number of pieces or chromosomes varies from species to species. In the mouse there are 40, in man, 46. The chromosomes vary in size and in other properties and can be identified individually under the microscope. They are randomly arranged in the cell, but microscopic study shows that they can be grouped in pairs, one member of each pair being of paternal origin and the other of maternal origin. Because of this paired arrangement, we often refer to the mouse as having 20 *pairs* of chromosomes and man as having 23 pairs (Figure 4.2).

The genome, or gene string, is remarkable in the encyclopedic volume of information it carries in a structure that is physically minute. No computer chip in existence can compare in storage capacity. I say physically minute and it is, because it can all be packed, along with assorted proteins and other compounds, in a cell nucleus that measures four ten-thousandths of an inch in diameter, far too small to be seen by the unaided eye. Yet because it is in the form of a slender thread, its length is impressive.

The genome of a mouse contains 4.7 billion nucleotides. If stretched out to its full length, it would measure 74 inches. The human genome contains 5.6 billion nucleotides and is correspondingly longer (Figure 4.3). I per-

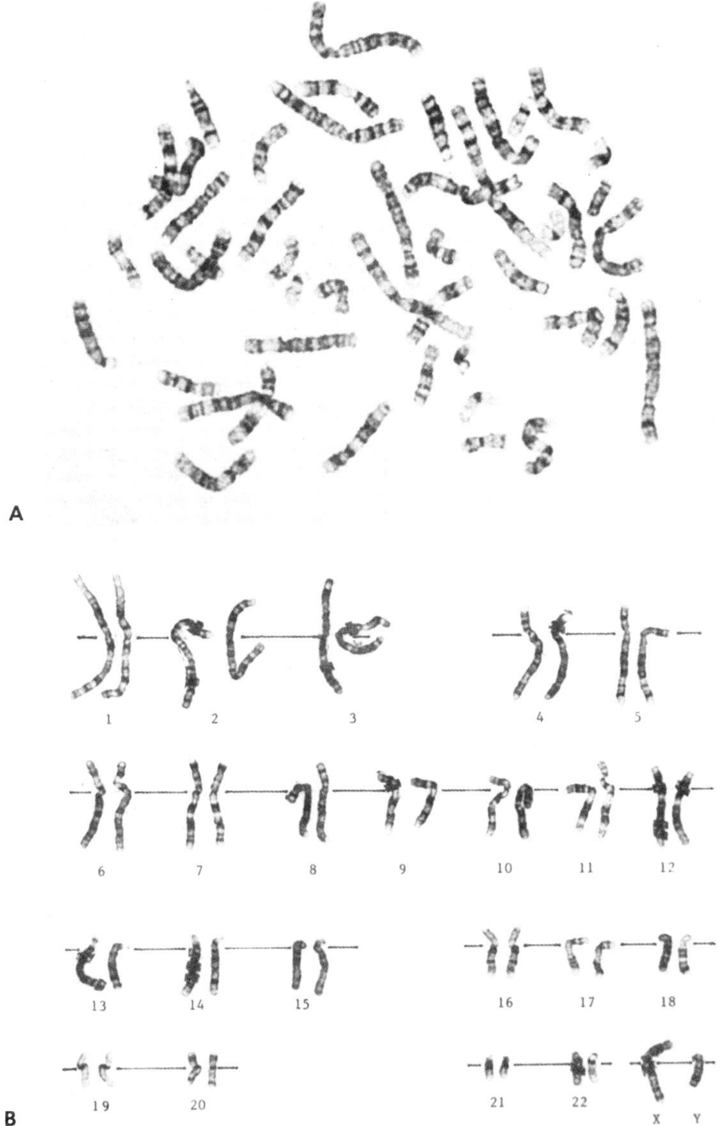

FIGURE 4.2. The 46 chromosomes of a normal human male. The DNA is packaged into these neat bundles shortly before cell division. Each chromosome is actually double, though this does not show in the figure. The two halves will be pulled apart when actual cell division occurs. (A) The chromosomes arranged at random as they actually appear under the microscope. (B) The chromosomes taken from an original photograph and arranged in pairs. One member of each pair comes from the father, one from the mother. The horizontal line passes through a point at which traction is applied when separation of the two halves occurs. [Photographs kindly provided by Dr. Uta Francke.]

FIGURE 4.3. An electron micrograph of approximately two five thousandth of the DNA from a single human cell. The cell was in the interphase between cell divisions. It was treated to free the nucleus, from which most of the material other than the DNA was then extracted. The remaining DNA was spread on the surface of water and photographed at high magnification. The treatment eliminated the second-order coils, leaving the third-order coils clearly visible. [From McCready el al., 1979. Electron-Microscopy of Nuclear DNA from Human Cells. *Journal of Cell Science* 39:53–62. Reproduced with the permission of the *Journal of Cell Science* and the kind assistance of Dr. McCready.]

haps can give some idea of the dimensions of the mouse genome by assuming it magnified into a piece of twine. The green twine that I sometimes use in my garden is approximately one-tenth inch in diameter. If the gene string of the mouse were enlarged to this size, it would be 1,530 miles long. If laid out along appropriate major highways in the United States, it would reach from Bar Harbor, to Minneapolis, with a little to spare. The cell nucleus, magnified to the same scale, would be 43 feet in diameter.

As we have noted, the genome is present in duplicate in all cells except germ cells. For many purposes, therefore, it is appropriate to think of the gene string as consisting of two threads about 37 inches long, on our magnified scale, 780 miles long.

Probably few people have escaped the annoyance of a snarl when trying to manipulate a long piece of thread or string. Nature's success with the gene string is awesome by comparison. Division into separate chromosomes shortens the length of the individual pieces to be manipulated. At the appropriate stage in the cell cycle, however, each of these chromosomes and its enclosed double helix must divide and the daughter products be neatly packed in a newly formed cell nucleus, where they can engage in their basic function of protein production. In growing tissue, the chromosomes must disengage themselves from the nucleus and repeat the cycle at frequent intervals.

The steps in the process are manifold, but the neatness of nature is well manifested in the way the gene string is packed in the nucleus. It is done through coils within coils within coils, with the secondary coils neatly wrapped around some millions of spools formed of a special protein, two turns to a spool (Figure 4.3 and 4.4).

Proteins, and correspondingly the genes that code for them, vary considerably in size. There is no accepted typical size, but a reasonable figure might be a length of 300 amino acids for the protein and hence, of 900 nucleotides for the gene. Given this size, the mouse genome could encompass 5 million genes. Actually, there are not nearly that many. This discrepancy is explained by one of those surprise discoveries that adds spice to the life of the researcher. A substantial part of the genome is made up of *repetitive sequences,* in contradistinction to the *unique sequences* that comprise the genes. It is as if the primordial printing press had a bad fit of stuttering. The gene itself may also be interspersed with repetitive sequences.

There are many guesses as to the reason or reasons for the existence of this repetitive DNA, but there are no firm answers. One attractive hypothesis is that it helps regulate the time and place of action of the structural genes. It has also been branded as *junk DNA.* Whether this uncomplimentary title is warranted is a matter of debate, but until we know better, it at least provides a useful handle.

In any case, we cannot calculate the number of structural genes from

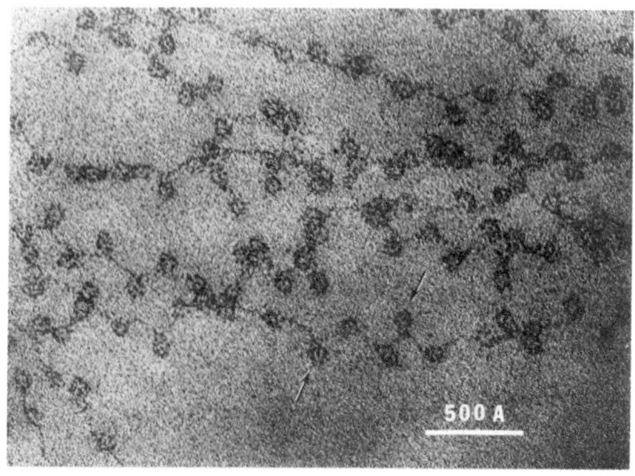

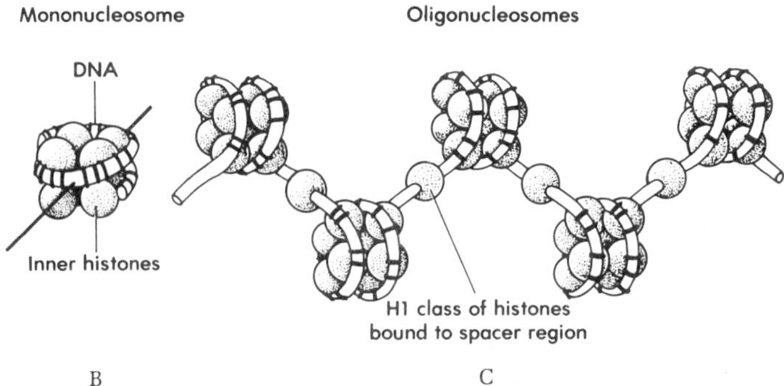

FIGURE 4.4. Photograph and drawings showing the second order coiling of the DNA from a chicken cell. (A) An electron micrograph of DNA prepared by a method that preserves the second order coiling absent in Figure 4.3. The magnification is about 20 times that of Figure 4.3. (B and C) Drawings showing the way in which the DNA is coiled around "spools" of histone (a protein), with two turns to the spool. [(A) from Olins A.L. et al., (1977). Nu Models for Chromatin Structure. *In* The Molecular Biology of the Mammalian Genetic Apparatus, Part A. P. O. P. Tso, Ed., Elseview/North Holland, Biomedical Press, Amsterdam, pp. 211–237. Used by permission of Dr. A.L. Olins (who kindly supplied the electron micrograph) and the publisher. (B) and (C) from Olins, D.E. and Olins, A.L., 1978. Nucleosomes: The Structural Quantum in Chromosomes. *American Scientist* 66:704–711. Used by permission of the authors and publisher. (The same combination of figures has been used by George Burns in his text, *The Science of Genetics.*)]

the total length of the genome. Indeed, there is at present no method that gives a firm answer to this important question. Estimates of the number for man range from 50,000–1,000,000. The mouse presumably has slightly fewer genes, but the estimates are no more precise.

Perhaps more important for our purposes than mere numbers is our growing appreciation of the enormous range of functions genes serve in development and in the operation of body and mind. We shall examine this area of knowledge from a number of viewpoints. We can usefully start with a discussion of the genes in the mouse that have been identified and characterized.

We first need to define a new term. We have already noted that when DNA reproduces, it sometimes gives rise to variant forms. If the change occurs in a structural gene, two similar but nonidentical genes can result, two alternative and mutually exclusive gene forms that occupy the same position in the gene string. Since two or more alternative gene variants can occur at the same site, we need a name for the site. The name we use is *locus* (plural *loci*). For specificity, the expression *genetic locus* is sometimes substituted.

The classic and, until recently, only method of discovering loci was through the observation of a variant form of life that could be shown to be inherited as a unit. Inevitably, in the study of the genetics of mice, the first variants discovered were those affecting some easily seen external characteristics such as coat color. Later, methods became available for the direct detection of chemical changes in some gene-determined body proteins such as enzymes, although these usually produced no visible bodily change.

Through the use of these methods, 1,200 loci had been discovered in the mouse by 1985 (80). The number is increasing rapidly, and even if old methods of locus detection reach the point of diminishing return, new ones will take their place. By means of techniques that we need not go into, 750 of these loci have been assigned to specific chromosomes.

Because genes with easily visible, external effects were the first discovered and because the search for hereditary variants has been going on at least since the last century, when European pet fanciers kept deviant mice as curiosities, we may assume that most of these loci are known. The prime example is genes affecting coat color, but other categories of genes with properties that favor easy observation have probably been almost as thoroughly explored.

We now examine some of these categories. Different sources disagree somewhat as to numbers; in every case I believe my figures are conservative. I draw primarily on a review by Dr. Margaret Green (29). Three intensively studied gene categories with visible effects are listed below with the number of loci placed in each category. The categories are arranged in what I judge to be the descending order of our understanding.

Color of coat or the presence of white spotting: 57 loci.
Texture of the hairs composing the coat and/or texture of the skin: 72 loci.
Head shaking or circling behavior and often deafness because of defects in
   the organ of balance in the inner ear: 41 loci.

The large number of loci affecting each of these relatively simple body structures gives some hint of the complexity of the total genome. A qualification, however, is necessary. The genes at some of these loci affect other structures as well as the one named. For our purposes, we should know the number of loci whose primary role is the building of the particular structure we are concerned with. There is no sure basis for eliminating intruding loci, and there are also some necessarily grey areas. For example, should we include as a hair-form gene one whose primary function is a role in the construction of the follicles from which the hairs grow, and which thereby influences hair structure?

With due allowance for the complexities, I believe it is reasonable to place the number of primary-function loci in each of our categories at 23, 33, and 24, respectively. The 23 coat color loci include some that produce pigment present in the skin and the iris of the eye as well as in hair. Some also affect the distribution of pigment rather than or as well as its production. The 33 loci listed as affecting hair texture do not include loci whose primary effect is clearly on the skin, but do include loci that may be operating via the hair follicle.

Even though these figures are not precise, they do give some conception of the number of loci required to organize the development of relatively simple structures. For most body structures other than the three specifically considered, the record ranges from incomplete to a total blank, but we do have some interesting glimpses of what may lie ahead. There are a substantial number of known loci in the mouse that, in their mutant form, cause specific hereditary diseases. Here again, it is common to find several loci operating on a single system. Thus there are six known loci that can cause obesity in combination with diabetes, each operating at a different stage in the regulation of carbohydrate metabolism (12).

Because the approach to genetics in man has usually been through medicine, the number of known, disease-causing loci is far greater than it is in the mouse. The uncertainties are also greater, partly because of the severe limitations on experimentation in man, but also because publication of data concerning human genes is often warranted even though there are many loose ends. A summary by McKusick of human loci reported before January, 1985 shows 1,789 fully confirmed loci and an additional 1,886 loci not fully identified or validated, for a total of 3,675. Probably at least 1,000 of these loci exist in forms known to cause disease. McKusick suggests as a reasonable assumption that three-fourths of all loci have pathological potential (58).

All these numbers will grow; ongoing research in both mouse and man is rapidly expanding our knowledge of the mammalian genome.

The three structures, hair pigment, hair form, and the inner ear, that we singled out for specific examination in the mouse are only a tiny sampling of the whole range of structures and tissues that make up the total body. These structures include everything from bones to brain and, at the microscopic level, from the membrane in which every cell is enclosed to a variety of white blood cells concerned with defense against disease. In the case of the white blood cells, we know enough to have some concept of their surprising diversity. There are five visibly distinct types of white cells and one of these, the lymphocyte, is divisible into six groups on the basis of function. Each cell type is also characterized by a certain combination of specific proteins in its cell membrane. The function of these proteins is uncertain, but one possibility is that they are labels that guide each type of cell along its own, appropriate, migratory pathway in the blood vessels, lymphatic vessels, lymph nodes, and spleen. Over 40 of these proteins, each genetically determined, have been reported, and the number is growing rapidly.

These cell-surface proteins are to some degree specific to particular white blood cells. They represent, however, only a tiny fraction of the gene products necessary to make such a cell and indeed are only a fraction of the loci required to produce the properties that make the various classes of white blood cells different from other cells.

Of all body organs, the ultimate in complexity is certainly the brain. Since we are interested in behavior, the brain is of particular importance in the context of this volume. Only a specialist can fully appreciate the intricacies of brain structure. Suffice it to say here that the brain and spinal cord contain something of the order of 100 billion neurons, or nerve cells, and billions of supporting cells. Microscopic studies of brain structure have indicated that there are several categories of neurons, but recent tests using highly specific antibodies directed against nerve cell proteins suggest that the visible differences represent only the tip of the iceberg. Every neuron links up in highly specific fashion with other neurons or with end organs such as muscle fibers, the organ of hearing in the ear, taste buds on the tongue, and rods and cones in the retina of the eye. As many as 100 trillion connections may be involved. We know little about how these linkages are established during development, but each neuron must carry instructions sufficient to insure that it scores a bullseye on its own particular target (41).

How many loci are required to produce these billions of specialized cells and their interconnections? A study by Chaudhari and Hahn of the chemistry of the mouse brain led them to conclude that the formation of the brain in this species requires at least as many protein-producing loci as are required for the formation of all other organs and tissues combined. Their evidence is of a quite indirect and uncertain nature, but it suggests 150,000 as the possible number of loci functioning specifically in the brain, and they believe that this may be an underestimate. Many of the individual genes at these brain-determining loci are probably active only in one specific type of nerve cell (7).

Variability in Human Chromosomes

The year 1959 was a turning point in the study of medical genetics. In that year, Lejeune, Gautier, and Turpin showed that Down's syndrome, or mongolism, is due to the presence of a specific extra chromosome, a chromosome assigned the number 21 and characterized, among other things, by its small size. Down's syndrome occurs in at least one in 800 newborns and is especially frequent in the children of older mothers and older fathers. Affected children have a low mentality and are likely to die at an early age from leukemia, heart disease, or infections. They make up about 10% of severely retarded individuals (68,70,75,95).

Because of the medical importance of Down's syndrome, the discovery of an easily demonstrable genetic cause immediately caught the attention of the medical profession. Could it be that other major developmental disturbances were due to chromosomal aberrations? Methods for the microscopic study of chromosomes had undergone major advances, and the time was ripe to apply them to genetic disabilities. In rapid order, other studies appeared linking changes in the normal chromosome pattern to a variety of afflictions. By 1983, a vast literature had developed linking hundreds of major disorders to visible variations in the genome.

Chromosome variants are in some ways less fundamental than are mutations at single loci, and certainly they are less significant for the genetics of social behavior. Nevertheless, they do provide a dramatic demonstration of the enormous variability of the genome, and for this reason we shall briefly summarize them.

Chromosome variants can arise in two ways. The first source of variants is a faulty distribution of chromosomes during cell division. One daughter cell can receive an extra chromosome, the other daughter cell thus being deficient by one chromosome. This is the way Down's syndrome usually arises. The other source of variants involves chromosome rearrangements.

It is not unusual for breaks to occur in the gene string. This could have a disastrous effect if the breaks were not repaired. To deal with this problem, cells contain enzymes specifically committed to repair. Usually the splicing restores the gene string to its original condition; occasionally, if two breaks have occurred simultaneously and in proximity, the junctions may result in a reordering. Several types of chromosome rearrangements are possible—there can be deletions, inversions, translocations, and the formation of ring chromosomes.

The mere rearrangement of the gene string is usually not harmful. It is when the rearrangement results in cells or individuals with a chromosomal excess or deficiency that damage occurs. Each specific excess or deficiency has its own specific pathological consequence, but, as a general rule, the extent of the damage increases with the length of the genetic material involved. Development cannot proceed normally if there is a major imbalance in chromosome structure, and in extreme cases death may occur very early in embryogenesis.

Numerous studies have documented the importance of chromosomal imbalance in spontaneous abortions in humans. Chromosomal abnormalities have been identified in approximately 50% of aborted fetuses.

Many embryos die at stages too early for the dead fetus to be retrieved and studied, so it is impossible to document the cause, but chromosome aberrations may well be even more important in the early than in the later stages of development.

At lower levels of chromosome imbalance, the fetus may survive to term but suffer from defects than can cover an enormous range in both severity and manifestation. Quite frequently there is growth retardation, mental impairment, and facial and bodily alteration. There may be quite specific changes in the ridges and creases of finger and palm. Pathologies, most of them lethal to the fetus, have been reported in association with deficiencies or excesses of all except two of the 23 chromosomes, and they are quite common in association with 9 of the chromosomes. The two chromosomes not associated with known pathologies probably are so essential that disturbances in which they are involved are consistently lethal (16). Twelve abnormalities in sex chromosome number have been reported, ranging from the presence of only one X and no Y to the combination of two Xs with three Ys. The presence of extra parts of one or more sex chromosomes can also occur. All of these cause disturbances in development (96). All told, about one child in every 200 is born with a congenital affliction caused by a chromosomal anamaly (75).

Chromosomal imbalance occurring in body cells but not in the germ line (cells in the lineage that gives rise to germ cells) can also cause disease. Approximately 20 types of cancer have been traced to this cause, and it now seems possible that chromosome disturbances are not only a factor in many types of malignancy, but can also serve to differentiate minor subtypes of the disease. Chromosome typing may therefore aid in diagnosis and treatment (17,35,108).

Recent research has demonstrated a large class of chromosomal variants in man not involving any chromosomal imbalance; indeed, our knowledge of the morphology of chromosomes has now become so refined that people's chromosomes appear to be as individual as their fingerprints. To cite just one study, McCracken and coworkers examined the chromosomes of 16 identical and eight nonidentical twins. The chromosomes of the two members of each identical twin pair were exactly alike, but all 16 nonidentical genotypes present in the sample were demonstrably different from each other in their chromosome architecture. This category of chromosome variation appears to reside in considerable part in the repetitive or "junk" portion of the DNA, so its genetic significance is uncertain. However, it is a striking manifestation of the variability of the total gene string (55).

The Variability of the Mammalian Genome

The impression of most people who have seen mice caught in a home or barn or grain bin is that, except for the difference between the sexes, they all look very much alike. This is a valid judgment with respect to superficial appearance, but it can be very misleading. Actually, mice are extraordinarily different one from another. A specialist in the immunogenetics of white blood cells could demonstrate, on the basis of the cell-surface proteins of these cells, that practically every wild mouse anywhere in the world is recognizably different from every other wild mouse. A specialist in enzyme chemistry could do the same thing by analyzing the structure of enzymes found in the liver and other organs. These variations are all strictly gene determined, so a high degree of genetic diversity in wild populations is implied.

What fraction of the genetic loci in mouse and man exists in more than one form we do not know. One study indicated that about one-third of the loci determining the production of enzymes in man are polymorphic. Every refinement in technique, however, holds the potential for the detection of new variants. I suspect that we will ultimately find that variants—some rare, some common—exist at the great majority of mammalian loci.

The effect of these variants on the whole organism varies from zero to damage so severe as to cause death in an early stage of development. This is a huge subject, but there is one point that should be made here. Several studies have indicated that there are many variant genes whose effect depends on the environment. In some environments they may, in greater or lesser degree, be beneficial, in others detrimental. The advantages and disadvantages may stem from many factors, but a well established effect of some variants is an alteration in the susceptibility or resistance to specific diseases. Some of these variants may be advantageous if a disease is common, disadvantageous if the disease is rare.

For our purposes, the important area of variation is in those genetic loci that influence brain structure and thereby human faculties and human behavior. This subject is in its infancy; at the moment some of the most specific information comes from the mouse.

In the mouse, 28 loci have been detected that influence brain structure. In all instances, the effects on the brain are of a rather gross nature, and in all instances except one they were detected because they produce obvious behavioral disturbances. The nature of the disturbances is suggested by the names applied to the mutant mice. Some examples: agitans, jittery, lurcher, reeler, sprawling, and trembler. In eight instances, microscopic studies have shown abnormalities in the Purkinje cells. A Purkinje cell is a structurally specialized type of nerve cell occurring in the cerebellar cortex. In another 11 of the mutants, there are abnormalities in the myelin, a fatty substance that forms an insulating sheath on some nerve cells. Although 19 of the 27 mutants with behavioral expression fall into these two classes, all show individual differences. Thus the Purkinje cell variants range from one in which the defect is detectable only with the use of the electron microscope to others with obviously underdeveloped parts of the brain. The one brain variant unaccompanied by behavioral disturbances consists of the absence of the corpus callosum, a large tract of nerve cells connecting the right and left halves of the brain. This was discovered accidentally in a microscopic study of mouse brain sections.

The significance of these observations for brain genetics is that 28 variants of a gross nature, all belonging to a few very specialized categories, had been detected in the mouse by 1981. This suggests that in the totality of the brain, an enormous number of genetic variants must be possible. Nobel Prize winner Roger Sperry has drawn the same conclusion from an entirely different sort of study. In his words, there "is a growing recognition of, and respect for the inherent individuality of the human intellect. The more we learn, the more complex becomes the picture for predictions regarding any one individual and the more it seems to reinforce the conclusion that the kind of unique individuality in our brain networks makes that of fingerprints or facial features appear gross and simple by comparison. The need for educational tests and policy measures to selectively identify, accommodate, and maximize the differentially specialized forms of individual intellectual potential becomes increasing evident" (93).

While genes, at least in terms of the evolutionary time scale, are constantly undergoing minor changes, there is also a substantial element of gene conservation. As in the evolution of languages, the spelling of words may change, but the basic vocabulary remains relatively stable. The chromosomes of man and his nearest relative, the chimpanzee, are remarkably alike as seen under the microscope, and biochemists have estimated that the genetic material of these two species is 99% identical (74,107). Comparable figures are not available for man and mouse, but we do know that many clusters of loci have been identified in the two species that share essentially the same genes with the same functions and often the same order in the chromosome (52).

DETECTING THE EFFECTS OF HEREDITY AND ENVIRONMENT

In studies of the development of personality, the social sciences tend to concentrate on the role of environmental factors, the biological sciences on the influence of heredity. It is not surprising, in view of this dichotomy of the life sciences, that a polarization of opinion on the nature-nurture issue has developed. Sociologists, anthropologists, psychologists, and psychiatrists tend to emphasize the importance of culture in the making of man, evolutionists and geneticists, the importance of genes. And in the 100 or more years that these compartments of the life sciences have existed, the public view of the issue has fluctuated with the eloquence of the opposing advocates, aided at times by such political factors as concern over immigration policy.

Despite the emotion that the nature-nurture issue always has and still does engender, the voices of reason are beginning to stake out common ground. Joint meetings of social and biological scientists have been held specifically to integrate relevant data, and researchers in ethology, neurophysiology, and the hybrid science of behavioral genetics are building a factual background for a middle view.

But what are the relative effects of heredity and environment on human variability? There is no one answer to this question. Human traits have to be examined individually, not as a group. No two mixes of the inner and outer factors are just alike. And in studies of man, the variables cannot be controlled as they can in experiments with laboratory animals. Despite these complexities, there are valid ways of getting answers to our question. We shall examine studies based on several different methods, but it will be useful first to examine the ideal experimental situation.

In the ideal experiment, groups of genetically identical individuals would be placed in diverse environments, and genetically diverse individuals in identical environments. Each such combination would be repeated as many times as necessary to make sure that the observed results were statistically valid. If the trait under study were quantitative, as are, for example, height or weight, the decrease in spread between the extremes produced by like genes as compared with like environments would then provide a measure of the relative roles of nature and nurture. Any one set of comparisons is fully valid, however, only for the specific degree of diversity existing within the particular total environment under study and the particular total gene pool under study. A study of the role of nature versus nurture of a human trait carried out in Sweden might not be fully applicable to the more diverse situation prevailing in South Africa. This precise type of experiment is, in its essence, possible with laboratory mice; it can be approximated with men.

There are now many strains of laboratory mice that have been inbred, brother with sister, for many generations. Close inbreeding reduces and in due course virtually eliminates genetic diversity. All the individuals of any one inbred strain are thus a standardized product. Inbred mice can also be mass-produced and hence made available in whatever quantity a particular experiment may require.

In man, the counterparts of inbred strains are identical twins. The ultimate experimental situation consists of pairs of identical twins separated at birth and reared in widely different homes, and of fraternal or nonidentical twins reared together in the same home. In any large population both these situations can arise, but even in the most favorable circumstances, the student of nature-nurture in man must be content with a less suitable population in a less controlled environment than he could expect to achieve with mice.

BEHAVIOR STUDIES WITH INBRED MICE

Mice, given the appropriate surroundings, display a variety of behavior patterns that are amenable to study. In dozens of experiments extending over a span of more than 40 years, behavioral geneticists have tested these patterns in various groupings of inbred strains. The behaviors investigated include building nests, hoarding food, consuming food following a brief fast, selecting plain drinking water versus water plus 10% alcohol, climbing ropes, avoiding a small electric shock, attacking crickets placed in the mouse cage, exploring a large unfamiliar cage, and fighting. All of these behavior patterns were found to vary significantly between inbred strains.

In our presentation, we shall limit ourselves to fighting behavior. This shows interesting diversity, has been extensively researched, is of particular social and ethical significance, and exemplifies effectively the joint roles of heredity and environment. One word of caution is necessary. While studies under laboratory conditions tell us with some accuracy the genetic potentials of mice for aggressive behavior, they do not tell us the frequency with which these patterns of behavior are displayed under natural conditions.

Before turning to the work with inbred mice, we should note that the importance of genotype in aggressive behavior has been tested and confirmed in selection experiments. Starting with a noninbred and hence genetically heterogeneous stock, it has been possible to alter aggressiveness either upward or downward by appropriate selection of parents over a succession of generations.

Five types of fighting in mice have been described. Type one is a self-defense reaction elicited by any sudden pain stimulus. A mouse will turn and attack a pair of forceps used to pinch its tail. Type two is defense of a nest by lactating females. While these two behavior patterns have been well established, they have not been extensively studied and strain differences have not been described. The other three fighting behaviors have been found to show strain differences. These three are attacks by females on lactating strangers, fighting among males presumably associated with the defense or claiming of territory or with the establishment of dominance, and fighting by either males or females over food.

I preface further discussion of this subject with a brief word on the designation of inbred strains. There is a standardized nomenclature for all strains. For the convenience of the reader, I have used instead a single capital letter for each strain mentioned. At least some of these will be recognized by mouse geneticists.

Haug and Pallaud noted that jointly reared females would sometimes attack an introduced female stranger (34). They tested this further with three inbred strains, B, S, and Z. Females of strain B attacked the stranger only if she were lactating. Strain S females attacked both lactating and nonlactating strangers, but the former much more savagely than the latter. Strain Z females did not attack in either situation. It was shown that the incitement caused by the lactating state was due to an olfactory factor present in the urine.

The genetic component in the difference between strains B and Z was tested by cross-fostering young of one strain on the other. Female young were exchanged between two unlike mothers which had given birth within the previous 12 or 24 hours. The fostered females, as adults, behaved like their natural, not their foster parents. We shall see that this is the usual, but not the universal, result of fostering experiments.

Fighting between male mice, at least if they are strangers, is common and easily observed. Fighting males bite, wrestle, squeal, and rattle their tails against the cage floor. The fighting can be vicious, and if allowed to continue, sometimes results in the death of one of the contestants.

Laboratory mice are pampered creatures. The conditions under which they are reared and the specific modifications of these conditions employed in most behavioral experiments are quite different from those encountered in the wild. There is therefore some uncertainty in attempting to equate laboratory-induced fighting with any form of natural aggression. A territorial reaction is suggested by some of the experimental situations that induce fighting. Males under essentially natural conditions defend territories, although of very limited area. Fighting to achieve dominance is another possibility, particularly in some experimental situations.

DeFries and coworkers carried out experiments using a design in which the fighting observed was probably of the dominance variety. Adult mice were placed in three, interconnecting cages, each cage being about one-half square foot in area. The subjects were observed at intervals over a period of several days, and dominance order was determined by the number of bite wounds on the tail. The investigators used three strains, B, C, and D, and also hybrids produced by crossing the strains. Various groups of four males were placed in the three pens. The order of dominance achieved varied somewhat with the pairing, but the hybrids, probably because of the well-known phenomenon of hybrid vigor, usually headed the list. Among the purebreds, the Bs were clearly dominant over the Ds. The addition of females to the group increased the fighting. The dominant male sired 90% of the litters in groups containing females and that were observed over a long period (15).

Valzelli and Bernasconi tested relative aggressiveness in six different mouse strains, using a quite different experimental design. Males were grouped for short periods, and the tendency to initiate an attack was scored. Enough groups were used so that strain differences could be clearly established. The percent of animals in a given strain showing aggression ranged from 95 to 18. Four representative scores were 95% for strain S, 84% for strain B, 52% for strain D, and 20% for strain C (97).

Many tests of fighting tendency have been carried out, and in all of them strain differences have been found. The order of dominance or aggressiveness of different strains has not been consistent in these experiments, but this almost certainly reflects the complexity of the interactions between experimental design and genotype and not any fundamental flaw in the tests. As an example of the complexities, Fredericson and Birnbaum found that C males, which are generally less aggressive than Bs in daytime pairings, killed their B partners in eight out of ten pairings if left overnight.

The phenomenon of fighting over food was discovered and thoroughly studied by Fredericson. Like self defense, but unlike the other forms of fighting we have described, this type of aggression is shown equally by males and females. The tendency to fight can be increased by prior fasting (21).

The typical experimental design consists of placing a single pellet of food between two, like-sexed mice that have been fasted for 24 hours. The result depends on the mouse strain under test. Adult Bs fight promptly and actively for possession of the pellet. Zs are slow to fight, but after some time, struggle actively for possession. Cs share the pellet with only a slight indication of competition. Interestingly, if the pellet is anchored so that possession by one mouse is impossible, the tendency is for all strains to share it. Bs that have fought in encounters over food when hungry, soon learn to fight over the pellet when well fed.

Another modifying influence described by Fredericson and coworkers, in a study with strains B and C, was the presence of a female mouse with a pair of competing males. The aggressiveness of the males was first elevated by a series of encounters. Like true gentlemen, two such trained B males, who would normally compete vigorously over a food pellet, became relatively peaceful if a lady mouse were present. The female could be either a B or a C. An added male did not have this effect, and the effect was not seen with paired C males (22).

There have been a number of studies of the effect of prior victories or defeats on the aggressiveness of strange paired males in the absence of food. Not surprisingly, victories increase aggressiveness and defeats decrease it (47).

These experiments by Fredericson and others demonstrate that fighting behavior can be modified by prior fighting experience, by hunger, and by the presence of a female. There are many other circumstances that can modify aggression, but before we examine these, we should note one test of environment that has given largely, though not entirely, negative results.

The early environment of mice can be altered by fostering them on a foreign strain. We have already noted one such test. There have been many others and almost always, the young have retained the behavior of their natural parents. Occasionally there seems to have been a slight effect of fostering *per se,* with the modification not always being in the direction of the foster parent. Southwick fostered unaggressive young on aggressive parents and noted a change in the direction of the foster parents, but only two strains were used in his study (92).

Fredericson, in a test of the effect of fostering on fighting over food, found that the behavior of the parent, not that of the fostered infant, was altered. When the possessively aggressive Bs were fostered on non-possessive C mothers, the mothers became more aggressive over possession of a food pellet, while the fostered Bs were unchanged (21). It is interesting in this context to note the results of studies in hospital maternity wards that suggest that maternal feeling is permanently enhanced if nature is given precedence over sanitation, and the newborn infant, instead of being kept temporarily in a separate, sterile room, is given immediately to the mother (79). Apparently, the behavior of the infant can modify that of its parent just as the behavior of the parent can modify that of the infant.

The most striking environmental influences on fighting have been observed in the intermale competition we have classified as territorial or dominance-seeking. One of these influences concerns the effect of prior isolation on aggressiveness.

As early discovery in studies of male fighting was that a period of isolation increases aggressiveness. Consequently, previously isolated males have often been used in aggression experiments. Likewise, females that have been isolated also show an uncharacteristic tendency to attack and an increased restlessness and hyperactivity (101). An interesting qualification of the effect of isolation has been noted by Scott. Young mice isolated from parents and littermates during a critical period between 20 and 30 days of age show reduced aggression. Apparently social contact at this particular stage of development makes mice less fearful of—and hence more aggressive toward—strange animals (85).

The effect of isolation on aggressiveness in mice may have a chemical basis. Several studies have indicated that the metabolism of serotonin, and perhaps of other substances that regulate nerve cell activity, is altered in mice deprived of normal contact with other members of the species (97,100).

Prior fighting experience is another source of major change in male aggressiveness. When males are paired with carefully selected opponents so that they either win a series of victories or suffer a series of defeats, their subsequent fighting behavior is altered in quite the way we might expect from human experience. The frequently beaten mice become timid; the frequent victors become bolder and quicker on the attack. There may be strain differences in this response; according to one account, naturally unaggressive mice do not easily acquire aggressiveness (27,43,88).

Another influence on fighting success is rearing mice of one strain with the opposing strain from 30–90 days of age. Levine and coworkers used two strains of mice, one of which was a consistent loser in interstrain pairings of unacquainted males. If members of this docile strain were reared with the more dominant strain, their subsequent fighting success in the interstrain pairing substantially increased (50).

Two studies of aggressiveness in mice have revealed environmental influences that are contingent on the recognition by inbred mice of a difference between their own and a foreign strain. It is quite possible that mice recognize color differences, but the most firmly proven form of recognition is based on smell. One type of olfactory factor, or *pheromone*, has been shown to be the product of a genetic locus, or possibly a small group of loci, on mouse Chromosome 17. The locus is polymorphic, existing in many alternative forms, so that there is a considerable probability that its product in different strains will be different. A basis of olfactory recognition is thus well established (106).

The role of a pheromone present in male urine in stimulating male aggression has been reported by Kessler and coworkers. Strain D males of proven aggressiveness were tested against castrated D males that had been treated with water or male urine. If the target males were treated with strain D male urine (urine of their own strain), they elicited a prompt attack. If treated, however, with water or with male urine from either of two foreign strains, the attack was delayed and less vigorous. Evidently, male fighting in mice is incited by a male odor that may differ from strain to strain (46).

Another study showing differences in male aggressiveness dependent on the strain of the opponent has been reported by Bauer. Two strains were used, the B and the C. Bauer employed a method for testing aggressiveness not used in any of the experiments so far described, though it has been successfully employed on many occasions. One of two paired males was held by the tail and bumped against his opponent, which was unrestrained and free to attack.

When the paired males came from the same strain, the Bs, as expected, were more vigorous attackers than the Cs. If the dangled male came from the foreign strain, the intensity of the attack was altered. Both strains attacked foreign males more vigorously than males of their own genotype (3). It is interesting to note that this reverses the pattern of Kessler's experiment, in which foreign urine proved less provocative than native urine. Different strains were used in the two tests and in all probability the contrasting results merely reflect the complexity of strain interactions.

All the environmental effects of fighting tendencies that we have described so far have been postnatal. Von Saal and Bronson have described a prenatal effect. Because of the large size of litters in mice, most mouse fetuses have intrauterine neighbors on both sides. Von Saal and Bronson found that females whose intrauterine neighbors were both males are more aggressive toward other females than are females gestated with female neighbors. They are also less attractive to males. A variety of evidence indicates that this increased aggressiveness is due to an excessive level of male hormone at this early developmental stage (99).

These studies with mice show very clearly that both genes and experience or environment can influence aggressive behavior in this species. Mice differ in their aggressive tendencies and both nature and nurture pay a role in determining these differences. If this is true for mice, a certain presumption is established that it is true for social behavior in general and for mammals in general. To apply this to man, however, we need more evidence from other species, with the most relevant evidence coming from man's near relatives and most of all from man himself.

SOME OBSERVATIONS SUGGESTING GENETICALLY DETERMINED BEHAVIORAL VARIABILITY IN CHIMPANZEES

Investigators who have worked with apes have been impressed by individual differences in their behavior and in their teachability. Nissen, who worked with 151 chimpanzees at the Yerkes Laboratory over a 25-year period, concluded that their behavioral variability was even more striking than their rather considerable anatomical and physiological variability (65). Since the animals were raised in a comparatively uniform environment, these differences presumably were mostly genetic in origin. The observations, however, cannot have the precision of those made in experiments with inbred mice.

Observations such as these, and also studies of the early appearance of variability in human infants, including variability in the moral sense, which we shall cite later, all point to a genetic component in behavior. If there is any one area of study that has swung the thinking of psychologists and psychiatrists in this direction, however, it has been brain research.

THE BRAIN AND BEHAVIOR

A few examples will have to serve to illustrate our growing knowledge of the relation of brain structure to behavior.

One method of determining the function of specific brain areas is to plant an electrode in one of them and then observe the effect of mild electrical stimulation. The effect is not lasting; a response occurs only while the electrical current is on. The techniques involved are so well developed that tests of this sort can be carried out and repeated at will with very little apparent disturbance to the subject.

John Flynn at Yale used this method to study brain areas in cats that control aggressive behavior. The animals used were normally unaggressive; some of them had lived with rats and had not molested them. When the lateral hypothalamus of one of these cats was electrically stimulated in the presence of a rat, the cats would attack viciously and lethally. The kill would be quite precise; a bite in the cervical region of the spinal cord in the manner typical of felines in general. In other experiments, Flynn tested the effect of electrodes in the medial hypothalamus. When this area was stimulated, the cat would ignore the rat and promptly attack the experimenter (44).

Robinson and coworkers carried out somewhat similar experiments with rhesus monkeys. When the anterior hypothalamus of these subjects was stimulated, there was an effect only if another male was present.

Under these circumstances, a small male, low in the dominance order, would violate all the rules and attack a much more dominant male. The attack was so vicious that the dominance order was often reversed (44).

These experiments indicate that very specific patterns of behavior are determined by very specific brain areas. The term *instinct* has gone out of fashion, but it is tempting to revive the term and to say that we can now relate instincts to detailed brain structure.

Other revealing experiments concerning what, again, we might call instincts have been carried out with birds. The behavior involved was not tied to specific brain areas, but the experiments are relevant because they reveal the sort of fascinating interactions between nature and nurture that can occur in the development of behavioral patterns. The studies were carried out with the white-crowned sparrow by Peter Marler and his students at Rockefeller University (53). They found that the song of these sparrows is learned and that the learning can occur only between the chicks' tenth and fiftieth days, a so-called *critical period* (86). If the chicks do not hear their species' song during this period, they can thereafter produce only crude, regularly patterned vocalizations. If they do hear it, they learn the song, including minor regional dialects. Songs of species other than the white-crowned sparrow are neglected. Observations by Marler on swamp sparrows indicate that the within-species song carries key sounds in the presence of which, and only in the presence of which, the learning mechanism is activated.

TWINS

Students of the effects of heredity and environment on humankind have found in twins a uniquely favorable material. Identical twins, derived from a single ovum and hence with identical genes, can be contrasted with fraternal twins derived from two ova and genetically no more alike than two nontwin siblings. Differences in identical twins must be due to differences in their milieu at some stage of their life history after the division of the egg into two incipient individuals. Fraternal twins can manifest both these environmentally determined differences and differences attributable to their genes. If, with respect to some measurable trait, the disparity between fraternals exceeds that between identicals, there is a presumption that the excess is largely or entirely caused by heredity because the environment has been the same.

There is a catch in this argument. The childhood environment of identical twins may in fact be more alike than that of fraternals. Identicals are more apt to be dressed alike, to play together, and to have the same teachers and friends. It can thus be argued, and sometimes is argued, that any excess difference between fraternals is mostly or entirely environmental in origin.

This issue has been extensively debated, but only recently has it been addressed experimentally. The results of several types of experiments all justify the conclusion that the greater differences between fraternals as compared with identicals is not, to any substantial degree, due to a greater difference in their environments (51,70). The validity of the twin method is confirmed.

Three approaches have been used; we can mention only one. It is not uncommon for parents of twins and for the twins themselves to be mistaken as to the twin class to which the pair belongs. This of course does not happen when one twin is a boy and one a girl—such twins are necessarily fraternals—but it happens rather often when they are the same sex. Two studies have been carried out of the effect of mistyping on the degree of measurable difference between members of a pair. The results showed little effect of mistyping on twin difference. This makes it unlikely that disparities between fraternals are due to differences in the way they have been treated.

There is also a theoretical reason for discounting the significance of environmental differences for the validity of the twin method. Environment certainly does to some extent shape the personality, but personality also shapes the environment. If and insofar as fraternal twins have environments less alike than those of identicals, it is because their unlike genes make them choose these unlike environments. Unlike infants inspire different treatments by their parents, and unlike twin adults choose different friends, different careers, and different marriages. Any excess environmental disparity of fraternals as compared with identicals may exaggerate the disparity between the members of a fraternal pair, but the disparities all start in the genes. (See Scarr (83) for a good statement on this same thesis.)

What do twin studies tell us about the relative roles of nature and nurture in personality formation? We have only begun to tap this potential, but such studies have already told us a great deal. This is a huge subject and I can scarcely do more than mention a few results and summarize the conclusions, although I shall go into some detail concerning twin and adoption studies of criminals. Because of the cursory nature of this review, it is in order to note that there are still harsh critics and to call attention to some of the shortcomings of their arguments.

The favorite whipping-boy of the environmentalists is Sir Cyril Burt, a psychologist by training, who during his many years as professor at Liverpool University and University College in London made important contributions in this area. After his retirement in 1950, although living in somewhat straited financial circumstances, he became interested in studies with identical twins, and between 1955 and 1966 published three papers on twins in which he reported a close correspondence in IQ test scores between members of identical twin pairs. His results were widely quoted.

About ten years after the last of Burt's papers was published, Professor Kamin at Princeton noted that the level of agreement between Burt's separate studies was unusually close and questioned the validity of his results. Burt was no longer living, but the medical correspondent of the London *Sunday Times* and subsequently Hearnshaw, who has written a very careful biography, found evidence from Burt's records that most of his results were purely fictional.

This unfortunate episode in the history of twin studies has been used by some writers to tar and feather twin work in general. This is totally unjustified. There are several dozen studies that, despite methodological flaws in some of the earlier investigations, are generally sound and some recent ones that are remarkable in their thoroughness.

Plomin, DeFries, and McClearn list 18 studies (Burt excluded) involving a total of more than 6,000 identical and fraternal twin pairs, in which the subjects were given various IQ ("general cognitive ability") tests. In every study except one, which involved 62 twin pairs, the scores of the identical twins were substantially more alike than were those of the fraternals. In the one exceptional study, there was almost no difference between the two twin types (70). Twins also have been used in studies of such matters as smoking habits, drinking, schizophrenia, homosexuality, and personality characteristics. In these, too, the identical twin similarity has quite consistently exceeded that of the fraternal twins.

Even with due allowance for the possibility that the two twin types are subject to different imposed environments, the conclusion is inescapable: there is a significant genetic component in the traits investigated. Loehlin and Nichols conclude from their study of 850 pairs of twins "that the genes and environment carry roughly equal weight in accounting for individual variation in personality" (51). It is my impression that the same 50:50 split is a reasonable rule of thumb to apply to IQ and other measurable qualities. To this we must add two significant qualifications. First, while I personally believe that IQ tests, in the appropriate place, can be very useful, there is room for considerable question as to just what they measure. Second, studies have suggested that the relative roles of heredity and environment vary significantly according to the specific aspect of mental capacity or personality under investigation (98). The 50:50 rule is only a rough generalization.

Occasionally, because of some family problem, twins are separated as small infants and raised in different homes. This provides a potential means of studying individuals of identical heredity exposed to unlike environments, an experimental situation otherwise obtainable only with inbred mice. Such twins have been the subject of a number of investigations going back to 1937 (63). There was a considerable period when no one was working in this area, but an extensive research project is now underway at the University of Minnesota under Professor Bouchard. All of these studies show that identical twins display remarkable resemblances even after long separation. Two British female twins in their late 30s who had been separated at the time of World War II and raised in quite different socioeconomic settings were noted, at the times of their first meeting with the study team, to both be wearing seven rings (39). In an earlier British investigation, two separated identical females, when interviewed in their homes, both were found to have open copies of Handel's *Messiah* on their pianos. Such instances might be charged off to coincidence were it not for the rather high frequency with which they occur.

Professor Bouchard, himself a psychologist, says that the similarities he and his team of investigators have noted exceeded his original expectations and have convinced him of the importance of the genetic component (5).

CRIME AND PERSONALITY

Our main interest in this chapter is a search for evidence concerning the genetic component in human social behavior. As usual, we have to be selective. Because crime is of great social importance; can be sharply defined, for our purposes, as breaking the law; and has been extensively studied, we shall confine ourselves to this area. Such minor crimes as smoking marijuana or exceeding a speed limit we largely rule out. We consider first some studies on criminal careers.

John Conrad, Simon Dintz, and their collaborators carried out an extensive study of criminal careers in Columbus, OH (13). The study was confined to males who had committed at least one serious crime during the years 1950–76. Qualifying individuals were those who had committed a murder, assault, or rape and had been recorded by the police in the "MARS" book, or those who had committed a robbery and were recorded in the "robbery" book. Robberies were treated as violent crimes because they directly involved the person robbed. The final sample contained 1,591 individuals, 61% derived from MARS records and 39% from those on robbery. This was not a random sample of criminals; the purpose of the investigators was to study the perpetrators of serious or violent crimes.

With these records in hand, the investigators next screened, with the generous assistance of the Columbus police and the FBI, records of all earlier arrests. They thus had a criminal history of their subjects, quite complete except for two significant omissions: Undetected crimes were necessarily excluded though probably quite numerous; juvenile arrests were excluded because the police withheld the records.

The principal finding from this study was that individuals picked from the Columbus police records because of a single crime classed as violent, generally had extended criminal careers. The 1,591 subjects were arrested a total of 12,527 times. The charges ran the gamut; these men in general were not criminal specialists. With the passage of time, more and more of the subjects ceased their criminal activities, but some persisted for years. The authors estimate that, "When all criminal activity is finally at an end, the average number of arrests will approach or exceed ten arrests per subject." As in other criminal studies, it was found that the long careers of crime tended also to be careers that began early.

It is evident from this work that a disproportionate amount of crime is committed by a small fraction of all criminals. Other investigators have reached the same conclusion. Wolfgang et al., in a Philadelphia study using a more broadly based youthful sample, found that 18% of their delinquent subjects committed 52% of the reported crimes (105). In a sample followed to age 30, the concentration of crime increased: 14.8% of the group committed 74% of all offenses and were particularly prone to violent offenses (30). Jan and Marcia Chaiken have reported similar results. They also noted the major role in all types of crime of persons who have committed violent crimes (6). Sheldon and Eleanor Glueck, in a classic early study of convicts in Massachusetts, likewise found a concentration of criminal activity. Thirty-two percent of their subjects persisted in crime during the 15 years that observations were continued (28).

Because of the length and personal detail of their study, the Gluecks were able to gather data on criminal proclivities shown by the children of their subjects. They estimated that more than a quarter of the children who had reached juvenile court age showed signs of delinquency. Obviously, these children had been brought up in unfavorable environments, but the authors nevertheless suggest a genetic component. They add the congent observation that bad environments may in part be generated by bad genes. Nature and nurture can interact in complex ways.

West and Farrington, who specifically studied crime in families, also found clear evidence of familial transmission (37).

Can anything be done to halt the careers of the chronic offenders? Many programs aimed at rehabilitation have been tried, but most observers conclude that benefits have been minimal. Perhaps the most authoritative summary has been prepared by the Panel on Research on Rehabilitative Techniques, operating under the auspices of the National Research Council of the National Academy of Sciences. The panel concluded that the magnitude of the task of reforming criminal offenders has been consistently underestimated and that there are no present grounds for offering recommendations for rehabilitation. Hope springs eternal, however. They believe that we should try again with "more intensive and extensive interventions" (69). S. Samenow and J. Wilson offer two specific but quite different suggestions as to how rehabilitation might be achieved (82,102). Despite the differences in their proposals, I see in both of them an indication that firm discipline must be a part of any successful program.

While with a few doubtful exceptions special rehabilitation programs seem not to have worked, it is possible that imprisonment as it is presently practiced is accomplishing more than is generally realized. A study carried out by the Bureau of Justice Statistics confirmed the finding of earlier investigations that only a fraction of all youthful criminals become truly chronic offenders. By age 40 a great majority are no longer lawbreakers. Perhaps, suggest the authors, there is a long-term deterrent effect associated with punishment (48). Increasing sentences for repeat offenders may also help. Fear of incarceration in many cases perhaps can induce reform.

The problem of rehabilitation is merely one facet of the broader problem of the environmental component in the socialization process. We can hardly doubt that there are environmental contributions—in fact, ethics would be pointless if the appropriate use of instruction, reward, punishment, and other social agencies were without effect on people's social behavior—but there are many differences of opinion as to what works.

With respect to the environmental component in crime, all I can do here is note a few reports that I have found interesting.

It is widely believed that rising unemployment increases crime. A review of the evidence by Wilson and Cook, however, did not reveal any association (103). Also, Hirschi found that juvenile delinquents are often employed and well paid, and a study in Racine, WI financed by the Justice Department shows a slight positive correlation between year-round employment of high school students, especially male students, and crime (37). This study also showed a link between car ownership or easy car availability and crime. This apparent contradiction of the rising unemployment hypothesis is perhaps explainable by the type of crime examined. It is probably mostly petty crime such as shoplifting that increases in hard times. Some drug pushers say that they have been forced into this occupation by the lack of jobs. The typical hard core offenders, however, may be largely uninfluenced by economic conditions. Another factor contributing to crime by teenagers with money to spend may be less dependence on parents and more dependence on peers.

The difference between the petty offender and the typical recidivist is further emphasized by reports that while rehabilitation programs are ineffective for the latter, they do work for the former. Initial accounts from a program established in Kansas City for shoplifters indicated almost 100% success. Three-fourths of the individuals enrolled in this program did have jobs and that may have contributed to the success.

Students of child behavior emphasize the importance of well-planned and consistent discipline in the socialization process (37). Sheldon and Eleanor Glueck, in their book on criminal careers, suggest that the lack of this kind of discipline in families with criminal records is one reason for recurrence in successive generations. They found attempts at discipline, but with a tendency to harshness rather than consistency (28).

As we noted in Chapter 3, violence in movies and television programs can increase the tendency to violence in young viewers.

We turn now to the question of the genetic component in criminal behavior. Much of the material already presented in this chapter strongly suggests that the genetics will be complex. Many aspects of personality must influence the likelihood that an individual will behave in an antisocial manner. Hundreds of genes must be involved, some of them acting singly, but more often, probably, in groups. The varieties of criminal personality should be correspondingly diverse. Examination of the genetic studies themselves confirms this impression.

There is one partial exception—one simple genotype that apparently predisposes, though to a slight degree only, to crime. The genotype is a chromosomal aberration: the presence in males of an extra Y chromosome, giving the sex chromosome constitution XYY. Useful reviews have been prepared by Plomin et al. (70) and by Omenn (68). The most thorough study has been carried out by Witkin et al. (104).

About one in 800 to one in 1,000 males is born with the XYY chromosome constitution. These men appear normal, but they are on the average several inches taller than XY males. A series of studies conducted between 1965 and 1969 indicated that, besides being taller than normal, they are present in disproportionate numbers in prison populations. It now appears that there is only a slight, though a significant, tendency towards crime. Witkin et al. found that their XYY subjects obtained relatively low scores on IQ tests. Since criminals tend to have a lower than average IQ, they believe that this may at least in part explain the slightly elevated incidence of crime (104).

For at least three centuries, persons interested in human psychology have been aware of one rather specific personality type with criminal or antisocial tendencies. In 1652, Thomas Abercromby, physician to King James II, wrote of a condition "in which all the upright sentiments are eliminated, while the intelligence presents no disorders" (76). Another writer in the early 1800s referred to individuals with "moral alienation" or "defective organization of the moral facilities" (13). This type is now well recognized and is usually referred to as the *sociopath*. It is sometimes also, especially in Europe, referred to as the *psychopath*.

Abercromby's description of the sociopath, as far as it goes, would be hard to improve on, but recent observers have added a number of details. Some characteristics, as described by Checkley and others, are: intelligent, poised, not given to delusions or irrationality, and often charming, yet given to lying, stealing, and the commission of other unethical acts, usually without any discernable reason or justification. The sociopath shows no shame or remorse for his acts and does not profit from experience. He is pathologically egocentric, and shows no real capacity for love. Often there is nothing in his family background or training to account for his behavior. He is the prototype of the black sheep (8).

Male sociopaths have attracted the greatest attention, but the condition is not confined to males. It often first appears in childhood.

Studies of brain waves carried out in several countries suggest that there may be a neurophysiological basis or bases for sociopathy. As noted by

Schulsinger, these studies show that the electroencephalograms (EEGs) of sociopaths show an excess of abnormal EEGs as compared with those of the general population. The more violent or impulsive the sociopaths are, the greater the number of their EEG abnormalities (84).

Because the sociopath, as thus defined with his intelligence and charm yet totally irrational unethicality, has inevitably attracted attention, there has perhaps been some tendency to overlook the frequency of types that only partly conform. Much of the material we have examined so far in this chapter would lead us to expect diversity rather than uniformity, and the sociopath apparently is not a homogeneous type. Two distinct subdivisions, the hostile or aggressive and the simple or relatively nonaggressive, have been reported by Conrad and Dinitz (13). The finding that the two give different cardiac responses to the adrenal hormone epinephrine emphasizes the validity of this classification.

Checkley has estimated that sociopaths may constitute from 1%–3% of the adult population. One investigator, in a study of inmates of Sing Sing prison, found that 19% were sociopaths. Another study indicates that they are among the most recidivistic and violent members of the prison population. Sociopaths also are often found as patients in psychiatric institutions. Of course estimates of incidence are contingent on the criteria of psychopathy that are used (8,60).

We shall look later into the genetic component in sociopathy.

Elliott has described a condition, significant for the etiology of crime, that has been designated the *dyscontrol syndrome*. Affected individuals have a history of recurrent episodes of uncontrollable rage. Elliott found that electroencephalography and other tests could be used to show that 94% of his subjects had developmental or acquired brain defects (19).

Mednick and his collaborators report a high correlation between tendencies to criminal behavior and certain physiological responses presumably originating in the autonomic nervous system. To study this they used a series of uncomfortably loud noises to induce a fear response and then measured the rate at which electrical conductance of the skin, which is altered by fear, returned to normal. Mednick cites studies by his own and other groups that demonstrate an association between slow "ectodermal recovery" and criminal tendencies. The slow recovery has been reported for sociopaths, adult criminals, and adolescent delinquents. Mednick believes that this test measures the rate of recovery from a fear stimulus and proposes an interesting theory linking rapid recovery to moral teachability (59). Whatever the merits of the theory, the test, if indeed it correlates with criminal tendencies, suggests a neurophysiological basis for and hence probably a genetic component in much criminal behavior.

A number of studies have indicated that criminals have lower than average intelligence (59). Low intelligence has therefore been suggested as a cause of crime. It is also almost certainly true, however, that many members of the Mafia who dominate organized crime in the United States and Italy have average or above-average intelligence. Some of them, in their own way, are very competent people. So probably also are most swindlers and white collar criminals. The men around Adolph Hitler had an average IQ of 130 (26). This suggests that evil tendencies are an independent variable, with no direct link to intelligence. Low intelligence, however, might conduce to the common forms of crime indirectly. This could happen in two ways. First, the more intelligent individuals with criminal tendencies would be likely to choose swindles and other white collar crimes that are less likely to be detected than common thefts. Second, individuals with below-average intelligence and perhaps some other handicaps are likely to develop frustrations that could lead them into crime. In fact, two groups studying the linkage of low intelligence to delinquency have concluded that it originates in the humiliation and resentment generated by their poor performance in school (37,71). A similar association has been noted between delinquency and dyslexia, and for similar reasons (4). It is a reasonable assumption that individuals who become involved in crime because of handicaps are less likely to become chronic offenders than are persons with sociopathic tendencies.

We have dealt so far with the sort of criminal who is often caught and convicted. Our examination of crime and personality would be incomplete, however, if we did not also include the organized crime families that are the source of some of the most socially damaging criminal activities in both the United States and Italy. These Mafia members, or Mafiosi, are numerically a tiny fraction of the criminal population; individually, they spend little if any time in jail. With the resulting very limited jail record, they never have been open to the sort of studies that are applicable to other offenders. Our information comes mostly from news sources or congressional investigations.

Like the typical sociopath, the Mafiosi, besides their lack of inhibitions against wrongdoing and their intelligence, are, if not exactly charming, at least capable of putting on an excellent front. They differ from the sociopath, as usually defined, in that their wrongdoing is purposeful instead of seemingly nonsensical. It is skillfully directed towards the accumulation of power and wealth. And whereas the sociopath often commits pointless, petty crimes, the Mafiosi are the recognized masters in successful extortion and murder.

With these personality characteristics, it is tempting to say that these men are, in the broad sense, sociopaths, though they certainly are uniquely different from the prototype. If the Mafiosi are in any sense sociopaths, this implies a major genetic component in their criminal tendencies. Intermarriage is common among organized crime families, both in this country and in Italy and, with the possible exception of the families centered in Naples, they all trace back to a common stock of Sicilian origin (90). These families could very well have genes predisposing them to crime that are not shared with the population at large. The most likely assumption in this case is that any genetic determination is complex, not simple.

EVIDENCE FOR A GENETIC COMPONENT IN CRIMINAL BEHAVIOR

Is the degree to which we behave ethically entirely or almost entirely the result of learning or does heredity play a substantial role? The answer to this question must be profoundly important in the formulation of any ethical philosophy. In the earlier sections of this chapter we have assembled some background information relevant to our question; in this section we approach the question more directly. We concentrate on criminal behavior because of the thoroughness with which the question of a genetic component has been examined.

The genetic component in criminal behavior has been estimated from studies both of twins and of adopted children. We start with the work on twins.

Christiansen has published an extensive study of criminality among twins and has also reviewed work done in this area from 1929 through 1977 (10,11). The review includes ten twin studies. (This includes Christiansen's own investigation. I also include in this two separate works on adults and juveniles by one investigator.) The basis of selection in all cases was a criminal record for at least one member of a twin pair. The pairs were classified as either concordant or discordant, depending on whether the second twin did or did not have a criminal record.

The ten studies involved 301 identical twin pairs and 361 like-sexed fraternal twin pairs for a total of 662 pairs, of 1,324 individuals. Both male and female pairs were included, but the majority were males. In all of the studies, concordance was greater in the identical than in the fraternal twins. In the five studies of adult criminals that were published between 1929 and 1962 and that were based on 30 or more pairs, concordance of identicals averaged from 64.5% to 77.8%, and that of fraternals, 11.1%–53.5%. In a 1942 study of 67 pairs, all juveniles, concordance was 97.6% for the identicals and 80.8% for the fraternals. In the two latest and most extensive studies, there was a notable drop in the concordance of identicals. Dalgard and Kringlen found 25.8% concordance in 31 identicals and 14.9% concordance in 54 fraternals (14). Christiansen found 35% concordance in 71 identical male pairs and 13% concordance in 120 fraternal male pairs. For female pairs, his figures were 21% for 14 identicals and 8% for 27 fraternals.

While the basic agreement in all ten studies on the dissimilarity between identicals and fraternals is notable, the difference between the studies are significant and require comment.

The juvenile study was characterized by a high concordance for both identical and fraternal twin pairs and a relatively slight difference between the two twin types. These results suggest a limited genetic factor in juvenile crime and also, because of the high concordance, a substantial importance for those environmental factors that are commonly shared by teenagers. These results, perhaps, are not surprising in view of the permeating influence of peer pressure, a widely shared feature in the juvenile environment and one likely to be shared equally by both twins of a pair.

The 1976 Dalgard and Kringlen study of adult crime was carried out in Norway, the 1977 Christiansen study, in Denmark. In both Norway and Denmark, excellent birth and criminal records are maintained, and these were used to obtain and analyze the twin samples. Because in both studies the starting point was a list of all twins born during a specific period, the extracted sample of criminal twins should have been genuinely random. In the earlier studies, carried out in other countries, the samples were usually obtained directly from the police records. If both members of an identical twin pair have criminal records, this is likely to attract attention. As Christiansen points out in his review, this could lead to an overloading of the samples with concordant identicals and hence to the high concordance of identicals which distinguished these studies.

The earlier studies also may have been somewhat inferior to the last two in twin diagnosis and in the criteria of criminality. This may reduce the dependence that can be placed on the precise figures reported, but it by no means necessarily invalidates the results. Descriptions in the earlier accounts of rather striking similarities in criminal careers of some identical pairs adds weight to the emphasis on genetic factors. On the basis of size alone, however, the Dalgard and Kringlen and the Christiansen studies are the most significant, and we now turn to a further examination of these two.

The twins in Dalgard and Kringlen's sample of 85 male pairs were carefully tested to determine whether they were identical or fraternal and were thoroughly characterized through the use of official records and personal interviews. All the convicted twins had committed serious crimes. The investigators also had a list of twins guilty of lesser offenses such as violation of motor vehicle laws, and the inclusion of these nearly eliminates the difference in concordance between the identicals and the fraternals. Even with these excluded, the difference between the 25.8% concordance observed for identicals and the 14.9% for fraternals is one of the smallest recorded. The authors report observations indicating that the identical pairs had been "treated more frequently as a unit" and had "experienced more extreme closeness than" the fraternal pairs, and they suggest that these environmental differences may account for the differences in concordance. Their data indeed indicate that concordance in crime among the fraternals was greatest for twin pairs than had experienced concordant environments.

This interpretation of the results of the study needs to be weighed in the light of comments made earlier in this chapter concerning the role of genes in determining environment. Investigations of criminals suggest that propensities to crime are likely to show up as personality problems at quite an early age. If both of any pair of twins, whether identical or fraternal, showed such problems, this would inevitably influence parental treatment. The parents presumably would do the best they could to correct the problems, but it seems unlikely that they would try a distinctly different method for each twin of a pair. On the other hand, if one twin of a pair was the problem child, a situation most likely from the genetic viewpoint to occur with fraternals, efforts at separate treatment would be quite natural. It seems quite possible, then, that it is personality shaping environment rather than environment shaping personality that accounts for shared concordance in environment and crime.

We turn now to Christiansen's study of Danish twins (11). The 1977 report of these authors is labeled "preliminary" because there are plans to increase the number of twin pairs studies, but it nevertheless is based on extensive data. It includes a total of 232 twin pairs (41 of them female pairs) with at least one twin having a criminal record. This is more than twice the size of any other group. The crime sample was drawn from a register of all twins born in Denmark between 1881 and 1910; it should be a geniunely random sample. Christiansen and coworkers did not attempt individual study of the twins to the extent employed by Dalgard and Kringlen, but examination was sufficient to characterize the pairs as identical or fraternal with an adequate order of reliability. Crime diagnosis was based on the excellent Danish polish records.

The concordance rates among males (35% of identicals and 13% for fraternals) show a greater difference than the corresponding figures found by Dalgard and Kringlen (26% and 15%, respectively). The sample also is more than twice as large. The difference between identicals and fraternals in Christiansen's data is highly significant ($p < 0.001$). Christiansen's figures for females (21% for identicals and 8% for fraternals) are even more disparate, but these are based on a much smaller sample.

Christiansen concludes that both genetic and environmental factors play a role in determining criminal behavior. Data had been obtained on the year of birth and place of birth (urban or rural) of the twins and on the social class of the fathers. An analysis was carried out to estimate the role of these specific environmental factors on the crime rate. The results point to a considerable complexity of interactions.

Besides Christiansen's study of crime in twins, Denmark has been the site of an extensive study by Hutchings and Mednick on crime in adopted children (42). The starting point of their study was a list of 1,145 adoptess, 30–44 years old, and a list of an equal number of nonadoptees. The subjects were matched for sex, age, residence, and the occupational status of their adoptive fathers. The investigators were provided with the names of the biological fathers of individuals on both lists when known and of the

TABLE 4.1. Percent of criminals among 662 adoptees as related to the criminal records of both their biological and their adoptive fathers.[a]

Is adoptive father criminal?	Is biological father criminal?	
	Yes	No
Yes	36.2% (of 58)	11.5% (of 52)
No	22.0% (of 219)	10.5% (of 33)

[a]Reproduced from Table 8 of Hutchings and Mednick (42). The table includes only individuals, whether adoptees or their fathers, who are classed as either criminals or as without a police record. The omission in both generations of all individuals classified as guilty of only minor offenses accounts for the reduced size of the sample.

adoptive fathers. Not all the biological fathers of the adoptees were known, and the sample size was thereby reduced to 965 individuals. All names were checked against police records to detect any individuals who had either a criminal record or a record of minor offense.

The first finding of interest was that whereas among the nonadopted control group, only 8.9% had a criminal record, the incidence among the adoptees was 16.2%. Adoptees, it would appear, are almost twice as prone to crime as nonadoptees. With respect to minor offenses, the records of the two groups showed much less disparity. When the three groups of fathers were compared the results were: biological fathers of adoptees, 30.8% with criminal records; adoptive fathers, 12.6%; and fathers of nonadopted controls, 11.1%. Evidently, children put up for adoption come from homes where the fathers are likely to have criminal tendencies.

The point most relevant to the role of heredity in crime is the influence of a criminal biological father as compared with a criminal adoptive father on the criminal record of the adoptee. The critical data are presented in Table 4.1

If neither the biological nor the adoptive father was a criminal, 10.5% of adopted sons had criminal records. If the adoptive father only was a criminal, the crime rate was 11.5%, a relatively minor increase. If the biological father only was a criminal, the crime rate rose to 22%, or more than double that when both fathers were crime free. This suggests a substantial genetic influence of the biological father on the criminal tendencies of adopters. If both fathers had criminal records, the crime rate of adopters rose to 36.2%. This represents an increase of 65% compared even with the group with a criminal biological father only. A synergistic effect of the genetic and environmental factors predisposing to crime is indicated. Perhaps the youngster with less natural resistance to crime is more easily corrupted by an unfavorable home environment than a naturally more resistant child. This is an attractive but unproven hypothesis. In fact, Hutchings and Mednick warn that while a genetic role in the determina-

TABLE 4.2. Percent of relatives diagnosed as having mental illness.[a]

	Biological relatives[b]	Adoptive relatives[b]
Sociopathic adoptees	19.0%	13.7%
Normal control adoptees	13.0%	12.0%

[a] Data of Schulsinger (84).
[b] The relatives included were father, mother, and siblings.

tion of crime is strongly suggested by their data, their sample was not large enough for a statistically conclusive proof.

Table 4.1 omits 303 of the adoptees because either the biological or the adoptive father was on the records for a minor but not a criminal offense. In the case of these adoptees also, a record of a minor offense or offenses by the biological father appears to be a much more significant determinant of the criminality of the adoptee than a similar record for the adoptive father. Again, the data point to a major genetic role in the determination of criminal tendencies.

Danish government adoption records have been used in another study relevant to the problem of the genetic component of criminal behavior—a study by Schulsinger on sociopathy (84). Schulsinger uses the term "psychopathy," but this is sometimes used synonymously with "sociopathy" and his criteria are clearly those of the disorder usually referred to as sociopathy in the United States.

From a list of 5,483 adoptions in the Copenhagen area between 1924 and 1947, Schulsinger selected 507 individuals with a record of mental illness. These were then screened for sociopathy on the basis of criteria that gave a reasonable degree of concordance when used by different psychiatrists. This yielded a sample of 57 individuals (40 male, 17 female). From the total list of adoptees, he also chose a matched sample of 57 individuals without a record of mental illness. Data were then gathered on records of mental illness among biological and adoptive relatives of both groups, and the data analyzed as to the incidence of mental illness, and more specifically, of sociopathy. The results are summarized in Tables 4.2 and 4.3.

Table 4.2 deals with mental illness of all kinds. The data show that there was nearly 50% more mental illness among the biological relatives of sociopathic adoptees than there was among the biological relatives of

TABLE 4.3. Percent of fathers diagnosed as sociopaths.[a]

	Biological fathers	Adoptive fathers
Sociopathic adoptees	9.3%	1.9%
Normal control adoptees	1.8%	0%

[a] Data of Schulsinger (84).

normal control adoptees. A genetic component in the determination of mental illness in the families of sociopathic individuals is indicated.

Table 4.3 deals with the other extremes: the incidence of sociopathy in the four groups is compared. There are more than five times as many cases of sociopathy among the biological fathers of sociopaths as there are among each of the other three groups. Again a strong genetic component is indicated. The statistical significance of this one grouping is marginal, but taken together, the data pointing to a genetic component in sociopathy are highly significant. As further proof, Schulsinger found no evidence that the sociopathy in his 57 adoptees was caused by problems during pregnancy or birth.

The studies of crime that we have been examining provide, in their totality, strong (not to say overwhelming) evidence that there is a genetic component in criminal behavior. They do not permit anything approaching a precise interpretation of the genetic factors involved. Despite the uncertainties, it may be worthwhile to offer what perhaps can be regarded as informed guesses concerning some aspects of the genetics.

In our examination of studies with mice, we found that a variety of pathological conditions is caused by recessive genes and that, as a rule, a number of mutant or deviant genes, each at a distinct genetic locus, can cause very similar pathological states. Thus genes at six loci are known to cause diabetes and genes at 24 loci are known to cause problems of balance and often deafness owing to defects of the inner ear. At the other extreme, there is no evidence that single genes have a direct and pronounced effect on fighting behavior. A gene causing blindness could have a substantial effect, but its influence would be indirect.

I suggest that we are likely to find in criminal behavior in humans both extremes of genetic control: both the single gene effects and the conditions determined by multiple genes.

Sociopathy, because it is a sharply delimited type of antisocial personality, is a good candidate for single locus determination. If its development in a given individual requires only the presence of recessive mutant genes at one locus inherited from both parents, it could appear in families without prior record of the trait. This is known to happen. It could also appear in children of a sociopathic parent, and as we have seen, this also happens. This sort of inheritance is also compatible with the existence of more than one locus, each capable individually of causing sociopathy. By analogy with such pathologies in the mouse as diabetes, I suggest that this is the case with sociopathy. There easily could be six or more loci, each capable of causing a minor variant of this condition. It would not be surprising if fairly typical sociopathy were also sometimes complexly rather than simply determined.

The less sharply defined types of criminality also have a genetic component, but there the comparable situation in the mouse would appear to be the various forms of fighting behavior. In both cases, many interacting loci probably affect the traits in question. Genes at these loci can be present in a variety of patterns, each producing its own particular behavioral tendencies. In the human situation, if and insofar as there is intermarriage between families with criminal tendencies, there could be some fixation of specific genes and the familial character of crime could be perpetuated. The observations of Sheldon and Eleanor Glueck tend to support the assumption that this occurs (28). The various Mafia families in this country and in Italy could represent a particularly vicious example of this situation. It is, I suggest, plausible to assume that in these families there is a fixation of genes that reduce or eliminate those natural restraints against murder and deceit that help make human society function. In combination with the intelligence obviously possessed by many top Mafia, an extraordinary potential for evil is produced.

STUDIES OF THE PROSOCIAL SENTIMENTS

Because public records provide no abundance of data concerning humanity's prosocial sentiments comparable to the abundance available concerning crime, this aspect of our social repertoire has received a lesser share of attention from scientists. Recently, however, there has been a growing interest in this subject. Particularly informative are studies of the development of socialization and of a moral sense in children. There is also a valuable 15-year-old study of the development of social behavior in monkeys. These studies show that man and his near relations do have strong social bonds; that these are complex rather than simple; that during development they appear in a well-defined sequence; that they have a substantial genetic basis; and that, both as children and adults, people differ in the strength of their moral sense.

Perhaps the most revealing study of the nature and development of loving behavior has been carried out with monkeys. Affection is evident in all social mammals, yet for a long time it received little attention from psychologists. One team of psychologists stand out as the exception: Harry and Margaret Harlow and their coworkers have carried out a long and fascinating series of studies of the development of social bonds in rhesus monkeys *(Macaca mulatta)* (32). Their basic findings are that affectionate behavior goes through a series of developmental stages; that appropriate stimuli, sometimes quite different from anything that had previously been theorized, are necessary to call forth each of the stages; and that deprivation at any one stage will impair the development of subsequent stages. They see the link between stimulus and response as being genetically determined.

The Harlows distinguish five affectional systems or phases in their subjects: (1) maternal love; (2) love of the infant for the mother; (3) peer or age-mate love, a phase that continues through adolescence; (4) heterosexual love; (5) paternal or father love. Some of their most fascinating observations concern the first two phases.

Their basic method of studying the mother-infant bond consisted of removing the infant from the mother and providing a dummy or surrogate mother in her place. Infants without either mother or surrogate showed obvious distress. If an infant was placed with a cloth-covered dummy, the infant would cuddle up to it, apparently deriving considerable reassurance from this contact. A wire surrogate, even if containing a bottle of milk to simulate a breast, was not an adequate substitute for the soft contact provided by a cloth covering. The attractiveness of the cloth surrogate, however, was increased if it could provide milk or if it were capable of a rocking motion.

Clinging is a normal part of the mother-infant relationship, and apparently this contact comfort, at least in rhesus monkeys, is the most important factor in the development of the mother-infant bond. Nursing is less important than theorists concerned with human emotional development had postulated.

In the next or peer-interaction stage of affectional development, mother-infant contacts decrease and contacts between juveniles increase. Play is very important in developing attachments at this stage, though the affectional ties, as manifested through physical proximity and cooperative behavior, continue after the play diminishes. Contrary to Freud's assumptions concerning human development, the Harlows found in their subjects that successful passage through the peer stage was more important for adult social and sexual relationships than successful passage through the mother-infant stage was.

These and the subsequent stages follow a definite chronology. Probably the concept of critical periods for the development of specific types of behavior is applicable here as it was in some developmental examples cited earlier. The Harlows found that males were much more given to rough-and-tumble play than females, with the difference appearing rather precisely at 30 days. Harlow cites a study by Sackett showing a critical period in the development of certain fear responses. If infant monkeys isolated at birth were shown projected pictures of adults grimacing in a manner indicative of threat, they showed no special response until approximately 90 days, when a full-blown fear response appeared.

Though monkeys develop new bonds as they mature, the old bonds do not necessarily disappear. In natural habitats, rhesus mothers have been observed to come to the aid of their fully mature sons, indicating that the maternal urge to defend progeny is a lasting one. Paternal affection also is manifested in the defense of infants and, to a lesser degree, older young.

Primates in general probably are characterized by stages in the development of social bonds similar in many respects to those described by Harlow and his associates. However, a study by Kaufman of the effect on young monkeys of the removal of the mother shows that significant differences can be found even in closely related monkey species. Kaufman worked with pigtail and bonnet monkeys, both members of the genus *Macaca* to which rhesus monkeys belong. Both species were maintained in groups of one adult male and four or five adult females. If a pigtail mother of a youngster five or six months old was removed from the group, the youngster showed acute distress and agitation and did not fully recover until about a month later (24). In bonnet groups, on the contrary, a youngster whose mother was removed showed only a brief period of agitation and soon was interacting rigorously and happily with other members of the group.

In the last few years a number of investigators, with the full collaboration of the parents, have been studying the development of moral or prosocial responses in children. Most of these investigators lay particular emphasis on *empathy,* or the ability to sense and respond to the feelings of others. This first appears in very small infants. If a bottle is dropped on the floor of a hospital nursery, the children appear undisturbed. If something starts one infant crying, the air is soon filled with the cries of a dozen babies, all demanding comfort. Thus at a very early age children respond sympathetically to signs of distress in other children (1).

Kagan, who conducted extensive studies of moral development in children, is one of the child psychologists who agrees as to the importance of empathy, but he also distinguishes a number of other mechanisms that contribute to the development of moral standards (45). Among those that he mentions are the recognition of deviation from norms; the ability to generate or conceptualize ideal outcomes; the need for cognitive consistency among beliefs, actions, and the perceived demands of reality; and the capacity to identify with others. While not belittling the importance of parental training and other environmental factors, he also holds that "the child is biologically prepared to acquire standards . . . all children have a capacity to generate ideas about good and bad states, actions, and outcomes."

Kagan emphasizes that the development of moral capabilities is a process that continues throughout childhood. A definite sense of right and wrong, however, appears before the third birthday; children of this age know that it is wrong to hurt someone else. Kagan notes that in primitive tribes, children are commonly born about three years apart, and that the emergence in children at age three of inhibitions that would protect the newborn are therefore an appropriate adaptation.

Child psychologists who have studied moral sense in children generally agree that there are substantial individual differences. This subject has been discussed in a number of books and reviews. Rest cites striking differences in children in their spontaneous tendency to attend to the possible consequences of their behavior for the welfare of others (78); and Staub notes that while some children need reinforcement in addition to social teaching, others need only to be instructed (84). Marian Radke-

Yarrow and coworkers at the National Institute of Mental Health have found not only much diversity in types of helpful and sympathetic behavior, but also substantial and persistent individual differences (72,73). Nancy Eisenberg at Arizona State University finds that "Certain qualities of early empathic behavior appear to be associated with certain types of moral reasoning" (1).

The magnitude and early appearance of these differences in childrens' moral sensitivities strongly suggest that there is a major genetic component in their determination. As we have noted, Kagan holds that this is the case. A twin study carried out by Rushton et al. provides confirmation. A questionnaire designed to measure altruistic tendencies was filled out by 296 identical twin pairs and 179 like-sexed fraternal twin pairs, ages 19–69 + . The authors use altruism in a broader sense than I use it in Chapter 5, but it probably is safe to say that the questionnaire did provide a valid measure of prosocial tendencies. The degree of similarity of the responses of the two types of twins was compared and found to be much greater for the identicals than for the fraternals. This is the result expected if the qualities are substantially influenced by heredity. The authors calculate that genetic factors determine somewhat more than half of the observed variability (81).

The variability in moral tendencies revealed in these studies is evident also in everyday life. We have examined one end of the moral spectrum in our review of criminal behavior. There is no comparable wealth of information about the other end of the moral spectrum, but Sorokin, in his book *Altruistic Love: A Study of American "Good Neighbors" and Christian Saints* (91), has given a fascinating account of some truly benevolent people. It warms the heart to know that this order of goodness does indeed exist.

SUMMARY AND CONCLUSIONS

The studies we have examined emphasize the wide range of social interactions—ranging from affectionate to violently aggressive—of which man and his relatives are capable. As examples of the complexity, we noted the incidence of five kinds of fighting in mice and five separate affectional phases of systems in rhesus monkeys. Each behavioral system probably has its own underlying neurophysiological mechanisms. Experiments on aggressive behavior incited by electrical stimulation of specific areas of a cat's hypothalamus is one of the clear demonstrations of the extent to which the details of behavior are spelled out in the brain. In many cases this sort of association can only be hypothesized, but the Harlows specifically postulate that the affectional systems they have described in rhesus monkeys do have underlying neurophysiological mechanisms.

We have found evidence that there are critical stages in the development of behavior. This may mean that each neurophysiological mechanism first becomes functional at a specific time in the individual's development. This is typically manifested in the rather sudden appearance of the ability to respond to specific environmental stimuli.

The genetics of the central nervous system must be, in its own way, as complex as the system itself. Surely the brain, with its ability to generate appropriate responses to an infinitude of stimuli, both internal and external, and to modify these responses through learning, is the most complex and marvelous of all organs. As we have seen, even in the mouse at least 150,000 genes may be involved in its development and operation.

Finally, there is the question of genetic diversity. We found abundant evidence of genetic diversity in mice, including diversity in degrees of aggressiveness. There is certainly no less genetic diversity in man, and this diversity extends to the prosocial or moral proclivities. Studies of twins and adopted children provide convincing evidence that there is a major genetic component in criminal behavior and in sociopathy. There are no comparable studies of people at the other end of the moral spectrum, but there is evidence of genetic variability in the prosocial tendencies of people in general. Studies of social development in children reveal differences that appear so early and persist so consistently that investigators generally agree that they are partly genetic. One twin study of ethical perceptions in adults confirmed this conclusion. We shall, then, proceed from the hypothesis that man's moral tendencies are variable rather than uniform, and that a substantial part of this variability is genetic in origin.

If morality is partly gene-determined, this surely has significance for ethical theory. Before considering this relationship, however, we shall examine, in Chapter 5, some aspects of the evolution of our social and moral tendencies.

5

The Biological Background of Social Behavior: Evolutionary Factors

Introduction

Man, like all the earth's creatures, is the product of evolution. This was Darwin's great postulate in the *Descent of Man,* and subsequent research strongly supports his reasoning. Not only did evolution shape man's physical being, it also shaped his mental and moral characteristics. This again was Darwin's view, and in this, too, time has strengthened his conclusions (14).

Evolution made man a social animal, but it also created both an enormous number and an enormous diversity of other social species. Mother Nature has given social living her seal of approval. It works.

Animal societies can be examined from several viewpoints. Ethology, the study of animals in their natural state, can show in detail the nature and diversity of social structures. Evolutionary studies can tell us of the stages by which existing social species, and also earlier but now extinct social species, came into being and can indicate how the successive stages were produced by natural selection. Evolutionary studies, in turn, draw on such other sciences as paleontology. Our understanding of the details of natural selection can be enhanced by including modern genetics in our studies of the selective process. This introduces into our survey areas of research known as population genetics and the mathematical theory of natural selection.

The name *sociobiology* has been given to the sum of these scientific domains as applied to social species. This chapter is concerned with the contribution that sociobiology has made to our understanding of our social nature. We shall have a good deal to say about animal societies other than our own, but only because these can illustrate the ways the cooperativeness and the aggressiveness in our own nature have evolved.

Our survey will confirm the conclusion of the last chapter that man's moral nature, though flawed, is capable of producing strong bonds of love, affection, and helpfulness. Those who seek to build a better world, whatever their background or approach, cannot expect to find their pathway easy, but their route is not barred by insurmountable obstacles in human nature.*

Social Characteristics of Man's Near Relatives

The primates are a group of mammals, all rather closely related to man, that includes the lemurs, confined to Madagascar; the New World monkeys, distinguished by the presence of widely spaced nostrils and prehensile tails; the Old World monkeys including the baboons; the apes, man's nearest relatives; and man. All told, there are about 200 living species. The primates display a wide range of variation but, except perhaps for the lemurs and a few other relatively primitive forms, all show some rather obvious physical similarities to man. Many also foreshadow, in varying degrees, significant aspects of man's social behavior.

The great majority of man's primate relatives are tree dwellers and in the process of adapting to this habitat have developed two physical characteristics shared by and important to man. Their forelimbs have evolved in the direction of arms rather than legs, with hands capable of grasping the branches of the trees in which they live, and their eyes have moved to the front of the face and developed stereoscopic vision, enabling them to gauge distances and to move more easily through the three-dimensional world they inhabit. The apes, at least, have full color vision equivalent to that of man. This presumably helps them to recognize the various fruits that are an important part of their diets.

Most primate species live in troops containing adults of both sexes and assorted young. Troops vary greatly in size. In the monogamous gibbon, the social unit is the family—an adult pair and up to four young. In rhesus monkeys and baboons, troops may number over 100. Troops of most species contain more females than males, though in one species of lemur the reverse is true (7,12,18,25,74).

Most primate groups are *territorial,* living in defined areas they defend against other groups. Sometimes their foraging extends into overlapping areas that are not defended; territories with these undefined limits being referred to as *home ranges.* While battles in defense of territory, if infrequent, may easily escape notice, it does appear that in some species territoriality is not well developed. Gorillas have home ranges with consid-

---

*Two writers who have examined human evolution and drawn similar conclusions are Edward O. Wilson in *On Human Nature* and Sydney Mellen in *The Evolution of Love* (64,113).

erable overlap and they are generally regarded as peaceful, but Fossey has reported one instance of group fighting (1,18,25,29,47).

Primates are predominantly herbivores, but insects are widely eaten, especially by some of the smaller species, and it has been estimated that 37% of all species variously consume eggs, crustaceans, and small vertebrates (34).

Apes approximate man more than most primates, so it is appropriate to give special emphasis to this group. There are five genera of apes, of which three—the chimpanzee, gorilla, and orangutan, referred to as the great apes—are particularly notable for their human resemblances. The resemblances are greatest in the chimpanzee and gorilla, which live in Africa, somewhat less in the Southeast Asian orangutan. These resemblances are reflected in the number of pairs of chromosomes characterizing these species. Man has 23 pairs, all the great apes, 24. The two lesser apes have 22 and 25, respectively. More significant than the number of chromosomes is the degree of similarity or difference of the actual bearer of heredity within the chromosomes, the DNA. Recent developments have made it possible to compare quite accurately the DNA structure of different species and to estimate the time when different lines diverged. According to evidence from this source, chimpanzees diverged from the human line about 7 (6.3–7.7) million years ago, the gorillas about 9 million years ago and the orangutan about 14.5 million years ago (91). De Waal's *Chimpanzee Politics,* and especially the photographs in this volume, show how remarkably like humans chimpanzees can be (16).

Two of the great apes, like man, are at home on the ground. Gorillas are primarily ground dwellers, though within the limitations imposed by their great size, are also at home in trees. They are knuckle walkers, traveling on all fours though with the fingers folded rather than extended. Gorillas can walk upright but seldom do. Chimpanzees are equally at home on the ground and in trees. On the ground they are also typically knuckle walkers, but not infrequently, walk upright. Orangutans are highly specialized for tree living, but do occasionally travel on the ground.

Although they subsist primarily on fruit and other plant food, chimpanzees, unlike most primates, also include some meat in their diet. Gorillas occasionally eat insects, slugs, and snails but not birds or mammals, although rare instances of cannibalism have been observed.

Among chimpanzees, most of the hunting and meat eating are done by males who may stalk or pursue their prey in groups of up to five individuals. They hunt birds, young baboons, and occasionally monkeys and bush pigs. To catch them the hunting males may use strategies such as a cautious surrounding of their intended victims. There is some sharing of the kills but only in response to a stereotyped form of pleading. Females are adept at catching insects and employ elementary tools in the process but seldom stalk mammals. It is interesting to note that in the hunting-gathering tribes that were the basic human social group for thousands of years, a sex specialization in the search for food also occurred (18,25,30).

One of the most striking similarities of chimpanzees to humans is a capacity, clearly evident even though primitive, to use readily available and even somewhat modified objects as tools and weapons. Tool-using has been reported in many other vertebrates, including birds, but we shall confine ourselves to the chimpanzee. Jane Goodall has provided an excellent review of this subject (28). Chimpanzees frequently employ tools to assist in the gathering of honey and insects. Stalks of grass, twigs, or fibers from bark, usually carefully selected and often modified, as by the removal of leaves, are used to obtain ants or termites by thrusting them into a nest and licking off any clinging insects. The tool may be carried from one nest to another over a considerable distance. Sticks are used in a similar way to obtain honey. Chimpanzees have also been observed enlarging the opening to a bees nest by prying and pushing the opening with a heavy stick so that they can reach for the honey with their hands. Much more striking uses of tools have been noted in captive chimpanzees under experimental conditions, but these tell more about the intelligence of these animals than about any natural tendency toward tool use.

With respect to the use of weapons, there are many reports of chimpanzees using sticks in the manner of clubs or of their throwing stones or other objects at another animal. It is almost always males who engage in these activities. The article by Goodall includes three photographs of males brandishing sticks in a threatening manner. In one case, the stick was used in a charging display, a maneuver used to increase dominance. In another case the threat was directed at baboons. Stone throwing has been observed on numerous occasions. Chimpanzees are not particularly good marksmen—the throwing often serves merely as a bluff or threat—but in one case a thrown stone hit a mother pig, which ran away, exposing the young to capture.

The significance for human evolution of these observations in the chimpanzee is the implication they convey concerning tool and weapon use by man's early ancestors. Once humans started making scrapers, spearpoints, and other implements from flint or other types of stone, they left artifacts whose durability provides conclusive evidence of tool and weapon making. The chimpanzees teach us that such developed use of implements may have been preceded by a long period in which perishable sticks and unfashioned stones were used in food gathering, food preparation, hunting, and fighting. Wood objects could very well have been substantially modified, and pieces of bone, perhaps the product of cracking to obtain the marrow, could have become part of the tool and weapon repertoire.

A final characteristic of chimpanzees and other apes, of interest in connection with the evolution of man's social nature, is the relatively wide birth spacing and long childhood in these species. It is characteristic of most primates (man is an exception) that they do not conceive while still caring for the last-born infant. In monkeys, which mature rapidly, this still permits a relatively high rate of reproduction, but in the apes, and especially the great apes, childhood is long and the reproduction rate low. Old World monkeys give birth every two to three years, chimpanzees only every five or six. Adolescence in chimpanzees extends from about age nine to 14, and the lifespan is only about 40 years. This means that female chimpanzees must live until they are nearly 20 merely to hold their own reproductively. They have a relatively small margin of safety beyond this (18,29,46,85). This low birth rate is one reason for the small populations of apes living today.

We can only speculate as to the cause of the wide birth spacing that distinguishes the great apes from the monkeys. Probably it is one of a group of interrelated traits that include stronger social bonds, slower development, an increase in the complexity of learned behavior, and a larger brain.

The Bones of Our Ancestors

Our knowledge of our early human ancestors comes almost entirely from fossilized bones. The fossil record has grown substantially in recent years, but there are still many gaps. It is not necessary to go into details, but a few points revealed by the record are important.

Most and perhaps all early Old World primate evolution occurred in Africa (50). Like other successful groups, the early primate line threw off many branches. This provided an opportunity for extensive trial and error; only a relatively few successful lines survived (51). The branches that gave rise to the great apes and to man probably diverged during a period when some of the heavily forested areas of Africa were giving way to mixed environments consisting of dense forests, savanna woodlands, and more open areas (18,55). These changes in the environment virtually necessitated a substantial degree of adaptation to ground dwelling. The human line made this adaptation by developing bipedalism and an upright posture. There is a long gap in the fossil evidence, but Johanson has described leg bones from Ethiopia dated as 3 million years old that are clearly in the human line and indicative of bipedalism almost as well developed as our own (46).

These bones belonged to a *hominid,* a term applied to the primate line that gave rise to humans, as distinct from *hominoid,* the broader group including both man and apes. Other bones have been discovered from the same period that probably were derived from the hominids or a closely related species. Richard Leaky and his mother and their coworkers found hominid bones in East Africa and Johanson and his group found them 1,000 miles away in Ethiopia. Johanson's best known find is Lucy, a relatively complete skeleton of an adult but small female from the same area as the leg bones.

Lucy is remarkable because of the completeness of the skeleton, but another find by Johanson was equally remarkable though in a quite different way. In the same region of Ethiopia where Lucy was found, he discovered a single group of numerous bones that examination showed had come from at least 13 individuals. The assumption is that they were killed by some natural catastrophe, perhaps a flood. Men, women, and at least four children were represented. The find provides strong evidence that at this early stage of human evolution, some 3 to 3.5 million years ago, man was living in social groups (46).

One of the most striking discoveries was made by the Leaky group. In a layer of volcanic ash that fell about 3.7 million years ago, which was then wet by rain and later covered with more ash, searchers came across fossilized footprints of two individuals walking side-by-side. The individuals that made them had human-like feet, clearly walked upright, and, judging by the length of stride, were of somewhat unequal height, but both considerably shorter than people today (55).

Johanson has given the group of upright *hominids* that left these bones and footprints the name *Australopithecus afarensis*. These creatures were short but heavily built and well muscled. Their brains were small, more like those of apes than of men. Their teeth were intermediate between those of apes and man. However, whereas apes, and particularly male apes, have large canines that make effective weapons, their canines were not conspicuously large.

The foremost conclusion to be drawn from these creatures is that upright posture preceded any major development of the brain. Bipedalism was well developed in Lucy and her relatives. This change in habit required changes in the shape of many bones, changes in musculature, and probably changes in the sense of balance and in those areas of the brain that control locomotion. Changes of this order of magnitude require time, perhaps several million years. All this occurred without significant changes in size.

No worked stone tools have been found with the *A. afarensis* remains. Any use of wooden or unworked stone tools by these creatures must be conjectural. Nevertheless, as argued by Washburn, their upright posture and well-developed hands should have predisposed them to tool use (105). Also it seems an almost necessary assumption that simple, unworked tools were evolved before the acquisition of the techniques necessary to produce worked tools.

The relationship of *A. afarensis* to the other more recent hominids that have been identified from skeletal remains is still a subject of debate. I shall follow the interpretation of Johanson, which I find particularly plausible.

Beginning about one million years after the earliest indications of *A.*

afarensis, two other major types appear in the fossil record. The first was originally discovered in South Africa in 1924 by Raymond Dart. Subsequently, in both South and East Africa, skeletal remains were found that clearly belong to the same general group but that display a sufficient range of variation so that at least two species, *Australopithecus africanus* and *Australopithecus boisei,* are now recognized. These forms, like *A. afarensis,* walked upright while retaining many primitive characteristics. Their brains were approximately the same size as those of the great apes, and their jaws and teeth had some apelike properties. In *A. boisei,* the more widely distributed species, the molar teeth were much enlarged compared to the incisors, suggesting a diet of coarse plant food. How the three species of *Australopithecus* are related is uncertain, but Johanson believes that *afarensis* gave rise to *africanus* and *africanus* to *boisei* (46).

The other form found in Africa and dated to this same period shows more development in the direction of modern man. Most notably, the brain had begun to enlarge. The original specimen was found by Louis Leaky in East Africa and named *Homo habilis.* The name means "handy man," and was given because bones of this species have been found in association with modified stone tools. The earliest such tools and the earliest *H. habilis* go back at least 2.5 million years.

While the increase in brain size in *Homo habilis* is the most obvious mark of humanness, there are indications of other brain changes in the human direction in both this early *Homo* and the contemporary *Australopithecus.* By studying the inside of well-preserved skulls, Ralph Holloway has found evidence that some details of structure in both these species were becoming manlike (38). Thus, along with upright posture, other changes foreshadowing the coming of man were taking place.

Beginning with the emergence of *H. habilis,* skeletal remains have been found that provide a fairly clear picture of subsequent human evolution. The human line at this point entered a period of extraordinarily rapid change. Most notable, in evolutionary terms, was what we can only regard as an explosive growth in brain size. Probably somewhat more than a million years ago, our ancestors moved out of Africa into Europe, where winters could be harsh but where, in season, food could be abundant. At a later date they moved into Asia (55). The demands of different climates propelled further evolutionary developments.

A. boisei, the most recent and most robust of the *Australopithecines,* coexisted in Africa over a considerable period with the more manlike *H. habilis.* The two seem to have lived at times in close proximity. This raises interesting questions, since it is a rule that when two species exploit the same resources, the resulting competition is likely to result in the extinction of one of them. As Schaffer has suggested, the *Australopithecines* and the earliest *Homo* may have evolved different dietary needs partly because of the pressure of this competition. The large molar teeth of *A. boisei* suggest that it relied primarily or entirely on bulky plant foods. *Homo habilis* is the earliest species in the human line that we definitely know to have been a hunter. Pressure from *Australopithecus* could have been one factor in this dietary development (84).

Hunting-Gathering Societies

Johanson's discovery of a single group of bones of *A. afarensis* representing at least 13 individuals indicates that even before the appearance of *Homo,* our ancestors lived in tribes. The details of tribal structure are much more difficult to deduce, but an important clue comes from studies of primitive societies that have survived into recent times and have been investigated by anthropologists. These all share or approximate a life pattern that has been called the hunting-gathering society.

More than 100 of these societies have been studied (114). They are widely scattered over the globe, but generally live in remote and inhospitable areas such as tropical rain forests, Patagonia, or the far north (55). They are, in a very real sense, relic societies, representative of a stage of man's social evolution that preceded agriculture. Despite their wide separation in space, which presumably also indicates a wide separation in time, and despite many differences in lifestyle, they nearly all share one important pattern in common. They are, within the limits permitted by their environment, omnivorous in their diet. The men are the hunters, although sometimes the women catch small animals or assist in game roundups. The women gather plant food, but in this, also, both sexes sometimes collaborate.

Eskimos, by virtue of their northern home, which in most of the areas they inhabit is entirely plantless, live at one end of the food spectrum— they subsist almost entirely on meat (45). Near the other extreme are the Tasadays, a single tribe that lives in isolated, mountainside caves in a Philippine rain forest. The wild yam is their dietary staple, but they also eat berries, flowers, and wild bananas. Before some recent contacts with the outside world, they had no weapons and hunted no game animals, though they did eat tadpoles, frogs, crabs, little fish, and grubs from rotten logs which they broke open with ultracrude stone axes (62). More nearly in an intermediate position are the !Kung of the arid South African Kalahari region. !Kung men possess weapons and hunt local game in addition to gathering plant food (55,114). Until the encroachment of civilization disrupted their lifestyle, the Indian tribes of Patagonia hunted the guanaco, which provided them with meat for food, wool and hide for clothes and shelter, sinew for sewing, and images for their mythology (26). Some tribes make considerable use of fish, and in some cases, as with the Tasadays, insects form a minor part of the diet. A tabulation by Hayden indicates that although plant food is the single largest item in the diet of the typical tribe of hunter-gatherers, the various sources of animal protein contribute more than half the total (35).

The size of hunting-gathering bands varies considerably, but Hayden cites 25–50 as a representative number. The bands show varying degrees of fluidity. Sometimes they break up into separate families; in some cases several tribes congregate to hunt big game. Intermarriage between tribes is common.

Besides their life in groups, surviving hunter-gatherers resemble their relatives the apes in another respect: they tend to be territorial. An excellent review of territoriality in primitive societies has been written by Elizabeth Cashdan. An important principle that emerges is that territorial defense increases in proportion to the density of resources. Dense resources mean that a tribe can subsist in a relatively small and hence easily defensible territory. The tendency to territoriality can thus be shown to be based on what are essentially economic considerations—defense of small but not of large territories is economically feasible. The same principle is found to apply to social mammals in general. Cashdan notes that exceptions occur in some inhospitable areas where population densities are very low, but she believes these exceptions can be explained (8).

Since most existing or recent hunting-gethering societies live in marginal areas with a low concentration of resources, territories tend to be large and hence territoriality relatively undeveloped. There are, however, many parts of the world where resources appropriate for hunting-gathering should have been abundant. It is a reasonable assumption that in these areas territoriality would have been the rule.

While the lifestyle of hunter-gatherers resembles that of most of their primate relatives in that it is social and to some degree territorial, it also differs in important respects. Most of these center around the relation of the sexes.

Among primates in general, there is a great diversity in sexual groupings. Some 80% of all primates are polygamous (42), and among these the social group typically consists of one or a few adult males and a larger number of females. Surplus males live alone or in male groups.

In hunting-gathering societies, the numbers of males and females are approximately equal. Chance inequalities in the sex ratio within one small tribe are compensated for by marriages between tribes. Family groupings of one male and one female are relatively permanent, and while varying degrees of polygamy occur, usually in the form of multiple wives, so also do strong family bonds. In the Tasadays, the primitive, isolated, Philippine tribe already referred to, there were ten men and only five women, but there was no wife sharing (62). The family, as noted by Sahlins (82), is an important economic unit. Food and other useful products derived from hunting by males and gathering by females are shared, and the males share in the support, and to some degree, in the care of children.

The reason for the evolution of this family pattern, which results in social groups with approximately equal numbers of males and females, has been the subject of considerable debate. According to one theory, this social structure would have been favored because male group hunting would have made large game available as a food source even when weapons were still quite primitive. Recently, there has been a tendency to question this theory. There are, in any case, several plausible alternatives. We draw extensively on a review by Hayden in discussing this subject.

Among existing hunting-gathering societies, there is a great diversity of hunting patterns, but there are several circumstances in which group hunting has been noted. There are reported cases of large and dangerous game, such as bears, being hunted by groups. Eskimos form large parties to hunt caribou, which is available in abundance, but only during a short season. Much meat, therefore, must be obtained, prepared, and stored within a short time. In areas with sparse resources, communal game drives, in which women and children may participate, are sometimes found. Both theoretical and observational studies of animal foraging suggest a positive relation between scarcity or unpredictability of resources and cooperative exploration. Individual search accompanied by sharing of finds should be particularly profitable. Bees use this method, individual bees communicating to the hive any discoveries of plants in bloom. Hunting in parties, with the individuals spread out to broaden the area covered, should be another effective strategy.

Hayden has noted one other advantage in the exploitation of game that could accrue from multimale tribes, but that does not involve group hunting. The success of any particular hunt is quite unpredictable. Sometimes the bag is substantial, sometimes the hunter or hunting party returns empty-handed. The resulting ups and downs in meat supply can be at least partly cancelled out if the yields of multiple, separate hunts are shared. Apparently such sharing does occur. The use of this method requires a considerable degree of mutual trust; one nonsharer could disrupt a group (35).

Besides his social tendencies, man differs from his near relatives in another respect: his greater fertility owing to the shorter spacing between births. Chimpanzees and gorillas, as we have noted, give birth only about every 5 or 6 years. The human female has a higher reproductive potential. Two studies of primitive societies are relevant. The Yanomamö Indians of Brazil complete a pregnancy about every three to four years, and approximately 85% of these pregnancies terminate in a liveborn child permitted to live. This birthrate and the subsequent survival rate is sufficient so that the population is growing. The !Kung bushman of the Kalahari have a similar or perhaps slightly shorter birth spacing. In both tribes, mothers nurse their infants for two or more years. Both also practice abortion and occasional infanticide to prevent too close child spacing. Among the Yanomamö, male infants are most likely to be spared, and among both tribes, also, physically normal infants are favored (52,69).

These data indicate a birthrate for primitive tribesmen at least 50% greater than that of the great apes. Evidently tribal mothers would have had more progeny to care for than their lowly cousins. Evidently also, primitive man had a potential for quite rapid population growth. We are led to ask what selective forces led to the birthrates characteristic of the different species and how tribal wives could have cared not only for more small children than female chimpanzees or gorillas, but also children with an even longer dependency.

A possible factor in the ability of humans to support a family of growing children could be the establishment of relatively stable home bases where children and at least some of the women could stay for weeks or months. To some extent this does occur in hunting-gathering societies today. Among the Eskimos, a dozen or more families will congregate each winter in a small community from which the males sally forth each morning to harpoon seals (45). It is also true, however, that many societies are predominantly nomadic. !Kung women, according to one estimate, walk with their children and their few belongings about 2400 kilometers per year.

It would be expected, and has in fact been observed, that the amount of nomadism increases as the density of food resources decreases. It is quite possible that in the past, many tribes were blessed with environments so favorable that little nomadism was necessary.

We leave for now the question of how our ancestors adapted to a reproductive potential substantially greater than that of their relatives, the apes, but shall return to it again.

WHAT ARCHAEOLOGY CAN TELL US ABOUT HUMAN SOCIAL BEHAVIOR

Archaeological studies cannot provide us with anything approaching the detailed evidence concerning the social structure of tribal man that we can get from observing existing societies. Many of the facts they do provide, however, are important because they apply to early man. The evidence takes us back to *Homo habilis,* some 2 million years ago, and comes primarily from twenty or more sites in East Africa (55). The evidence has been reviewed by Isaac and Crader (43).

One of the most informative sites is located in Tanzania and was excavated by Mary Leakey and her colleagues. In one small portion of a single stratum that once had been a lakeside, numerous stone artifacts and animal bones were found. The site has been dated as 1.8 million years old. Near the center was a small area bare of remains that the excavators interpret as the living quarters, probably originally provided with some kind of a shelter. This interpretation was suggested by the discovery elsewhere of a ring of stones, apparently man-made. There was some clustering of remains around this bare spot, but there were also remains more widely scattered. All told 2470 stone artifacts were found and 3510 animal bones. Most of the artifacts were unretouched stone flakes, but there also were worked flakes, shaped tools, and stones that apparently had been used for pounding. The bones were from both small and large mammals, and about 8% from turtles. Most of the large animals were members of the family Bovidae, which includes antelope and cattle. The bones were usually broken, presumably so that the marrow could be extracted. This suggests a use for the pounding stones. Fragments of a *Homo habilis* skull were found, but there was also a skull of *A. boisei* nearby.

What information can we extract from the excavation of this site? On the basis of what is known from other sites, the assemblage probably should be attributed to *Homo* and not *Australopithecus*. The presence of the australopithecine skull confirms the coexistence of the two species, but otherwise raises more questions than it answers.

The number of remains suggests the work of a group of people and hence a tribal structure. Isaac and Crader do not interpret the site as a permanent one, but it certainly was occupied for some weeks and perhaps on several occasions (43).

Meat made up a substantial part of the diet of these people, and it came from numerous species. The nearby lake provided the turtles. We may safely assume that the other small animals were hunted and captured, but there is a possibility that the larger beasts were scavenged from kills made by lions or other predators. Some recent studies of tooth marks and tool marks on the bones of the larger prey tend to confirm the scavenging hypothesis (89). If some or all of them were scavenged, there is again the possibility that man drove away the original possessor. Bushmen in South Africa have been known to obtain fresh carcasses from cheetahs in this fashion, and scavengers often try to steal food from other scavengers or from the original predator. It is of course also possible that man both scavenged and hunted the larger game. Another important point established by this excavation is that primitive man carried animals or animal parts back to a home base.

One other conclusion that is indicated is that *Homo habilis* had reached a stage in which a substantial cultural tradition was transmitted from generation to generation. This is suggested by the presence of worked stone tools and by the whole hunting complex, including the carrying of meat to a home base.

If we try to go backward in time from this period of East African prehistory, we encounter a paucity of hard facts. However, as Isaac and Crader state, "It is entirely reasonable to envisage a prestone tool phase of *hominid* evolution in which meat eating was an important component" (43).

Looking forward, we find evidence of a gradual refinement of stone tools and weapons. At the same time there was an increase in the ability to kill large game, though the variety and quantity of such game must have varied with the environment (55).

The ability to use fire must have been a very important addition to our heritage. When this occurred we do not know, but it probably was at least one million years ago (55). The invention of bow and arrow, another milestone in human development, came much later.

The Evolutionary Sources of Aggressiveness and Cooperativeness

The basic source of fighting in animals is competition for the possession of resources in short supply that are essential for survival or essential for the production and rearing of progeny. Three of these essential resources can be identified: food, the availability of a mate or mates, and the conditions necessary for the successful rearing of young. Since all food comes from the earth or its waters, access to territory can be the key to access to food. Competition for food, therefore, can take the form of competition for territory. In social species, competition for territory usually is competition between groups rather than individuals. There may, however, also be intragroup competition both over specific items of food and over mates. This intragroup competition, rather than being directed at the objects wanted, may take the form of competition for a high rank in a dominance hierarchy. Thus while food, mates, and offspring are the basic resources, competition to secure or protect these may be disguised as competition for territory or for dominance.

Because of the importance of territory, it is one of the commonest sources of conflict. Many factors, however, can influence the frequency of such conflict. Since food and territory are closely linked, anything conducive to a food shortage should increase territorial demand. As we shall see, both drought and population increase, either of which can put pressure on the food supply, are sources of increased fighting.

Within social groups food is usually shared; but if the supply is short, there may be competition, especially for choice items. In this situation, access is usually determined by rank in the dominance hierarchy.

Competition for a mate or mates is a major source of intragroup conflict. As Wilson has put it, sex is "an antisocial force in evolution" (112). In mammals, this competition is usually between males. The number of young that a single female mammal can produce and rear is severely limited. The mother also invests substantial energy in each potential offspring during pregnancy, and additional energy and much loving care throughout lactation. Under these circumstances, the mother has a high stake in the continued survival of each of her infants. The male mammal, unlike the female, can potentially sire a very large number of young. From the evolutionary standpoint, therefore, a likely strategy for the male is to acquire as many mates and sire as many young as possible and to leave the care of the young to their mothers. Special counterforces are necessary if fatherhood is to extend to a share in the care of young.

Because of this difference between the sexes, harem formation or *polygyny* is rather common in social mammals. Another common pattern is promiscuity, with access to females in estrus going to the dominant males. Monogamous matings also occur, but are the exception rather than the rule. Because of the great evolutionary advantage accruing to males with multiple mates in species where this mating pattern occurs, there is a marked tendency towards modifications in the male that will help him win fights. It is, in fact, an accepted rule that in polygynous species, the male is bigger than the female and may have special armaments, such as the antlers of male deer or the large canine teeth common in male primates. In some species, this sexual dimorphism assumes truly extravagant proportions.

Competition for the survival of offspring in social mammals usually takes the form of the killing of infants not one's own. Both males and females have been seen to practice this form of infanticide. The advantage gained is either a greater chance of survival of the killer's own offspring or, in the case of infanticide by males, an earlier opportunity for mating with the mothers. Mothers often fight strenuously to protect their young, but subordinate females may submit rather passively to killing by a dominant female.

Because of the significance that a pecking order in a dominance hierarchy can have in access to food or mates or in the survival of young, the establishment of the hierarchy is an important part of social life in many species. The process often begins when animals are quite young. Animals may be assisted by close relatives in achieving rank. Some serious fighting often occurs. However, there is no advantage in starting fights almost certain to be lost, and fighting is inimicable to cooperation. Limiting tendencies, therefore, have evolved. Lorenz, in particular, has called attention to a variety of ritual behavior that inhibits combat between unequally matched individuals. Once a dominance order is established, it can become a stabilizing influence within the group (59).

The sources of cooperation or mutual helpfulness in social animals are less obvious and more difficult to analyze than the sources of conflict. They all trace back, however, to the inherent advantages of group effort. Even where the group structure is very primitive, a group working together can often accomplish what the members of the group could not accomplish individually.

A number of specific situations can be identified in which group action is profitable. Most species are subject to predation by one or more other species. Where this is the case, group defense against the predator is likely to be more effective than individual defense. If a species is territorial, groups can be more effective than individuals in defending or conquering a living space. As we shall see when we examine cases, the number of cooperating individuals can be the determining factor in victory or defeat. If a species is omnivorous or carnivorous, the chance of success in a hunt can be enhanced by group effort. A hunting pack can catch game too large or too dangerous for an individual to catch alone. In the pursuit of small but elusive game, a group can often succeed by virtue of strategy. Because hunting is a daily activity in carnivorous species, the selective effectiveness of this agency in producing cooperation should be particularly high. It is no accident that some of the most closely knit mammalian societies occur in carnivorous species.

The need for cooperation in the gathering of plant food is less obvious than in hunting, but there are situations in which it can be helpful. If food is unevenly distributed over a wide area, sharing of information about the best sources can be beneficial.

Insofar as the members of a species are capable of using experience to choose a favorable course of action, the discovery and transmission of these favorable patterns will be enhanced by the diversity of input possible only within a group. There are cases in both monkeys and chimpanzees in which some new use of a tool or other advantageous practice adopted by a single individual has spread throughout the troop.

Finally, cooperation in the care of young may be helpful and perhaps necessary in assuring their survival. The cooperation may be extended by females other than the mother or it may be provided by the fathers. Patterns of paternal care are likely to evolve only if the males can be sure that the young they nurture are their own. Promiscuity, therefore, is not likely to be conducive to helpful fatherhood.

Aggressive and Cooperative Behavior as Seen in a Few of the Social Mammals

There are hundreds of species of social mammals, and in no two are the social patterns exactly alike. The diversity is extraordinary. All these species, however, display both aggression and cooperation. Most of the aggression is between groups, but some intragroup aggression always occurs, though with much variation in degree. From the ethical point of view, factors increasing or decreasing this aggression, both internal and external, are of particular interest and will be one of our major concerns.

Violent aggression is rare, but it does occur, and probably occurs in all species. Because of its rarity, it was not seen in some of the early studies of social species in their natural habitats. This led to a widespread view that most social mammals are more gentle by nature than their human cousins.

As the hours of observation increased, however, instances of intergroup and intragroup killing were almost always seen. We should remind ourselves that, except on the TV screen, violence in man is also infrequent. The proverbial observer from Mars, unless he landed in the midst of a war, probably would have to watch for hundreds or thousands of hours before he saw an example. If man is unique, it is because he lacks natural armaments, and through his intelligence, has been able to substitute crafted weapons, originally crude and primitive but endowed with ever increasing sophistication and lethality.

The sample of social species we shall now examine is necessarily small and imperfect. It does, however, serve to illustrate the various sources of aggressiveness and cooperativeness we have been discussing.

We have noted that reduction in the food supply owing to drought or other environmental fluctuations can cause an increase in aggression. This is well illustrated in a study by Errington of muskrat populations in a group of marshes, lakes, and streams in the Midwest. In the summer, the study shows, the muskrats are widely dispersed, maintaining relatively small home ranges; but in the winter there may be considerable sharing of lodges and burrows. In normal seasons there may be some sexual fighting in the spring, or attacks on animals trying to move into an already occupied body of water, but serious injury is infrequent. In a year of bad drought, however, Errington found that fighting was greatly increased. A trapper reported that 90% of the pelts he took from a marsh shrunken by the drought were damaged. Not many of the muskrats were killed in the fighting, but losers who were driven from the marsh faced almost certain death from predators (20).

The effects of hunger are complex, but studies of other species also tend to show that it is a cause of aggression. Since really bad seasons are infrequent, we may expect that episodes of serious fighting will be correspondingly infrequent.

Like bad weather, population increases can also cause food shortages. In a study of population cycles in small rodents, Krebs et al. showed that male aggression peaked at the same time as population (53).

Elephant seals illustrate some of the consequences for aggression of a polygynous, harem-type mating system. There are 18 species of seals, and their mating patterns range from monogamous to highly polygynous, with up to 100 females per male. Seals come out of the water to give birth and to procure mates, and the nature of the breeding area largely determines the breeding pattern. If space is limited, so that a single male has easy access to numerous females, the development of polygyny is favored (48). Elephant seals give birth and mate on small islands, and the breeding season was studied by LeBoeuf and Peterson on Ano Nuevo Island off the coast of California (56).

Harem bulls are three to five times heavier than females and require about twice as long to reach breeding age. Besides their large size, the males are characterized by a short trunk from which they get their name. They come ashore on Ano Nuevo Island in December. The researchers marked individually 93 of the 115 males that landed; those unmarked were immature transients. Competition broke out immediately, but whereas in many species of seals it is for territory, in these seals it was for rank in a dominance hierarchy. Most of the competition was in the form of threats, the seals rearing back and grunting. A total of 6000 such threat displays was counted, but some undoubtedly were missed because observation was not continuous throughout the period. Usually one male retreated, but in 90 instances, both stood their ground and began to fight, biting and slashing with their teeth. Fights lasted anywhere from a few seconds to 15 minutes, with one male almost always being the clear victor. The fighting went on throughout the breeding period, and the investigators were able to arrange the males in a clear and nearly linear hierarchy. There was only one exception: two males, despite two bloody fights, maintained an equal status. Some changes in the dominance order occurred, but only following fights.

The females came ashore in January, a month after the males. Two hundred twenty-five were counted, nearly twice the number of males, and the observers were able to mark 45. LeBoeuf and Peterson suggest no reason for the unequal sex ratio, but according to Jouventin and Cornet, the polygynous species of seals show an excess of male mortality. The late maturity characteristic of male elephant seals would also have the effect of reducing male number.

The females gave birth soon after landing and suckled their young for about four weeks. At the end of this period they mated and then returned to the water with their pups. The males stayed on the island for about another month and made no contribution whatever to the care of their progeny.

A major purpose of the study was to determine the role of dominance in mating success. The individual tagging and the observations on male confrontations and individual matings made a clear answer possible. Four percent of the males inseminated 85% of the females, and access to the females was clearly a function of rank. Mating success went to the males who won fights or whose threat displays intimidated the challenger. Studies of other highly polygynous species have resulted in similar findings.

Social species that live entirely or in substantial part by hunting are a particularly interesting group, both because of their shared behavior patterns and because man is one of them. The hunting in these species is typically a group activity and it appears to favor strong social bonds. The principle mammalian groups other than man notable for this life pattern are lions, wolves, African wild dogs, hyenas, dolphins, and killer whales. Like all other species, the hunters, despite their internal cohesion, do display internal as well as external aggression.

In my summary of hunting species, I have drawn on summaries by

Wilson (112) and Leaky and Lewin (55), but have also made use of other sources, espcially with respect to lions and African dogs.

Lions are unique among members of the cat family in that they are a social species. They are polygynous, the social group, or *pride*, containing 2–18 adult females but only about half as many adult males. The females are usually closely related and show a high degree of cooperation. Infant cubs may nurse on several females besides the mother. The adult males in a pride often include nonkin as well as close kin (70,71,87).

In keeping with their practice of polygyny, lions show sexual dimorphism. The males are somewhat larger than females, and their abundant mane protects the neck, which is the usual point of attack when males fight. There is considerable intragroup fighting among the males over females in estrus, with victory increasing the opportunity to sire young.

Young males and occasionally young females leave the pride as they approach maturity, and live alone or in small groups with no defined territory. They later seek access to prides either peaceably or by force.

Hunting occurs at dusk and is primarily the province of females a reverse of the pattern we noted in chimpanzees. It is notable for the high degree of cooperation involved. The lionesses often stalk prey by fanning out and then rushing simultaneously from different directions. Schaller's data show that several lionesses stalking together are generally twice as successful as are solitary hunters. Group hunting also enables lionesses to catch such large and dangerous prey as adult male buffalos. In one instance, a male buffalo attacked by a single lioness was the undisputed victor. There are indications that the number in the hunting group adjusts to the nature of the prey (87).

The prey is shared, but seemingly more on demand than willingly. Adult males have the first claim. Schaller describes a case in which three males from one pride took a gazelle from four females of another (87).

The males patrol and defend pride territory, which usually is sizable, since it has to provide an adequate number and variety of prey animals. Occasionally one coalition of males may control several prides and their territories, so the ratio of females to breeding males is higher than the figures for single prides would suggest. Schaller found clear evidence that the male coalitions sometimes wage major battles, occasionally over a fresh kill but more often over territory. Of 22 subadult or adult lions whose deaths he recorded, he estimated that five died as the result of fighting. Six of 92 lions that he observed were blind in one eye. Numbers give a significant advantage in these territorial battles, and this doubtless has been a factor in the evolution of multimale prides (87).

In some cases, intergroup battles lead to the death or eviction of all males of the losing coalition. The females who had been paired with these males are then taken over by the victors. This is usually followed, as an expression of the competition for the production of progeny that we have noted, by the killing of all cubs under a year and a half to two years of age by the incoming males. Females often vigorously defend their cubs and may be wounded or killed in the process. Older cubs are evicted by the newcomers, and young born to females who were pregnant at the time of the takeover usually do not survive.

Besides the defense of young, female lions may have evolved more subtle strategies for increasing their own reproductive rate. For several months after a takeover, the pride females are frequently in estrus and encourage mating but seldom conceive. There have been several suggestions as to how this might favor the survival of subsequent offspring. Packer and Pusey believe that by remaining attractive to any potential incoming males for several months, the females increase the chance of takeover by a relatively large male coalition. Since a large coalition is likely to remain in possession of a pride longer than a small one, this would increase the chance of a long, uninterrupted period of breeding (71).

Wolves and African wild dogs are pack hunters. Both sexes participate in the hunts, but the puppies are quite helpless and must be tended by the mother, so some females may remain at a den prepared at the time a litter is born.

In wolves, as in lions, the number that engage in a particular hunt is adjusted to the nature of the prey. Ten or more wolves may join in a moose hunt, but fewer are required when sheep or caribou are the target. The skillful cooperation of wolves hunting Dall's sheep in the difficult terrain of Mount McKinley National Park was described by Murie. On an uphill run, the sheep can outdistance the wolves, but in a downhill chase the wolves have the advantage. When two or more wolves participate, they can usually use their numbers to control the direction of the chase and thus increase their chance of a catch. Even under favorable circumstances, and whatever the nature of the game, the victims are likely to be old or ill, though immature animals also may be captured (65).

Hugo and Jane van Lawick-Goodall, who have studied African wild dogs, have reported packs ranging from four adults to one of approximately 40 that subsequently split into three. Both sexes were about equally represented. In a pack of 12 adults that they observed over the longest period, all except the mother of a recent litter participated in hunts. Their prey included gazelles, wildebeests, and zebras. These were seen to outrun their pursuers somewhat more than half the time; success in the hunts seemed to depend on singling out the ailing or aging animals that could be most easily caught. The van Lawick-Goodalls believe that considerable skill goes into this selection process. One strategy employed by the dogs was to break temporarily into separate packs and to reunite as soon as there were indications that one chase was going better than the others (102).

The helpless young of wolves and wild dogs must be fed, and these species have solved this problem by the evolution of the capacity to regurgitate food. Thus these species, like human hunters, though in more involuntary fashion, carry food to a home base and share it with mothers and young. Males as well as females participate in the feeding process.

African wild dogs have a very large home range and have not been observed defending a specific territory. The conflicts that have been noted in this species appear to center around food, access by males to females, and the survival of offspring.

Some observers have failed to see indications of a dominance order, but Hugo and Jane van Lawick-Goodall found that this is present, though the precise ranking was clearer in the females than in the males. Much of the time, rank was displayed only in subtle behavior and gestures, but serious fighting did also occur. In one case a bloody battle, leading to a reversal of rank order, took place between two females. The advantages of dominance were indicated by only a few observations. Apparently dominance in males gave access to females in heat and perhaps in some cases preference in access to newly killed game. Low-ranking females were frequently harassed and in at least one instance also chased from fresh kills. More important, in certain circumstances dominance determined the life or death of a litter. Usually only one female littered in any given year. In one case in which two litters were born in close succession, the dominant mother killed the litter of the mother that ranked below her (103). The function of infanticide in this instance presumably was to insure an adequate supply of food for the surviving pups. The problem of adequate food for a wild dog litter is exaggerated by the large size of litters in this species. Litters have been observed numbering from 8–16 individuals, and there is a report of a litter of 19 pups, three of them still-born, delivered in captivity. Even with the whole pack except the mother providing food, a litter in this size range might well be about all that could successfully be sustained.

It is evident that the need for group hunting has produced strong social bonds in both wolves and wild dogs. Surely, it is no accident that the domestic derivative of the wolf has been called man's best friend. The devotion of wolves to their young becomes, in the dog, directed to children, leading Lorenz to note that children are perhaps safer with a dog than with other children (59). It has been reported that wild dogs will sustain even sick animals, unable to hunt, with regurgitated food, an act of altruism unusual in animals other than man (55).

Hyenas, although only remotely related to the dog family, have developed a somewhat similar social patern. They form sizable clans or packs with equal numbers of males or females (87). A major difference from wolves and wild dogs is that hyenas rely to a considerable extent on scavenged animals. With their strong jaws, they can crack the bones of animals that have been dead for some time and extract the marrow but, acting in packs, they also steal game from wild dogs or even single lions. Scavenging is not their only food source, however. They often do their own hunting and their packs can bring down game as large as the wildebeest (55).

Killer whales, which have been called the wolves of the sea, are also pack hunters. Their prey includes other species of whales (which may be much larger than they are), sea lions, and porpoises. An extraordinary degree of cooperation has been observed. One pack of 15–20 killer whales succeeded in surrounding about 100 porpoises that they had been pursuing. They gradually constricted the circle and then took their turn, one by one, in eating the trapped animals.

The porpoises are another group of carnivorous marine mammals that have evolved a high level of sociality. They travel in groups that vary greatly in size but that may number up to 100. Schools contain adults of both sexes, but these may break up into a variety of subgroups. It is not easy to study the ethology of such a marine form, but from what is known we can surmise some of the methods they use in catching fish. Two things we do know: They detect schools of fish through their highly developed sonar, and they are capable of a diversity of underwater sounds that should permit communication over considerable distances. This suggests that they fan out to locate fish and communicate any discovery to the rest of the group.

Two quite distinct forms of cooperation have been described. Hoese saw two dolphins capture small fish by making waves that drove them up onto the muddy shore but still within reach of their captors (37). Another form of cooperation, observed many times, is the support of injured comrades, giving them the opportunity to recover sufficiently from the shock of injury so they will not drown from too prolonged submersion. This is obviously related to reports, some ancient and some possibly authentic, of dolphins having saved mariners.

What clues can we draw from this body of information concerning social carnivores that may be relevant to the role of hunting in human evolution? Most of the groups we have examined derive two rather distinct advantages from group hunting. First, it enables them to capture large and dangerous game and the size of the hunting group tends to be adjusted to this need. Second, group hunting enables them to use strategies in capturing small game that would be denied to the individual hunter. It is generally agreed that primitive man exploited the first of these advantages of group hunting, though there is debate as to how early in his evolution big-game hunting began. Hayden, in his review of hunting by existing hunter-gatherers, does not, it seems to me, adequately consider the strategic advantages of group hunting. Surely when we find these advantages exploited in quite ingenious fashions by other social predators we can safely attribute a similar use to primitive man. And we may go a step further and postulate that the potential for such use was a contributing factor in the evolution of man's brain and the capacity for speech.

The social patterns found among the primates are of particular interest because the primates include man and his nearest relatives. We cannot begin to convey the diversity of social structure in the approximately 200 species in this group, but we can select genera that illustrate the factors influencing conflict and cooperation or that are important because of their close relationship to man. The groups that we have chosen are the macaques, or more specifically, *Macaca mulata,* the rhesus monkey of India, and its Japanese relative *M. fuscata;* the baboons; the gibbons; the gorillas; and the chimpanzees.

Rhesus monkeys were studied by Carpenter on Santiago Island, a part of Puerto Rico, where they had been introduced and had become well established. A group Carpenter regarded as typical contained six adult males, 32 adult females (25 of which had infants), and ten juveniles, making a total of 73 individuals (7). Troops of the Japanese macaques, studied by Sugiyama, show the same unbalanced sex ratio but tend to be larger, perhaps because of a long history of artificial feeding. The sex ratios at different ages in the Japanese troops indicate that males begin to leave the troop as adolescents (at 3–4 years of age). Fighting increases during the breeding season and most of the departures take place at this time, but Sugiyama did not see evidence that the adolescents were forcibly driven out (98).

Carpenter made observations on dominance relations in one rhesus troop of 85 individuals. He found that the seven adult males could be arranged in a linear dominance order. If the animals were provided with a tray of food, the top ranking male had exclusive possession until satisfied (7). Subsequent work has also shown that while several males may have access to any one female during the breeding season, it is the dominant male that has access during estrus when conception is most likely. A dominance order is evident in quite young animals and is established in females as well as males. Success in fights is a major determinant, but this in turn is influenced by or correlates with the dominance rank of the mother. Once established, threats usually are sufficient to maintain the order, though serious fighting resulting in injury may occur over females in estrus (27,82).

The juvenile males that leave the troop usually live in all male groups. During the breeding season many of these exiled males gather on the periphery and try to lure away or mate with troop females. Apparently they achieve some degree of success (98).

The mating system of macaques is thus a mixture of polygyny and promiscuity, and as is usual in polygynous species, there is considerable dimorphism. The males are bigger than females and are armed with large canine teeth.

Another accompaniment of a polygynous structure in primates may be the killing of infants by formerly exiled males who succeed in penetrating a troop. This increases the chance that the immigrant will promptly have progeny of his own. This phenomenon has been noted in macaques, but has not been studied in detail. It may be infrequent in this species. In the *Hunuman langurs* in which it has been specifically studied, it is quite common (39,40).

While both Carpenter and Sugiyama noted that macaque troops occupied definite ranges, neither saw evidence of territorial fighting. In both these cases the animals were being provided with supplemental food, and the resulting abundance would favor peaceful relations. In other circumstances, especially where there was crowding, fierce fighting resulting in serious wounds has been reported (27,95). Investigations on Santiago Island subsequent to Carpenter's original study did indicate intergroup fighting. Thus Altmann, while noting an overlap of ranges, also saw encounters that resulted in threats and sometimes overt battles. Transgressions of home ranges and hence the risk of battles increased when the usual supplement of food was withheld (1). Another study showed a higher mortality among males than among females, especially after age four (98). Presumably this was the result of fighting, though whether intergroup or intragroup is not clear.

While the males of macaque troops may fight among themselves, they tend to unite in intergroup conflict. Even here, however, dominance remains important. Thus the removal or illness of a dominant male reduces the territory available to a group (7,27).

If macaques are capable of aggression, and sometimes of lethal aggression, they also are capable of warm affection. This is best revealed in the laboratory studies of the Harlows cited in Chapter 4, but it is also seen in the native environment. The touching, grooming, and embracing that macaques display goes a step beyond the manifestations of sociability seen in mammals other than primates. Maternal affection is highly developed. Carpenter reports two cases in which rhesus mothers carried dead babies until only skin and skeleton remained. Females other than the mother may also share in the care of young (7).

Baboons, like macaques, are Old World monkeys, but are ground dwelling instead of tree dwelling. In all four species of baboons, the males are much bigger than the females and are armed with particularly vicious canine teeth. We shall consider only two species, the *savanna* and the *hamadryas*.

The *savanna* baboons live in open grassland south of the Sahara. In this habitat, they are exposed to attack by lions and several other species of predators. They escape predation at night by sleeping in trees, but when a troop is ranging its territory in the daytime in search of food, it may have to rely on self-defense. In response to this need, the *savanna* baboon has evolved social groups that are not only large but, unlike most monkey species, contain approximately equal numbers of males and females. The males develop a clear dominance order, but even after it is established, may erupt in noisy and vigorous threat displays that sometimes have to be quieted by the top male (27). One of the privileges of high rank is access to females at the height of estrus. Despite this system of privilege, the males band together if predators threaten and can put up a defense that frightens off most attackers. If approached by lions, however, they do sometimes seek to escape by climbing trees (63,112).

Hamadryas baboons live in semidesert areas where predators are few but where food, in unfavorable years, is a limiting resource. The basic social unit in this species is the harem, consisting of one adult male, several females, and their offspring. The males keep constant watch over the females and harass them if necessary to keep them from straying. As males mature, they are driven from the harem and may form all-male groups, but even before reaching full maturity they may try to form harems of their own. At night, harems congregate on cliffs, which provide protection from leopards that sometimes prey on this species (63,112).

While there are aspects of baboon social structure, like that of macaques, conducive to conflict, the baboons, again like the macaques, display tactile bonds that help keep the groups together (63).

Of all the primates, gorillas and chimpanzees have attracted the most attention because of their obvious similarity to man. And because of this similarity, their social and aggressive tendencies can be particularly informative.

Our ideas on this subject have gone through three stages. Accounts brought out of Africa by nineteenth-century explorers and hunters tended to depict the gorilla as a monster that beat elephants with clubs, killed men, and carried off women. The chimpanzee fared somewhat better, but by some accounts was also a brute. The first major scientific studies of the gorilla in its native habitat were carried out by Schaller and Fossey. These studies totally reversed the picture, painting the gorilla as harmless toward man—so long as it did not feel threatened—and gentle with its own kind (24,85). Fossey wrote, "After more than 2,000 hours of direct observation, I can account for less than five minutes of what might be called 'aggressive' behavior." Further research, however, changed the picture again and showed that lethal aggression against other gorillas, both young and old, is a very real part of these animals' behavioral repertoire (25). The concept of chimpanzees conveyed by Jane Goodall's studies passed through a similar transformation. As in so many other species, including man, lethal attacks are infrequent, but they do occur.

Dixson has given us a thorough review of the social nature of the gorilla. There are slight differences between the two subspecies included within the genus *Gorilla,* but these are generally minor. Typical troop size ranges from 6 to 16 (18). Schaller, in ten gorilla groups, counted twice as many adult females as males (85). Juveniles and infants outnumbered their mothers. The excess of adult females is largely due to some of the males spending some or all of their time in isolation. A small fraction of the excess may be due to high male mortality. There is a dominance hierarchy, though reports differ as to its rigidity. Fossey noted in one troop that the aging, dominant male had far more opportunities to sire offspring than a fully mature son. She also noted that when the male leader of one troop was killed by poachers, the group nearly disintegrated (25).

Groups have definite ranges, but there is considerable overlap and some intermingling of groups. However, mutual avoidance has also been seen. It is possible that the reduction of the gorilla population that has been caused by poaching has reduced the tendency toward territorial defense (18,25).

While there does not seem to be territorial conflict, Fossey has described an instance, during an "interaction" between groups, in which a male of one group killed the male of another (25).

The aging male that Fossey described and named Beethoven ultimately came into conflict with his son Icarus. This conflict led to two of the lethal attacks observed in this study. A six-month-old infant sired by Beethoven was found killed and partly eaten. Icarus was suspected as the killer. Soon afterwards, the mother of the dead infant mated with Icarus and in due course presented him with his first offspring. Somewhat later another killing occurred and in this instance was observed in detail. Icarus attacked an aging mate of Beethoven, pounding her and jumping on her, and continuing the assault even after she was dead. The adaptive significance of this act is not entirely clear, but apparently it hastened the replacement of Beethoven as dominant male by his son.

Fossey saw several other instances of infanticide. All told, of 38 babies born in 13 years, infanticide claimed six. There is no certainty as to the evolutionary reason for these killings, but Fossey believes that they can be explained on the ground that they provided a reproductive advantage for the killer (25).

Gorillas, like other primates, groom one another and derive social cohesion from personal contact. There also appears to be somewhat more paternal care of infants than is normally seen in macaques or baboons. In all these groups, males play a protective role, but Fossey saw dominant gorilla males go beyond this, taking infants from their mothers' arms and grooming them. In another case she saw a grim old male tickle an infant with a flower, apparently in fun (24). Schaller described gentle play between juveniles and infants (85).

Of all the primates, the chimpanzee, in both appearance and behavior, is most reminiscent of mankind. This is particularly true of a pygmy species that lives in rain forests in Zaire and, because of its isolated location, has only recently been studied in detail (76). This species has a flatter face than other chimpanzees; it often walks on two feet when on the ground; it shares none of the aversion to water common in other primates; it mates throughout the estrus cycle; and it often copulates face to face. Whether the special similarity of the pygmy chimpanzee to man is due to convergent evolution or to less divergence from our common ancestor than that seen in the common chimpanzee, or to some combination of these two, is at present unknown. At least in its small size, the pygmy chimpanzee probably resembles the common ancestor.

There have been a number of studies of chimpanzees in their native habitats, but few can compare with the two decades of observation carried out by Jane Goodall and her associates in Tanzania. Wilson has summarized various aspects of other studies (112).

The basic social unit is a group of perhaps 30–80 individuals, with adult males and females about equal in number. This group has a defined home range; Goodall found that the range of the group she was studying fluctuated between 5 and 8 square miles (29). While chimpanzees inhabiting this range are a definite unit, they spend most of their time in separate groups that may vary substantially in composition. Because the chimpanzees are primarily arboreal, an organization designed for defense against predators is not needed. Organization is dictated instead by economy in foraging and especially by the food needs of the females and their young (30).

Whereas baboon males are continually on the alert for possible predators, chimpanzee males are concerned with territorial defense. The adult males in the unit studied by Goodall, acting in groups of three or more, regularly patrolled the territorial boundaries. They patrolled silently, sniffing and studying the ground or climbing tall trees to detect any sign of strangers. If two groups met, they usually exchanged threats and then retreated, but if a lone individual or a mother with an infant was encountered, the patrolling males often gave chase and attacked (29).

An instance of intergroup competition followed by Goodall for several years was particularly devastating in its outcome. In 1970, the main unit under study divided to form two groups. Seven males and three females with their offspring split from the original unit and took possession of the southern portion of the original home range. For several years, aside from the usual border encounters, the two groups lived peaceably. Then, for reasons not easily defined, the main (northern) group went on the offensive. Over a one-year period, the patrolling northern males encountered three males and an old female of the southern group near the boundary, foraging alone. All were caught and so brutally beaten that, although they were left alive, they did not survive. After another peaceful interval, two more southern males were killed. Together with two deaths from natural causes, this left the remainder of the southern group defenseless, and the northerners soon reclaimed their territory. Compared to the intergroup fighting, relations between adults within the group were found to be relatively peaceful (29).

Male chimpanzees have a clear dominance hierarchy, and this is a source of rather frequent minor confrontations. DeWaal and Roosmalen noted that these often ended in a friendly embrace (17). The most serious dominance fights seen by Goodall were between the alpha male and competitors for his top spot. Successful aggressors often used strategy. The easiest victory was won by a small male who in a few months time cowed all competitors by accompanying his charging displays with bangs on empty kerosene cans. Another male enlisted the aid of a brother, but still had to climax his threats by jumping on the alpha from above while the alpha was reclining in his tree nest. The enlistment of relatives in the search for status apparently is rather common. Goodall saw one case of unaided takeover, but the displaced male in this case was weakened by age (29).

Although chimpanzees strive actively for rank, it is associated in these particular primates with only limited advantages. The sex structure of chimpanzees is promiscuous rather than polygynous, and there is no clear evidence that opportunity to sire offspring is significantly correlated with dominance. There is no breeding season, but females have a definite estrus period and usually mate with several males at this time. Occasionally a male persuades a female in estrus to leave the group with him, but what qualities in the male induce the female to form this temporary monogamous pairing is not clear. As an evolutionary consequence of the access of multiple males to females in heat, chimpanzee males have unusually large testes. It is hypothesized that if there is multiple insemination, a particularly generous production of sperm will increase the chance of parenthood (90). High rank probably does confer some advantages in access to food or favorable feeding areas and it certainly frees the alpha male through most of his period of dominance of any need to devote time and energy to personal defense.

Goodall found, much to her distress, but only after several years of observation, that there is a second area beside territoriality in which chimpanzees display lethal aggression. Two females, a mother and a daughter, were observed killing and eating the infants of other females in the group. At least three such instances were actually observed and a number of others were suspected. In each case the mother was actively attacked, but in only one instance was there fierce fighting before the parent yielded her newborn baby. Males were sometimes seen protecting mothers and their infants (29).

What is the significance of this particular form of infanticide in chimpanzees? It contrasts with the infanticide occurring in lions and gorillas, which is male-perpetrated, but resembles that in African wild dogs. Since infanticide by females requires overpowering or overaweing the mother, it is presumably limited to dominant females. In African wild dogs, it is believed to be due to the need to adjust the number of young to the available food supply. It is, I think, a reasonable speculation that this also is the case in chimpanzees. In the area studied by Goodall, there is a dry season when food is scarce, so food is probably a limiting resource. Perhaps this is particularly true of high protein foods; hence the cannibalism. Goodall, it should be noted, suggests the possibility that the two females practicing infanticide were psychopathic (29).

Besides the instances of infanticide by female chimpanzees seen by

Goodall, each of two observers has reported an instance of infanticide by males (6). In one of the two cases, there is reason to believe that the mother, her infant, and an accompanying female who were attacked, and the five aggressor males, were from different troops. The relationship is uncertain in the other case. The infants in both cases were eaten by the males. The cannibalism of killed infants in these cases also is compatible with protein deficiency as at least a partial explanation.

If food is a limiting resource, then territory is also a limiting resource, and this could account for the territorial aggressiveness seen in chimpanzees.

As Goodall notes, one of the striking similarities of chimpanzees to humans is their capacity for both violence and affection. Probably in no other social animal except man are evidences of affection so easily seen. There is tender care of young by mothers, often aided by older sisters or other relatives, through a prolonged infancy. Though there is little paternal care, adult males will fondle and groom infants. There is solicitousness for ailing relatives. There is no clear dominance of males over females (83). There is friendliness and hugging between male rivals, and male cooperation in hunting and the patrol of territorial boundaries. There is play, including play between animals of quite different ages. There is food sharing by males, though only with individuals who do the appropriate ritual begging. All this adds up to a group structure with strong internal bonds.

The relative lack of paternal care in chimpanzees presumably is a correlate of the promiscuous mating system. Except perhaps in the infrequent cases of temporary pair bonding, there is no way that a male can tell that an infant is his own. In the monogamous gibbon, paternal care is more evident. In one species, even infants older than a year are carried by the male when the troop is moving (41).

What Animal Studies Tell Us About Man's Social Nature

The fourteen mammals whose manner of adaptation to social living we have examined illustrate the selective forces that predispose either to conflict or to cooperation. All these forces have operated to one degree or another in human evolution. What can we learn from these illustrations concerning the special factors in human evolution that have shaped our potential for conflict, cooperation, and love?

The high level of cooperation seen in social carnivores suggests that man's transition from herbivore to omnivore should have helped to strengthen his social bonds. Group hunting by tribal males has often been suggested as an important source of cooperative tendencies in human nature, and I concur in this point of view. Scavenging, insofar as this was used to obtain meat, probably also would have sometimes required cooperation. The demands of plant gathering and the care of small children should at the same time have strengthened the bonds of tribal wives.

The group defense against predators displayed by male baboons suggests that the same cooperative behavior pattern may have occurred in man. Likewise the relatively high proportion of males in groups of baboons in comparison with those of most other primates, which improves their defense capability, suggests that this has been a factor in the move toward monogamy and hence an approximately equal sex ratio in man. Simonds has suggested that the high proportion of males that he noted in the bonnet macaque of South India, another species subject to predation, serves here also to increase defense capabilities (92). The threat of predators may thus have been a second factor in strengthening male cooperation in man.

Along with the adoption of group hunting and group defense, man moved in the direction of monogamy, or at least of relatively permanent pair bonding. This caused or contributed to the replacement of periods of estrus by continuous receptivity on the part of the female; reduced the sources of male fighting we have noted in polygynous species; led to a demand for faithfulness in marriage partners, with much more emphasis, however, on the female than the male; removed the occasion for the killing of infants by males taking over a harem; decreased the significance of dominance order in the human social structure; led to an equalization of the adult sex ratio within the tribe; and contributed to the successful development of the hunting group and to group defense. It was also favorable to father love, paternal care of the young, and the transport of food to a home base, and it made the family an important economic subunit within the tribe.

We would expect, too, that the move in the direction of monogamy would decrease or eliminate sexual dimorphism. With respect to the canine teeth, which are reduced in size in human males as compared to their size in the males of many other primates, an elimination of dimorphism has occurred. In fact, it goes back at least as far as *Australopithecus afarensis.* Humans, however, still show some size dimorphism, and it has been argued that this invalidates emphasis on the importance of monogamy in human evolution. The answer to this is that human size dimorphism is a result of the specialized roles assigned to males and females in the economy of tribal societies. Big males make better fighters and hunters. Smaller females, so long as their child bearing and child rearing faculties are not impaired, require less food and hence increase the chances of tribal survival, especially in times of hardship.

Another consequence of the trend towards the monogamous family was a growth in the emphasis on mate selection. The offspring of any individual get only half their genes from that individual, the other half come from the other parent. The selection of fit mates who will transmit good genes has some reward under any mating system, but the premium is much greater under monogamy than under systems that promote promiscuity or polygamy. Mate selection may result in competition between suitors, but the competition is usually resolved by the mate sought, without resort to fighting. Compared to the overt aggression that can occur in the accumulation of a harem, mate selection is a peaceful process.

While the degree of territorial conflict differed substantially in the different social mammals we examined, some recognition of territorial boundaries, at least, was noted in all except the aquatic mammals and possibly gorillas. In two of the species, lions and chimpanzees, the conflict reached lethal proportions. We have yet to examine the evidence for the occurrence of tribal warfare in primitive human society. At this point, however, we can draw from the animal studies one inference significant for social evolution. In both lions and chimpanzees, one group was observed to essentially replace another as a result of intergroup competition. There has been much debate among evolutionists as to whether group selection has played a role in the evolution of sociality. Our two specific examples at least show very clearly that group selection does occur. Why, in these examples, one group replaced the other is not entirely clear, but sheer numbers appeared to play a role in each case. Numbers could, in turn, be a reflection of an ability to cooperate, but our knowledge of group selection in operation is far too scanty to permit firm conclusions. It is sufficient here to take note of the reality of selection at this level.

One other lesson conveyed by our survey is that intergroup conflict is influenced by two environmental factors: the weather or climate and population density. Since these factors fluctuate, we can assume that tribal conflict in man must have fluctuated. Periods of relative peace and periods of war may have alternated, with the peaceful periods probably being longer. Looking ahead, we can say that for a peaceful human future, two conditions are essential: population control and the use of the world's resources in a manner that will preclude periods of unmanageable hardship.

Warfare in Primitive Human Societies

Fighting over territory is widespread in the animal kingdom, occurring in species as diverse as insects, fish, frogs, lizards, and mammals (88). In social animals, it is typically a group endeavor, with success often going to the side that can martial the greatest number of contestants. Territorial conflict certainly has been common in man in historic times, but the farther we go back into prehistory, the more uncertain the record becomes. The question of its occurrence among our early ancestors can be approached from three viewpoints; the first, theoretical, the second based on the occurrence of warfare in existing primitive societies, and the third founded on the search for archaeological evidence.

Territory, as we have already noted, is a basic requisite for all life. If population outruns the food supply obtainable from a given area, individuals or groups either gain control of enough territory for self-support or starve. This principal applies quite as much to man as to any other species. In fact it may have been a more urgent problem for man than for his relatives the great apes because his shorter interval between births endows him with a greater potential for population increase. Of course accident, disease, and perhaps infanticide can act as limiting factors, but it seems altogether likely that in some human societies at some periods, population pressure would give rise to territorial conflict.

A case can also be made for the thesis that an effective combination of external aggression and internal cohesion—or to use Herbert Spencer's terminology, external enmity and internal amity (96)—should be, for social species, a formula for success in the evolutionary arena. The effective use of the formula by any group or tribe should favor the spread of its genes through the gene-pool of the species. I should emphasize that, so far as man is concerned, I am speaking only of the long period of human evolution during which man lived in small groups. With the advent of agriculture and the resulting growth of population, the situation began to change, and modern technology has transformed it totally.

In the early days of this country, as Europeans moved across the United States, they had an opportunity to observe Indian tribes that were still relatively unmodified by European contact. Most of these tribes practiced a limited agriculture, cultivating corn and sometimes beans and squash, but they also depended substantially on hunting or fishing. The additional food available from agriculture had resulted in a substantial increase in the size of the typical community (4).

Father Nicolas Point, a priest who lived for a considerable period with tribes of the Kansas-Missouri area of the United States before these tribes had more than very limited contact with whites, has given us a fascinating account of their lifestyle (75). Their culture had been substantially modified by the adoption of horses, which had been introduced by the Spaniards at a considerably earlier period and allowed to run wild. This enabled the tribes to hunt buffalo with greatly increased efficiency, so that although they apparently did not practice agriculture, they lived in communities of as many as 700 or 800 individuals.

Communities belonging to the same tribe generally maintained peaceful relations, but when groups belonging to different tribes met, fighting could easily erupt. Sometimes the fighting assumed major proportions. Father Point tells, for example, of a massacre of 80 Pawnee women and children by their neighbors the Kansas. Warfare also went on between Crows and Blackfeet and between Pends d'Oreilles and Blackfeet. One slaughter of

Blackfeet by Pends d'Oreilles is specifically linked by Father Point to a food shortage.

The capacity of the Indians for external aggression was more than matched by manifestations of internal affection and mutual aid. As Father Point writes, "Although the Kansas were vindictive and cruel toward their enemies, they were not strangers to the most tender sentiments of friendship and compassion. At the loss of someone close to them they were sometimes utterly disconsolate." Individuals were generally honest; the rewards of buffalo hunts were shared by the more successful with the less successful; and their leaders were expected to show high standards of courage, vigilance, and generosity.

Early French explorers in the New World told of the massive stockades with which the Iroquois surrounded their villages (72). Archaeological studies have revealed a string of similar fortifications going at least as far west as Tennessee. These are assumed to mark the route of a pre-Columbian Iroquois migration, perhaps from as far west as the Ozarks to their ultimate home in New York State. Presumably the Iroquois brought the practice of corn agriculture with them. The Iroquois were known in historic times as a warlike people, and their fortifications are a clear indication that their acquaintance with warfare antedated any contact with the white man (5).

The cliff-dwellings of the Mesa Verde Indians of Colorado, to which they moved from a less protected site in about 1200 A.D., provides a further indication of intertribal conflict (4).

Similar but more convincing evidence comes from the discovery by Larry Zimmerman and his colleagues in Crow Creek, SD, of a mass burial of at least 486 persons in what had been a walled village. Both sexes were represented, but there were few females of ages 15–19 years. Marks on the skulls clearly indicated that a massacre had taken place, with only young women and small children being spared. At the time of the massacre, about 1325 A.D., these Indians cultivated corn, and also hunted buffalo. Despite this dual source of food, careful analysis of the bones indicates that the villagers were seriously undernourished, again suggesting a link between want and warfare (119).

Evidence suggesting warfare among North Americans at an earlier date comes from an excavation by Lovejoy and coworkers of a site in Ohio. An analysis of 1,327 well-preserved skeletons buried between 800 and 1100 A.D. indicated that mortality was considerably higher among adult males than females. These were primarily a hunting-gathering people, though they may have practiced marginal corn agriculture during part of the occupation. They were, however, a well-nourished people with a relatively low infant mortality. This find suggests the occurrence of sporadic, but not totally lethal warfare, in a period of relative abundance (60).

Loy, using specialized techniques, identified human blood as well as the blood of game animals on weapons ranging from 1000–6000 years old unearthed in British Columbia. He offers no interpretation, but tribal warfare certainly is one possibility (61).

Few problems in anthropology have generated more dispute than the question of the extent to which fighting occurs in those hunting-gathering societies that have been subject to study in recent times. Some authorities have seen these societies as predominantly peaceful, others have reported a wide range of aggressive behavior. Leaky and Lewin argue that the practice of warfare was rendered more likely by the introduction of agriculture. Crops, cropland, and a settled habitat, they suggest, are valuables to be defended at all cost. With respect to offense, once a crop is harvested, it provides the resources to support an aggressive venture. They cite a study of South American Indians that showed that the members of agricultural societies are much more likely to cannibalize their victims than are hunting-gathering people (55). It should be noted that, while suggestive, this study was based on a sample not really designed to test the point at issue. Wilson sums up his view on this problem in the statement, "Throughout history, warfare, representing only the most organized technique of aggression, has been endemic in every form of society, from hunter-gatherer bands to industrial states" (113).

Several Indian tribes living in the jungles of Brazil, which have remained largely untouched by civilization well into the twentieth century, have been studied by anthropologists. One of these, the Yanomamö, is notable for its warlike tendencies (9-11,66,67).

The Yanomamö Indians live in villages of about 65 individuals, hunt with bow and arrow, fish in the nearby rivers, and raise some crops by a slash and burn method of cultivation. They have occupied this same area for at least 200 years, and, aside from the use of the white man's metal fishhooks, appear to have generally retained their original lifestyle. They are well nourished and have a low natural infant mortality; hence the population is growing despite prolonged nursing of infants and some practice of abortion and infanticide to reduce the live birth rate. Marriages are polygynous, and with food abundant, most competition seems to be over women. The dominant males father a high proportion of the children. New villages occasionally have been formed, with most of the emigrants belonging to a single family. A consequence of this method of village formation is that, as shown by detailed genetic studies, the people of adjacent villages may show a surprising degree of genetic disparity.

The Yanomamö are a warlike people. Intervillage conflict may occur within the tribe as well as between tribes, though it usually involves remote rather than adjacent villages. According to one estimate 25% of male deaths, in some areas at least, may arise from warfare. While there doubtless are multiple causes of the fighting, the Yanomamö themselves attribute it to competition over women, and observation tends to confirm this view. It has been suggested that the importance of women as a source of competition reflects the principle that in animal societies where food is abundant, it is females that become the limiting factor in the struggle of males for biological survival. The practice of polygyny by the Yanomamö is, however, at the very least an intensifying factor. In a monogamous society there might be competition for particular mates, but there would be little incentive to acquire women by war. Though the Yanomamö practice war, they do not necessarily enjoy it. They have been heard to say, "We are tired of fighting. We don't want to kill anymore. But the others are treacherous and cannot be trusted" (9,11).

Chagnon did not find the Yanomamö pleasant people to live with. He describes them as "sly, aggressive, and intimidating" and contrasts them with another South American tribe with which he spent some time and found helpful and friendly. It seems possible that the unpleasant nature of the Yanomamö traces back, at least in part, to their practice of polygyny and its concomitant tendency to male competition. It should be noted, also, that they practice a limited agriculture—they are not true hunter-gatherers. Some authorities might argue that this accounts for their aggressiveness.

The Eskimos have been cited as a totally peaceful people. According to Eibl-Eibesfeldt, this view of Eskimo society originated with Nansen, one of the early Arctic explorers who lived with the Eskimos and came to love them. Eibl-Eibesfeldt believes Nansen spoke from inadequate observations and overstated the case for Eskimo peacefulness (19). Some observers, while agreeing that Eskimos usually settle their disputes without bloodshed, have found obvious conflict. According to these authorities, the tribes of Siberia, Alaska, Baffinland, and Northwest Greenland, settle their disagreements by wrestling bouts in which a combatant occasionally is killed. Jenness, on the other hand, states that families sometimes engage in lethal blood feuds and that these may last for several generations (45).

With respect to the !Ko-Bushmen of the central Kalahari, another people sometimes cited as lacking aggression, Eibl-Eibesfeldt concludes on the basis of his own observations and that of other anthropologists that while, again, harmony may be the ideal, fighting does occur. Children were found to be at least as given to fights as children in our own society. Intragroup fighting of adults was seen to arise from a variety of sources ranging from fits of anger to adultery, and bloodshed or death occasionally resulted. Eibl-Eibesfeldt reports no cases of intergroup warfare, but both individual bands and a larger, intermarrying group referred to as the ban nexus, which apparently corresponds to the Indian tribe, do recognize specific territories. The nexus territory is separated from adjoining territories by a no-man's-land that is generally avoided by all nexus members (19).

While Eibl-Eibesfeldt is doubtless correct that the Eskimos and !Ko-Bushmen show some aggressive tendencies, it nevertheless is true that in terms of intertribal warfare they are notably peaceful. It could be, however, that this reflects special factors in their environment.

The point has been made, and I think is valid, that existing hunting-gathering societies are relic societies. A few live in habitats of relative abundance, but it is probably fair to say that in no case are their territories as attractive as those where agriculture and civilization has replaced the more primitive way of life. We cannot necessarily extrapolate reliably from one area to the other. The Eskimos live in a habitat so harsh that it may be impossible to sustain life and wage warfare at the same time. On the other hand, the Kansas and some of their neighbors of the western plains of the United States who, according to Father Point, were abundantly provisioned by their buffalo hunts without recourse to any agriculture, lived in large communities and waged serious warfare. It is true that they had recently modified their culture by the adoption of horses, and this lends some uncertainty to the applicability of their behavior to an interpretation of other thriving hunting societies. Nevertheless, there is a certain plausibility to the hypothesis that hunting-gathering tribes inhabiting favorable territory would be staunch in defense and would have a potential for population increase conducive to offense.

Any attempt to assess the occurrence of fighting between pre-agricultural human groups in prehistoric times must rest on scattered archaeological evidence. This evidence has been summarized by Roper. Of the many human crania that have been unearthed, a small fraction show injuries that suggest that the owners met a violent death. These fractured crania represent many different stages of human evolution. It is Roper's judgment that these findings support the conclusion that tribal warfare, and perhaps warfare between different hominid species or subspecies, did occur during much of the two million years of human evolution. Evidence of a different nature indicating the occurrence of fighting during the later stages comes from cave paintings in France depicting men pierced by arrows (80).

Since there is evidence that unfavorable seasons tend to increase aggressive behavior in human and other animal societies, it is relevant to examine the record concerning climatic changes during the period of human evolution. Much of human evolution took place during the two million years of the Pleistocene, an epoch in the earth's history marked by the occurrence of four great ice ages. During the four periods of glacial advance, not only was much of Europe and North America covered by ice sheets, but other more southern areas were affected as well. So much water was tied up in the ice that the oceans receded, exposing large areas of the continental shelf. Patterns of rainfall were altered, erasing deserts and causing major and widespread changes in vegetation. During each interglacial period these processes were reversed. Darwin took note of the drastic effects, including extermination in some cases, that these climatic fluctuations must have had on plant and animal species.

The human inhabitants, no less than the other species, were affected by the climatic changes and had little choice but to migrate. At the very least, nonmigrants would face difficult problems of adaptation. In West and South Africa, the effect was less drastic than in Europe, but man was alternatively forced from the lowlands by flooding and back down again from the highlands by drought (15). The emigrants must at times have penetrated territory of other tribes not yet prepared to move. Conflict almost inevitably would result, and it seems likely that, over the centuries, many tribal groups would have been wiped out in the resulting battles. These processes of tribal conflict would necessarily have a major selective effect, resulting in the multiplication and territorial spread of those genes conducive to tribal cohesion, communication, good planning, and the development of viable cultures. The evolution of qualities typically human would be accelerated.

Even if warfare were an important element in our past, this does not mean that it is inevitable in our future. Many divergent views have been expressed on this issue. Wilson, it seems to me, has taken a sensible middle position, and what I say in considerable part echoes his thinking (112).

There is, in our nature, a streak of competitiveness that sometimes develops into overt aggression. This is a heritage that goes back millions of years in our evolutionary history. But because fighting is dangerous, and excessive intragroup struggle hurtful to the tribe, we also have evolved inhibitory predispositions. The numerous protests that are staged, at least in free societies, against nuclear armaments and specific military actions, testify to the presence of a strong aversion to war. There are indications of this in primitive societies. If causes of war, other than any inherent aggressiveness, can be identified and removed, war ultimately should be preventable, but the obstacles in the way of removal are formidable.

The Driving Forces in Human Evolution

We have examined the aggressive aspects of human behavior and found that, at least with regard to territorial aggression, man is not very different from his near relatives. Even more important in the context of ethics than man's competitiveness is his capacity for helpfulness and love. These also can profitably be studied from the evolutionary point of view. Before we turn to this specific subject, however, it will be profitable to examine the driving forces that have shaped the human species in its totality and given it its unique place in the animal world.

Interpreting the selective forces that shape a species is of necessity a matter of speculation. In the case of man, the process is made more difficult by gaps in the fossil record. We know that some 12 or 15 million years ago, when the advent of a drier climate changed some forest areas of Africa to a more open grassland, a number of new apelike creatures appeared that probably spent some time on the ground. One of these,

Ramapithecus, has been regarded as a possible ancestor of man. *Ramapithecus* was only a fraction of the size of man today. Like man, but unlike existing great apes, it had thick tooth enamel, possibly indicating that its diet included grassland seeds of the sort that must be ground to be digestible. Its powerful jaws and large, flat-topped molars also accord with this interpretation. Unlike man, it presumably was not a meat eater. Also unlike man, it may have been dimorphic in both body size and the size of its canines, but this point has been disputed. *Ramapithecus* was extraordinarily successful, surviving for some 6 or 8 million years and spreading from Africa, where it probably first appeared, to Asia and parts of Europe (22,33,49,50,51,77).

Following this early proliferation of apelike forms, there is a gap in the record of some millions of years. The trail can be picked up again, at about 4 million years into the past, with the appearance in Africa of *Australopithecus* and later *Homo.*

In the meantime, our ancestors had assumed a fully upright posture. There have been many suggestions as to why man became two-legged instead of adapting to the ground as a four-legged creature. A suggestion by Rodman and McHenry is attractively simple. During the millions of years that our ancestors lived in trees, their extremities had become modified for grasping and were no longer well suited to walking on the ground. The long fingers of the forelimbs could be folded under, giving the knuck-walking gait of the chimpanzee and gorilla, but this is not really an efficient form of four-legged locomotion. Alternatively, tree-dwelling primates that returned to the ground could use their hind legs, less modified for an arboreal existence than the front ones, for bipedal locomotion. Rodman and McHenry argue that, from the point of view of ease of adaptation, there was very little to choose between these alternatives. The occasional use by chimpanzees of bipedalism adds plausibility to this assumption (79).

The upright posture would bring both rewards and problems. It would free the hands for grasping tools or weapons or for carrying food, but it also would make it more difficult for an infant to ride on its mother's back. As Leaky and Levin point out, this problem would be aggravated further because the feet of these hominoids, as they became increasingly adapted for locomotion on the ground, would lose their grasping ability. Since the infants would share this loss, they would thenceforth be able to cling to their mother's fur only with their hands, not with their feet. The natural response would be for the mothers to start carrying their children in their arms. This would reinforce the upright posture and set up a feedback loop that would accelerate the trend to bipedalism (55).

Besides being upright walkers, our ancestors of four million years ago were also social. We have noted the evidence for this in Johanson's find of at least 13 skeletons of *Australopithecus afarensis* who apparently had been killed by some catastrophe while traveling as a group. In view of the social nature of most anthropoids, this characateristic of the human line may have antedated the appearance of *A. afarensis* by millions of years.

Another human trait that may have been present at this time, although we lack firm evidence on this point, is the short interval between births as contrasted with the longer internal characteristic of the great apes. Monkeys have a shorter birth interval, so it is possible that it was always present in the human line. Lovejoy, whose ideas have been clearly presented by Johanson and Edey, has laid great emphasis of the more frequent births and used it to develop some interesting concepts concerning this period of human evolution (46). I follow these rather closely.

At the time our ancestors first appear in the fossil record as grounddwellers, they already had developed a level of intelligence and a brain size exceeding that of monkeys and approximating that of modern apes. Along with this went a lengthened childhood. The apes solved the problem of caring for slow-maturing infants by having fewer of them. If our ancestors were not to take this route, mothers would have to be able to take care of more than one child at a time. This need touched off a chain reaction. A home base made it easier for the mothers to cope with this added responsibility, but care for a family restricted opportunities for the search for food. The solution for this was to enlist the aid of the fathers. With their hands freed by bipedalism, males could cary food, but to carry enough, the food value had to be high. Hence the shift to a partial meat diet.

Another problem created by the move to open ground was an increased exposure to predators. Konner has noted the numerous predators that might endanger the lives of !Kung infants (52). This problem was enhanced for our ancestors by their total lack of natural weapons. The single most definite clue that we have as to how this was solved is a study by anatomists Susman and Stern of the finger bones of *Homo habilis*. They found that the hands of this earliest *Homo* still had some of the grasping characteristics of a tree climber. A run for the nearest tree may have been a common method of escape (99). It is also reasonable to assume that the use of clubs and perhaps spears as weapons appeared at an early date. If chimpanzees can brandish sticks, certainly the early hominids could have done it too, and they might soon have learned to increase the effectiveness of their weapons through careful selection and perhaps some shaping with sharp bones or stones. The threat of predators would also increase the selective pressure for group living. Half a dozen of these small but well-muscled creatures armed with clubs might deter even lions in search of prey.

While male hominids might be able to defend themselves, females and children would be more vulnerable. Sleeping in trees might be a possible recourse at night, but to get several children into a tree in the face of a sudden threat might be impossible. Ultimately some form of shelter building might develop. It is likely, however, that some children were lost to predators in the early stages of human evolution. This would increase the selective value of frequent births.

With the advent of *Homo habilis* some two million years ago, a concatenation of circumstances was in place favorable to an acceleration of the evolutionary process. Some of the circumstances resided in the physical and mental nature of *Homo habilis,* some in the environment of the Pleistocene epoch that began at about this time.

Evolutionists generally agree that the division of a population into small, partially isolated groups, which, however, from time to time undergo mixture so that new combinations of genes can be produced and submitted to the crucible of natural selection, is a condition favorable to rapid evolution (23,116,117). Man's tribal structure met the first part of this condition almost perfectly. This had been in place for a long time, but with the advent of the Pleistocene and the appearance of the Ice Ages, new elements were introduced. As we have already noted, the advances and retreats of the ice sheet forced extended migrations and brought tribes into contact with other tribes previously remote. Instead of gene exchange being limited to near neighbors, usually closely related, it would sometimes take place between groups of wide genetic disparity. Diversity would be increased, and this would occur at a time when the changing climates of the Pleistocene imposed a particularly rigorous level of natural selection.

These same forces were at work on many social species during the Pleistocene, but *Homo habilis* had within himself qualities favorable to what, in evolutionary terms, was a great leap forward. It was the combination of all these factors working together that changed *Homo habilis* into *Homo sapiens.*

Homo habilis was endowed with an upright posture, hands capable of grasping and manipulating, the capacity to fashion primitive tools, the practice of male group hunting and female gathering of plant foods, prolonged infancy but also frequent child-bearing, and the need for a home base to which food had to be carried. Pair bonding probably was already present, and substantial food sharing and other forms of mutual aid were common practices. The brain had begun to enlarge. These characteristics of *habilis* set the stage for the development in interacting processes analogous to those we described in Chapter 2 in discussing cultural revolutions.

A list of the reacting traits would certainly include the capacity of our ancestors to create, to communicate, to learn, to reason, and to love. Progress in any one of these traits would increase the fitness of the group and also would increase the need for progress in some one or more of the others. Each in turn would demand an increase in the size of the brain. A mutually reinforcing system of positive feedbacks would be in place that would lead to a relatively rapid replacement of the less modified tribal groups with those that were more advanced. The end result of this process was man as we know him today.

The Evolution of Man's Moral Nature

Of all the social mammals, man is the most capable of love—of those thoughts and feelings favorable to life in social groups. Modern societies, in which many people live together in relative harmony, are a testimony to the strength of these bonds. Our moral nature has its flaws. As we saw in Chapter 4, individuals differ in their endowments of sociality. There are discords in even the best society. The social insects, guided solely by instinct, have a perfection of social design that we cannot match. Dogs and chimpanzees display a degree of faithfulness and affection that warms our heart and stirs our admiration. But the bonds that bind mankind together are unique in their strength and their warmth.

In the scientific literature, the sentiments that bind are usually referred to in coldly scientific terms. *Sociality,* for example, is a popular word. I shall often borrow from this technical vocabulary, but I also shall not hesitate to use words like *love* and *morality.* We are dealing with feelings that run warm and deep, and it is not inappropriate to use words with this connotation. Mellen, in his *The Evolution of Love,* has made the same choice of vocabulary (64).

I shall use love in a broad sense, similar to the biblical sense, but more inclusive. It means, as here employed, all the sentiments and motives that lead people to comfort, help, or give pleasure to one another. These include affection, sympathy, feelings of responsibility, the impulses that have made people risk their lives to save another life, pleasure in the company of others, conscience, mother love, and romantic love. The term also appropriately can be extended further to include love of nature; of other, nonhuman creatures; of sights and sounds. These sentiments also can redound to the benefit of society.

The principle of natural selection, as set forth in Darwin's writings, has been shown in innumerable instances to explain the patterns of animal behavior, including animal social behavior. Man should be no exception. It is one of the triumphs of sociobiology and its component sciences that they have confirmed and reconfirmed this relevance of evolutionary theory, not only to animal anatomy and animal physiology, but also to animal conduct. If there is any one area where sociobiology's application is most difficult, it is in the explanation through natural selection of man's capacity for social behavior and for love. It is relatively easy to explain our aggressive tendencies; it is less clear that we can attribute our social tendencies to the action of natural selection. Darwin was aware of this problem. In *The Descent of Man,* he devotes a considerable part of two chapters to discussing the origin through natural selection of man's "moral sense" or "moral faculties" (14). Some subsequent writers, overlooking Darwin's arguments, used natural selection to justify a diversity of predatory behavior. In recent years, however, evolutionists and sociobiologists, in an outpouring of papers, have sought, with a substantial degree of success, to explain the prosocial aspects of human nature. Some of these papers are highly technical. Genetics and mathematics have been added to selection theory to produce an area of study appropriately called the *mathematical theory of natural selection*. There are still areas of disagreement, but the trend has been towards the successful explanation of man's substantial capacity for love and goodwill. I shall not attempt to abstract these papers in any detail, but shall use their conclusions to reinforce a common sense approach.

Before turning to a more detailed analysis of man's cooperative tendencies and their evolution, we need to consider how natural selection works. We need also to carefully define terms. Natural selection is an abstract concept, and in disucssions of abstractions, verbal confusions are common.

Natural selection is a process that goes on in nature whereby the individuals that are best adapted to the environment in which they live will, on the average, produce the most surviving offspring. In more technical terms, it is the differential survival of genotypes. Man may practice selection in the propagation of his domestic animals and plants, but such selection is excluded from natural selection.

Biological survival is the basic value in evolutionary theory. By biological survival we mean not only survival of the individual, but also the production and rearing of offspring. For the sake of brevity, we shall use *survival* to mean biological survival where ambiguity will not result.

The emphasis placed in this chapter on biological survival as a value should not be taken to mean that there are no other legitimate value systems. We shall consider some of these in the next chapter.

Natural selection can usefully be divided into several different catagories. A first distinction is between *handicap selection* and *competitive selection*.

Some individuals die or fail to reproduce solely or predominantly because of genetic handicaps. Two rather extreme examples are lethal genes and genetically determined sterility. Handicap selection increases in times of environmental deterioration when just staying alive becomes a problem for many individuals.

Competitive selection is selection determined by the degree of success in competition with other members of the same species for a resource in short supply. Handicap selection and competitive selection are not mutually exclusive. There can be substantial overlap between the two, but severe genetic defects are common and in much handicap selection there is little if any involvement of competition.

Competitive selection in turn may be subdivided into *individual selection* and *group selection*. Handicap selection is a form of individual selection, but I shall use the term *individual selection* to refer only to the competitive variety. The groups subject to group selection can be either social groups or groups demarked by geographical or ecological boundaries. These latter are referred to as *demes*. The frogs in a single pond are an example of a typical deme, but demes also are not uncommon in situations in which the boundaries are much less obvious.

The importance of group selection in evolution, and particularly in the evolution of man's social nature, has been the subject of intense debate. Darwin believed in its importance, and Sewall Wright, one of the founders and most able developers of the mathematical theory of natural selection, has also supported it. Darwin wrote in *The Descent of Man:* "When two tribes of primeval man, living in the same country, came into competition, if . . . the one tribe included a great number of courageous, sympathetic and faithful members, who were always ready to warn each other of danger, to aid and defend each other, this tribe would succeed better and conquer the other" (14). And Sewall Wright, in a discussion of the evolution of species in general, wrote, " . . . success in intrademic selection may be neutral or injurious to the success of the population. . . Interdemic selection can correct in part the shortcomings of mass selection from the standpoint of the species as a whole" (117). Wright, in many of his papers, stresses the importance of the division of populations into small breeding units both for rapid evolution and for the protection of the population from the development of traits that confer a temporary advantage on the individual but damage the species. Both the weight of expert opinion and the evidence from a variety of sociobiological studies appear to justify the conclusion that group selection has been important in evolution and, more specifically, in the evolution of social species including humankind.

One of the problems in analyzing the effect of group selection, as applied to social species, is the complexity of group structure within these species. In many primates, siblings or mothers and their offspring form cooperating subgroups. This is true of some human tribes, and in man the family is also a distinct subgroup. Intermarriage may bind some adjacent tribes together into loose confederations. We shall note some results of these complexities.

The social bonds that bind mankind together in groups can be analyzed from several points of view. Perhaps the most fundamental would be their physiological and neurophysiological nature. This subject is still in its infancy, and in any case is beyond the scope of this book. For our purposes, the most important property of social bonds is the nature of the selective forces that have produced them. Darwin devoted considerable space in *The Descent of Man* to this subject, and sociobiologists in recent years have examined it at length. There has now developed a considerable, though by no means complete, agreement as to the factors involved. The presentation that follows is generally in accord with curent thinking, but I have not hesitated in places to use my own terminology or my own shades of interpretation or emphasis.

The first point to consider is the basic advantages and risks of *help-*

fulness. Help or aid is defined as any contribution or assistance to another individual that increases, however slightly, the chance of survival of that individual. The basic advantage of helpfulness is that groups in which the members in one way or another help each other can accomplish some of the tasks necessary to living more effectively than they could if acting alone. We have seen examples of this in some of the species of social mammals in which group members cooperate in hunting, in defense against predators, and in defense or acquisition of territory. The risk in helpfulness is that it will not be reciprocated. A *cheater* may use the help of others without repayment, or with inadequate repayment, and thereby gain a selective advantage. It has been difficult to show how natural selection can prevent the spread of cheating to the point where the group no longer functions as an effective social unit.

A great deal of effort has been devoted to the problem of explaining the evolution of helpfulness in view of the damage that can be done by the cheat, and we can now say that an explanation in fully Darwinian terms is possible. As Wilson notes, however, the selective forces producing helpfulness are complex rather than unitary (111).

For the purposes of our discussion, we divide helpful behavior into categories, distinguished by differences in their selective origin. The categories are kintruism, mutualism (also called reciprocity or reciprocal altruism), justice (usually called, in this context, moralistic aggression), numonism (a word that I have coined), and altruism. These categories are generally recognized by sociobiologists, but there is considerable variation in the way each category is defined. In selecting my own definitions, I had two major alternatives. The first was to define the categories so that they are mutually exclusive, with no overlap between them; the second was to define them so that there is some degree of overlap. I have chosen the second alternative. This to some degree complicates the discussion, but I believe it most accurately reflects the complexity of the selective forces.

My discussion will be directed primarily to the evolution of helpful or moral behavior in man, but in a few cases I shall use examples from other social animals.

The recognition of kintruism as a category of human helpfulness produced by natural selection is largely due to Hamilton. *Kintruism* is defined as the extension of aid to close relatives. By all odds the most frequent expression of kintruism in mammals and birds is parental, and especially maternal, care of young. Children in human families often do help their aging parents, but this is not necessarily the case, and in most species of birds there is no reciprocation of aid. The evolutionary reason for care of children is obvious: it is necessary for the perpetuation of the genes of the parent. But the genotype of an individual or parts thereof can also be perpetuated through the reproduction of close relatives. Brothers and sisters share parts of the genotype of both parents, and cousins parts of the genotype of two of four grandparents. An individual can therefore achieve what we may call *survival by proxy* through the survival and reproduction of close relatives. The degree of benefit to the individual extending the aid can be expressed in a formula that combines the biological cost to the individual of the aid extended, the gain to the relative or relatives aided, and the degree of relatedness.

Kintruistic social practices occurring in the Florida scrub jay have been studied by G. E. Woolfenden. This jay is confined to a limited area of Florida with a sandy soil and a "scrub" vegetation, but in this one area they are common. The birds do not breed until they are at least two years old. Pairs remain together for life and occupy permanent territories. Young unmated birds attach themselves to pairs and assist in feeding fledglings, defending territory, and defending against predators. Not all pairs have helpers, but the ones that do raise a higher percentage of young than unhelped pairs. The key finding from the point of view of kintruistic theory was that helpers attached themselves to close relatives. In 74 instances of helping followed through one breeding season, helpers assisted their parent pairs 48 times, a father and stepmother 16 times, a mother and stepfather twice, a brother and his mate seven times, and an unrelated pair only once. Thus kin, and especially very close kin, are strongly preferred. Besides profiting from survival by proxy, helpers may get some direct benefit in their own propagation from the helper relationship. A dominance order is established among the helpers, and if one member of the breeding pair dies, the dominant male or female, as the case may be, takes its place. If helper males do not achieve an opportunity to breed through this route, they may sometimes succeed in establishing their own territory where they will be joined by a mate (115).

A great diversity of other forms of help reasonably interpretable as kintruism has been discovered. The most striking and widespread manifestation is the help extended by the worker castes of ants and bees to the queen and her progeny. We confine ourselves here to a brief mention of probable cases of kintruism in primates, including man.

Tribes of primitive man and troops of monkeys and apes are composed of individuals, many of whom are likely to be close relatives, though as a means of avoiding the harmful effects of close inbreeding, mates often come from adjoining tribes or troops. If close relatives are present, random extension of aid may yield some return in the form of survival by proxy. Kintruism is more evident, however, when aid within the tribe or troop is preferentially directed towards relatives. We saw that in primates other than man, siblings or parents and offspring often form subgroups, the dominance of whose members is influenced by the rank of parents or the aid of siblings. In the Yanomamö Indians, also, close relatives form subgroups. An expression of this subgroup formation, in both monkey and man, is the occurrence of intragroup conflict, with subgroups pitted either against individuals or other subgroups. Chagnon describes an axe-fight between groups of relatives that occurred in one of the villages he was studying (10). It is a reasonable conclusion that the net effect of subgroup formation is a decrease rather than an increase in intragroup cooperation.

We have no way of knowing how common kin subgroups were in primitive human societies, but I suggest as a reasonable postulate that they were minimized in the most successful tribes. Families were a universal subgroup in human societies. In general, however, family groups would have more to gain by cooperating with one another than by competing. The fathers would join together for hunting and the mothers for the care of young or the gathering of plant food. If this view of primitive man is correct, kintruism based on subgroup formation was relatively unimportant in early human societies. Rather than kintruism, the dominant social pattern was mutualism, with aid being extended to unrelated individuals as readily as to related individuals who were not immediate family members.

Mutualism is defined as the exchange of aid between the members of a group. It is assumed that there is some approximation to equality in the interchange. There may be very little time lag in the exchange, a situation likely to result, for example, when individuals engage in joint hunting. Mutualism with prompt exchange of benefits of this sort is refered to as *cooperation*. The alternative is *delayed mutualism*. An example: A successful hunter gives meat to families not his own, expecting reciprocation at a later date. Mutualism also may be divided into three other subgroups. *Kin mutualism* occurs when the participating individuals are close relatives. Kintruism is not necessarily mutualistic; kin mutualism consists of those cases of kintruism in which mutualism is present. *Nonkin mutualism* occurs when aid is exchanged between individuals not closely related but who are members of the same species. *Symbiosis* is mutualism involving members of different species.

The current thinking of sociobiologists does not go beyond the forms of mutualism I have described, but a paper by Storer, a philosopher rather than a sociobiologist, makes a plausible case for the addition of another and broader class to this category of social response. Storer's evidence comes from an examination of language and custom. He finds that in many languages ranging from Anglo-Saxon to Chinese, the words "ought" and "owe" (this latter in a commercial sense) have a common root. Combining this fact with the observation that people tend to give loyal support to a great variety of social groupings, ranging from political communities to churches, environmentalist organizations, and other public interest groups, he concludes that we have within us a consciousness of an obligation, not only to individuals, but to the community. He uses as a name for the broadest category of this perceived obligation the term "debt of shared responsibility" (97). The behavior inspired by this same obligation might, in our terminology, be called *group mutualism*.

It seems to me that Storer makes a good case that a sense of obligation to the community as such is a reality and that some of our acts of aid and support are engendered thereby. This goes beyond the feeling that we have an obligation to practice simple reciprocity or to make repayment in kind to individuals for aid rendered. At an intermediate stage, this more advanced social sense might lead us to extend aid to members of the community at random with the expectation of comparable return, though not necessarily from the precise individuals aided. In a more advanced form it might be manifest in contributions to a common pool of resources. Schools and highways are examples of this which we see in advanced societies; in a primitive society an example might be a common storehouse of seeds or nuts established as a safeguard against periods of want. Conduct based on group loyalty might also be manifest in group defense against predators or enemies.

In the sense of helpfulness inspired by a loyalty to the group rather than by a sense of indebtedness to individuals, group mutualism, if it is a reality, is probably far more developed in man than in other primates.

Mutualism or mutual aid, in some one or another of its forms, is by all odds the commonest manifestation of help to others to be found in social animals (54). That it occurs not only between related individuals, but, at the other extreme, also between individuals of different species, is testimony to its adaptive value. Group hunting in social carnivores is one of its manifestations, but only one of the many forms in which it occurs. It is commonplace in man today and shows little if any tendency to be confined to close relatives. Many business ventures, for example, require the cooperation of unrelated individuals. More than any other type of aid, it is the foundation of human society.

Group selection has certainly played a role in the evolution of mutualism. The practice of helpfulness can make groups stronger in a variety of ways, and the stronger groups ultimately prevail. Symbiosis, which is a form of mutualism, is a likely product of selection at the level of the symbiotic pair or group. We should not underestimate the complexity and variability of the genome and hence the diversity at the individual level on which selection can act. Progress in any particular evolutionary trend may be accomplished through a variety of routes. The difficulty mathematical geneticists have had in constructing models that will account for the evolution of mutualism through group selection may therefore reside in the unrealistic simplicity of the models.

A major problem already noted in the appeal to group selection is the potential group helpfulness creates for the evolution of cheating. If selfishness spreads more rapidly as a result of intragroup selection than it can be eliminated by intergroup selection, then group selection will not work. How often the balance between the two selective processes is tipped in favor of the group is a subject of much debate, but recent theoretical studies tend to show that there are plausible postulates that can lead to successful selection for mutuality at the group level (104,109,111).

Studies that demonstrate, with the rigor expected of the mathematical theory of natural selection, that nature can surmount the problem of the cheat, began to appear in the 1970s. Two methods of circumvention have been suggested. The first is the preferential association of the naturally cooperative. The cheat is left out in the cold. The second is the practice of justice. This goes a step beyond preferential association, adding to it the principle that the cheat is punished and, in a still more developed form, that the especially cooperative are especially rewarded.

Boorman and Leavitt, Axelrod and Hamilton, Eshel and Cavalli-Sforza, and Rushton et al. have all championed, in slightly different form, the potential of selective association. All of them note that such association would require, or at least be greatly assisted by, the ability to recognize and distinguish between cheaters and noncheaters (2,3,21,181). This would imply a considerable mental faculty. We may call this variant of mutualism in which a distinction is made between helpers and shirkers *cognizant mutualism.*

As Eshel and Cavalli-Sforza note, preferential association might start through the formation of subgroups of close kin who, owing to their relatedness, share the same cooperation-favoring genes (21). Also, as Axelrod and Hamilton suggest, functional equivalents of what we think of as recognition might occur in many life forms. Axelrod and Hamilton therefore extend the concept of preferential association to the evolution of symbiosis and cite the remarkable case of the fig tree and of the fig wasp that pollinates it. At the time of flowering, the wasp carries pollen from tree to tree and at the same time lays eggs in the figs pollinated. The helpful wasp is rewarded by the successful development of its eggs. If the wasp lays eggs without depositing a sufficiency of pollen, however, the tree simply sheds the sterile and still immature fig. The cheating wasp is "recognized," and does not get the reward of surviving progeny (2).

The evolution of justice within any social group would provide an even stronger form of protection for the practitioners of cooperation or mutualism. We define *justice* as the dispensing of reward in proportion to service rendered and of punishment in proportion to disservice rendered. (Some refinements of this definition will be presented in Chapter 7). In sociobiological writings, what we call justice is usually referred to as moralistic aggression, but this expression conveys unfortunate overtones, and I see no reason for not using the term that we apply to it in everyday life.

Both Trivers and Leigh very clearly point out the importance of justice in the evolution of "reciprocal altruism" or mutualism (57,101). Trivers cites studies of hunting-gathering societies that suggest that justice is present at this level of human evolution. A clue to the ingrained approval with which we regard it can be found in the appeal of western movies in which the posse of good guys catches and punishes the bad guy. Justice, to be fully effective, has to be applied in such a manner that it contributes to the survival of the particularly helpful individual and to the elimination of the cheat. It may not be easy to specify how this occurs, but it is not an impossible requirement.

The application of justice requires a degree of cooperation in its execution together with observation, classification, and memory of individual social behavior that probably does not occur, except perhaps in limited form, in social species other than man. The growth of intellect that began with the advent of the human species should, however, have rendered it possible for our early ancestors. Substantial mutualism occurred at the prehuman level, but it is plausible to assume that with the appearance of man, justice was added to the human selective repertoire. The efficiency of group selection was thereby increased, and the trend towards mutualism was correspondingly strengthened.

The practice of negative or retributive justice requires subgroup formation and a form of conflict between the subgroup and the cheat, but whereas in other situations such processes are likely to weaken the group, in the case of justice they strengthen it. While it cannot be said too emphatically that there is no such thing as a simple division of people into the "good" and the "bad," there are, as we saw in Chapter 3, variations in the distribution of moral qualities. This being the case, society can gain by singling out and punishing or restraining individuals with disruptive tendencies. Justice, by definition, can be applied on a quantitative, not merely a qualitative, basis and this allows accommodation to man's moral diversity. But where law exists, even in a very primitive form, people can be divided into the law-abiding and the law-breaking, and subgroups can be formed on this basis. In the more organized societies, law enforcement can be delegated, but in primitive human societies, justice probably took the form of group members joining together to ostracize or otherwise punish individuals who violated the rules of mutualism.

Positive justice, or the reward of unusual service, can be more difficult to apply than negative justice; it is especially difficult to apply in forms that influence biological survival. However, the accounts of native American Indian societies in the United States in the early days of European colonization do suggest that leaders were selected on the basis of ability to serve the group and that they were respected and honored, and Neel's studies of the Yanomamö Indians of Brazil indicate that, at least in this tribe, headmen have had a reproductive advantage (36,68).

The practice of justice, like the selective association that characterizes cognizant mutualism, can bring together unrelated individuals. It thus both strengthens and broadens the cooperating group.

Nature is ingenious; there probably are more routes for the spread of mutualism-favoring genes than we have yet thought of. One type of route that perhaps deserves more emphasis than it has received is what I may call the *entering wedge*. One example of an entering wedge has been suggested by Axelrod and Hamilton. Cooperation or mutualism, they note, can begin among groups of close relatives and then be passed on to descendants living in more genetically heterogenous groups. It starts out aided by the special benefits of kintruism, but once established, can persist without them (2).

Another form of the entering wedge is the appearance of mutualism in the restricted form of cooperation, later evolving into the more general form that includes delayed mutualism. In cooperation, since the exchange of benefits is essentially immediate, the cheat is easily spotted and excluded. If the practice of exclusion of the cheat becomes established, mutualism can be extended to situations where more trust is necessary.

A third form of entering wedge is the appearance of mutualism in situations when the benefits are high and the cost low, and its later expension to forms involving a greater risk. Group hunting is a possible example of low risk mutualism. Hunting is an instance of our second form of the entering wedge, cooperative mutualism, so it starts with this advantage, but it possesses other advantages also, Since every hunt normally involves considerable travel, the potential cheat, if he wants to share in the kill, has to expend the energy necessary to accompany the group. In the actual execution of the kill, unless the game is large and dangerous, the would-be cheat has little to gain from hanging back. If he joins in the kill, he is on the scene to share in the first morsels. Once cooperativeness became established by this route, it could gradually spread to the higher risk situation of big game hunting or to other riskier forms of mutualism.

A final entering wedge is the possible existence of social bonds that overflow to some extent into situations for which they were not originally selected. The adoption of entirely unrelated children is an example from modern society. The need to give and receive affection originally evolved for the benefit of close relatives, but if there was some spill over to nonrelatives that originally was nonadaptive, it ultimately could have become adaptive. From a neurophysiological perspective, it may be a more complex problem to restrict the genes conducive to love than to make them universal in their embrace.

We turn now to another category of aid to others, a category that overlaps kintruism and that I shall call *numonism* (nun-monk-ism). This is defined as aid to others by individuals who, for any reason whatsoever, cannot or will not reproduce. Since *aid*, in this context, is defined as a contribution to the capacity for biological survival, numonistic individuals cannot receive aid in a reciprocal form. If they extend aid to close kin, however, thereby practicing kintruism, they can receive a return in the form of an increased likelihood of survival by proxy. This cannot occur when they aid nonkin. In practice, numonists almost inevitably do receive some aid as gauged by other value systems, and it is therefore possible to assign values in nonevolutionary terms to the aid exchanged. If there is a net gain to the recipient, and if the recipient is unrelated to the aid donor, the numonism displayed can be said to be an instance of true altruism. It is worthy of note that such altruistic numonists in some cases do achieve a very real form of survival through contributions they make to man's cultural heritage. Some of the great composers are notable examples.

The most striking and widely recognized form of numonism is the aid extended by the worker castes of bees and ants to the queen and her offspring. In these insect societies, the function of reproduction has been separated from the functions of food gathering and the care of young, an extraordinary example of specialization. We have no really firm knowledge concerning the evolution of this type of social structure, but we do have one important clue. According to Wilson, societies with sterile castes have arisen on eleven separate occasions in the Hymenoptera, the group to which the bees and ants belong (110). This is an altogether unique situation. Also, in the Hymenoptera there is an unusual pattern of reproduction and sex determination. Males come from unfertilized eggs and hence have only one set of chromosomes instead of two. All the sperm produced by any one male are therefore identical, and his daughters share a higher proportion of genes than do ordinary sisters. They are halfway to being identical twins. It is generally believed that this reproductive peculiarity must be related to the propensity of the Hymenoptera to evolve in the direction of highly organized societies (32,93,112).

While no other animal group can compare with the Hymenoptera in the number of species with sterile castes, this caste structure has appeared also in the termites and in one species of mammal. Jarvis has described a species of rat living in underground tunnels in the manner of moles, in which breeding is delegated to a single female. Most of the other needs of the colony are met by the males and nonbreeding females (44). Since the underground mode of this species would tend to isolate the colonies, they probably are rather highly inbred. This would increase genetic uniformity and therefore increase the profitability of numonism.

The steps leading to the evolution of societies with sterile castes are doubtless complex and are not fully understood. Three probable prerequisites are a preexisting social structure, a high fertility in females so that breeding by some females is unnecessary and perhaps even undesirable, and enough genetic uniformity so that numonism is accompanied by a high level of survival by proxy. Early human societies may have met these conditions, although the last of the three imperfectly.

Does this mean that specifically numonistic individuals can have been produced in man by selective forces? Wilson answers this question in the affirmative. "There is," he suggests, " . . . a strong possibility that homosexuality is normal in a biological sense, that it is a distinctive beneficent behavior that evolved as an important element of early human organization. Homosexuals may be the genetic carriers of some of mankind's rare altruistic impulses." He goes on to point out that homosexuals in the United States may number 5–20 million and that they have occurred in all societies, including hunting-gathering societies, and have often occupied important places in these societies (113). Many of them also, it may be said, are exceptionally creative and dedicated.

While a selective contribution to the relatively high incidence of homosexuality is a plausible possibility, our understanding of evolutionary factors must remain imperfect until we know more about its neurophysiology and its genetics. If the genetics is complex, a considerable incidence, accompanied perhaps by a considerable diversity in type, can be accounted for on the basis of mutation pressure and other nonselective factors. The neurophysiology and genetics of human sexual preference may very well be complex, but it also is possible that an important component in its determination is a relatively simple switch mechanism that triggers either the typical female or the typical male preference pattern. Mutations in a trigger locus (or one of several such loci) could then be the common cause of homosexuality. There may also of course be environmental components.

Numonism in human society cannot reach the degree of perfection seen in the Hymenoptera because of the variability in helpful tendencies that occurs in mankind. Yet while there are few relevant statistics, I suspect that nonbreeding men and women generally give more to society than they receive in return. Of course some are primarily self-seekers, but among the numonists are also some who display the finest forms of altruism.

Altruism, the last of our categories of helpfulness, is defined as the performance of acts that yield a benefit to others in excess of any likely return. We include here returns in the form of an increased chance of survival by proxy. When the performing individual is a numonist, and hence someone who cannot be rewarded in terms of an increase in the likelihood of his or her biological survival, the exchange of benefits must be gauged by some value other than biological survival. An *altruist* is someone who practices consistent altruism and therefore is a net donor over the long term. Many authors, it should be noted, use altruism in a broader sense than is here stipulated.

Altruism can take many forms. The instances that catch public attention and that we are most likely to call altruistic are those in which an individual risks his life for others. Every year brings reports of this sort of self-sacrificing behavior. An example is recounted in an Associated Press item of January 1982. A commercial airliner crash-landed in midwinter in the icy waters of the Potomac River adjacent to the National Airport. A helicopter was soon at the scene, throwing out lines to people who had survived the crash and escaped from the plane. Five times the line was thrown to a middle-aged man clinging to a cake of ice. Each time, instead of taking it himself, he passed it on to someone else, and before the helicopter left, the pilot saw him slip below the fuel-blackened waters. Wars bring accounts of many instances of comparable self-sacrifice. Thornton tells the story from World War II of four chaplains who were on a troopship torpedoed on the way to Europe. When the chaplains found that there were not enough life jackets for everyone, they passed their jackets on to the soldiers and went down with the ship (100).

How can natural selection account for the evolution of this ultimate form of altruism? The lifesaving aid is often extended to individuals that are quite unrelated to the donor. On the battlefield, it often is extended by young men in the prime of life. It cannot be explained simply as an exaggerated form of kintruism.

Group selection almost necessarily is involved. One possibility is that the assortment of genes occasionally produces individuals with more than their share of genes conducive to helpfulness. But before this can happen, numerous prosocial genes have to become established in the population. I suggest the following as an example of a selective situation that might contribute to this establishment process.

The anthropological record makes it clear that intertribal warfare can occur in which the losing tribe is virtually wiped out. We have cited some evidence for this. In early American history, it is recorded that the Iroquois virtually exterminated the Hurons. Warfare of this degree of lethality may not have been a frequent occurrence, but it probably was not extremely rare, especially in unfavorable times.

If a tribe became involved in a conflict of which extermination was a possible outcome, the biological rewards for countermeasures would be substantial. If we assume a situation in which the losers would be exterminated, but in which the tide of battle could be turned by and only by the sacrifice of ten percent of the winning warriors, then there would be a clear gain from making the sacrifice. If we add to this the assumption that all the individual soldiers were taking an equal risk, then the risk clearly would be worth taking. These precise assumptions obviously are an oversimplification, but they do serve to suggest a selective mechanism that might give rise to the potential for extreme altruism. Kintruism and its special benefits may have been a factor in prehistoric warfare, but if the risks of battle or aid to fellow warriors in danger were undertaken on the basis of comradeship or duty to families at home and not specifically for the benefit of kin, the same basic motivations could easily lead to the sort of altruism exemplified in the examples we have cited.

Besides the ultimate altruistic act—giving one's life for someone else—altruism can take many lesser but by no means insignificant forms. Aid extended to handicapped people who cannot reciprocate is one of many possible examples. Such aid is not a luxury indulged in only by affluent societies. The bones of the Indians slaughtered in the Crow Creek massacre and buried in a mass grave (see the section in this chapter on warfare in primitive human societies) reveal that at least two of the villagers, one adult and one seven-year-old, had substantial handicaps and must have required special assistance to survive (119). Human sympathy is a deeply ingrained trait.

The hunting-gathering stage in human evolution enhanced man's moral nature, but the advent of agriculture altered conditions. Communities became too large for the favorable operation of intergroup selection, which works best with groups in the size-range of the typical hunting-gathering society. The conditions of intragroup selection also were changed. People had to deal much more with strangers. Selfishness could more easily go undetected and unpunished, and some societies came to be dominated by the selfish and the power-seeking. There were and still are differences within societies in the degree to which cooperation is rewarded. Cities are unlike small towns, and absolutisms unlike democracies. While the influences on human sociality of the ten millenia since the advent of agriculture have been diverse, and in some areas of human evolution doubtless beneficial, their net effect on man's moral nature probably has been negative.

Conclusions

The human species is the ultimate product of the evolutionary process, with power over nature and its fellow creatures totally without equal. But a home on the high ground does not guarantee man's evolutionary future. The extinction of species is as much a part of the evolutionary process as their creation. Our future is not guaranteed, but we have the power to shape it, the power to maintain and dwell in a beautiful, satisfying, and peaceful world. The still unanswered question is whether we have the wisdom and the will.

We come into the world with a mixed endowment of moral qualities. Compared to the bees and the ants, whose social behavior is directly programmed for the good of the nest, our social tendencies are flawed and contradictory. On a long-term evolutionary scale, there has been a downward trend in social perfection from the Hymenoptera and even more primitive orders to the mammals. Yet the primates have reversed this process, and man, as Wilson notes, has reached a new pinnacle. Our addiction to intergroup strife is much the same as that of our relatives, but the potential for love is greater. We are bound by social bonds and endowed with social potentials that differ from those of our primate ancestors not only in quantity, but also in some cases in quality (112).

What are these special endowments that set man apart? No two people would draw up exactly the same lists. I suggest the following as some of the qualities that put a livable social world within our grasp.

Unlike other primates that live in groups of multiple males and females, the sexes in human society are united in families that, at their best, are bound together by ties of affection and love. Sex has been changed, in considerable part, from a disruptive to a cohesive force.

The dominance hierarchy that is well established in most primate societies has been relegated in at least some human societies to a minor place. We declare all men created free and equal. Though we do not live up to this ideal, accepting as inevitable some stratification in our heterogeneous world, we can if we will keep the doors of opportunity open and avoid the threat of conflict inherent in a peck order of privilege.

Evolution has added the pursuit of justice to the human social repertoire and has given us the means to form truly effective social groups, unrestricted by the limiting factor of kinship. Justice provides a corrective force for the diversity in social and antisocial proclivities inherent in the genetic mechanisms of all social species except the Hymenoptera. Variability in the moral qualities presents serious problems, but they are problems that can be dealt with in a manner denied to other social mammals if we apply justice wisely and consistently.

Humankind is endowed with reason, a faculty that, certainly as much as any other human attribute, separates us from the rest of the animal world. We can look to the future and test out mentally the consequences of alternative courses of conduct with an acceptable degree of reliability, even if imperfectly. We can plan and choose and shape tomorrow to our will. It is this faculty that gives meaning to the expression "self-interest rightly understood." Reason, operating from a base of appropriate knowledge, can add a new dimension to our considerable but imperfect innate social tendencies. We will fall short in the service of our best interests, by whatever value system they may be measured, if we are not diligent in its use.

Man, like other social creatures, is guided in his social conduct by a major genetic component, but to that a broad cultural component has been added. We learn a great deal of our social behavior, but many other social species have the same potential, although to a more limited degree. Where we differ from our social relatives in kind as well as in degree is in the ability to formulate moral law and in the possession of a moral sense to enforce it. Of this aspect of human nature, Darwin wrote, "I fully subscribe to the judgment of those critics who maintain that of all the differences between man and the lower animals, the moral sense or conscience is by far the most important." Dogs, as Darwin noted, seem to possess "something very like a conscience," but only man formulates rules of conduct that are passed on from generation to generation and that give a clear basis for the moral imperative (14). If we know wherein lies self-interest rightly understood, application of this knowledge can be mandated by public opinion guided by appropriately articulated moral law diligently imparted to each rising generation.

And finally, man is a creator of institutions that can have an enormous effect for good or evil on his conduct. Reason certainly tells us that the wise design of our governments and other social structures is an important part of the attainment of the best interests of mankind.

6

What Is Ethics?

Some Definitions

Ethics is the study of moral law. This definition of ethics, one of many that have been proposed, has the advantage not only of conveying a meaning that is useful, but also of being simple, understandable, and of not leading directly into a morass of other terms that are difficult to define. The inclusion of the expression *moral law* is the major potential ambiguity. The general meaning of this can be stated quite easily, as we shall see, but much of the rest of this chapter is devoted to refining and sharpening this meaning.

Ethics also can be used in a somewhat broader sense to include widely approved principles of action that could be, but typically are not, formulated as injunctions with all or most of the properties of moral law. Much of the moral tradition of primitive tribes has always existed in this form (12).

All definitions of ethics lead ultimately to the problem of values and value systems. The first definition in Webster's Collegiate Dictionary is, "The discipline dealing with what is good and bad and with moral duty and obligation." Because this definition includes the terms *good* and *bad,* which have value overtones, it raises the values issue immediately. Any discussion of ethics that omits a consideration of values must be considered incomplete, and I shall not bypass this subject.

It should be pointed out that the term *ethic* (in the singular form) is often used to mean a specific system of moral law. We shall use it in this sense.

Most philosophical discussions, because they deal with entities that are abstract rather than concrete, require the use of terms not easily defined. We cannot point directly to examples of their meanings as we can when we are talking about bread, butterflies, or banks. Even as familiar a term as *bread* may have grey areas, but the problems involved when we are talking about right and wrong are enormously more complex. For clarity in discussing such subjects, careful definition is essential. A useful discussion of the principles of definition can be found in James MacKaye's *The Logic of Language* (13).

We turn now to the meaning of moral law. A *law* is an injunction or body of injunctions, typically directed at the members of a group rather than at a single individual and enforced by some form of group sanctions. The term *law* also is applied to natural laws or laws of nature, but as thus employed it has a very different meaning and is not our concern.

Laws, for our purposes, may be divided into two subgroups, *juridical laws,* or the laws of government, and *moral laws.* The Ten Commandments (Exodus, Chapter 20) are familiar examples of moral laws. We may take two as typical: "Honor thy father and thy mother," and "Thou shalt not bear false witness against they neighbor."

The first thing to notice about those statements is that they are commandments or imperatives, although the use of "Thou shalt not" rather than "Do not" slightly disguises the imperative form. An important property of imperative statements is that they cannot be said to be either true or false. Truth and falsehood are properties confined to declarative statements such as "He honors his father and his mother." This sharply differentiates ethics, which deals with imperative statements, from science, which deals with declarative statements. The test of truth is all-important in science; in ethics we must apply some other test or measure. The appropriate test for all imperative statements, I suggest, is the consequences of acting in accordance with the command or instructions conveyed. Recognition of the importance of the test of consequences goes back at least to Jeremy Bentham, writing two centuries ago, and indeed probably to Socrates (4). We use this test constantly. Our choice of recipes is an example. A recipe is a set of instructions or rules for making some item of our diet. Recipes whose end product pleases and the labor of whose use is commensurate with the result become our favorites. In this particular example, gauging the consequences is direct and easy, but it is important to note that in many cases, and especially where we have to rely largely or entirely on foresight, the gauging of consequences can be complex and difficult.

With this background established, we may define *moral law* as law that (1) is approved by the group and thereby gains some degree of enforcement through group pressure; (2) places some restriction on or demands some positive contribution through the acts enjoined; (3) benefits the group if generally observed; (4) has generality in the sense that it applies to a broad category of acts. All four of these properties of moral law can vary in degree. These variations are of considerable interest, but because of them it is difficult to draw a precise boundary between what is and what is not moral law.

Approved principles of action not specifically formulated as law, in terms of which we have suggested a broader definition of ethics, can also display the same four attributes, and again in varying degree.

I now consider some examples of variation in the four attributes, taking them up in the order in which they are stated. The injunction against killing, except in war, is almost universally approved. It stands at the high end of the spectrum of attribute one. Paying taxes or, more broadly, meeting all the obligations of a citizen of a functioning democracy, on the other hand, is an imperative high in the last three properties of moral law, but low in the first. The conviction that the obligations of citizenship, and especially the obligation of paying taxes, should be fulfilled is so low that evasion is common and often unconcealed. There are also categories of conduct strongly approved by some people and strongly disapproved by others and hence displaying an oddly mixed standing with respect to our first attribute. The use or nonuse of abortion under certain conditions is an example.

We may hypothesize, with reference to the foregoing examples, that moral principles that coincide with and reinforce our innate moral tendencies tend to be generally and strongly approved. Our approval, on the other hand, of principles of action not subjected to millenia of natural selection during human evolution and hence not part of our moral nature, even though these principles may be appropriate to the modern world, is apt to be either weak or inconsistent.

With respect to our second attribute of moral law, we may simply note that the more a person gives or surrenders beyond any likely return, the more moral an act is usually held to be. Thus giving one's life to save someone else would be viewed by most people as a profoundly moral act. Consistent giving in small ways is also esteemed.

Returning a benefit to the group is the very essence of moral conduct, and the great majority of precepts backed by group approval enjoin conduct with this property. There are, however, exceptions or partial exceptions. The practice, by Indians of the northwest coast of the United States, of *potlach,* or lavish distribution of gifts at a feast, with reciprocation being mandatory, is an example of a virtue carried to such an extreme that its utility is dubious. There thus is variation in our third attribute of moral law.

Generality is our fourth attribute of moral law. Those practices we call *customs* are examples of precepts that share the approval extended to moral law but that deal with conduct more specific than general. Driving on the right in the United States and on the left in Great Britain is an example. An older example is the specific modes of disposal of the deceased viewed as proper in different societies. Custom has sanctioned at least four forms of dealing with the dead: embalming, burning, burying in the ground, and exposing in the open for consumption by wild animals. Exposing in the open was right and proper to the Persians, but barbaric in the eyes of the Greeks, who practiced burial (24). Personal opinions would probably differ as to whether customs and the comparable taboos of primitive tribes should fall within the domain of moral law, and hence of ethics. My personal preference would be to leave them out, although there may be taboos that I am unaware of that should be included (22).

Our definition of moral law thus leaves some room for uncertainty, but it nevertheless will serve us well here. It does not, however, end our definitional problem. Thus the use of the word *benefit* raises again the problem of values, thus emphasizing the necessity of ultimately dealing with this subject.

Moral Law: The Cultural Component in the Determination of Our Prosocial Behavior

Like most aspects of our personality and our conduct, our social behavior is determined by both genetic and cultural components. In Chapters 4 and 5 we examined the genetic components in our social behavior. In this chapter and the one that follows, we examine the man-made moral laws that are the most basic and specifically formulated of the cultural components in the prosocial aspects of our conduct.

Our genetically determined social tendencies are in some respects more benevolent than are those of our primate relatives, but basically, they are similar. In fact, they have much in common with those of all social mammals. It is the possession of a social culture that sets us apart. Our ability to generate and apply moral principles and the extensive use that we have made of this ability has no counterpart elsewhere in the animal kingdom. This ability, it seems to me, deserves a place in human sociobiology that it has not generally received (20).

Perhaps the key to a common understanding of the nurture component in our moral behavior is two aspects of our nature that make the nurture component possible. These two natural endowments are reason and conscience. These natural endowments, quite as much as our heritage of moral law, set us apart from our animal relatives. The uniqueness of human reason is generally accepted, and as Darwin said, ". . . of all the differences between man and the lower animals, the moral sense or conscience is the most important" (8).

Reason enables humankind to foresee, with considerable success, the consequences of its acts, and hence to choose beneficial courses of conduct. This applies to social as well as to personal behavior. Primitive man could see that cooperating with other members of the tribe would serve him better than going it alone. It is quite likely that at an early date man went a step further and generalized the advantages of cooperation and perhaps of quite specific patterns of cooperation. Moral law, probably at first in a very unsystematized form, had appeared.

Moral law, if it was to work, needed enforcement. This role was assumed by man's evolving *conscience*—by his feelings of approbation and disapprobation and his sensitivity to the approbation and disapprobation of others. Some of the sentiments in our conscience may be inborn, but others are certainly learned. As we have already noted, totally contrasting behavior patterns may have their exponents. It is very unlikely that such contradictions would exist if they were not culturally implanted. These properties of conscience mean that conscience can absorb the teachings of reason and, as the enforcer, make them work. Reason and conscience are the perfect pair. They go far back in human history and must have contributed in a steadily increasing fashion to the generation of prosocial behavior.

While conscience is the all-important natural enforcer of moral law, we should not leave the subject without mentioning one other human attribute that can fill this role—our *aesthetic sense*. Our love of nature and its beauties can play an important part in making a conservation ethic work, and a conservation ethic is going to be increasingly important in today's world. Edward Wilson makes much this same point in his *Biophilia* (26).

It should be pointed out that an emphasis on the potential of moral law, rightly used, for human welfare is in no way incompatible with Darwinism. As Robertson pointed out in 1945 and Herman Muller in 1961, an appropriate culture can have survival value for the tribe possessing it (14, 19). And this applies quite as much to the moral component in our culture as to any other.

Moral Law Is Relative or Conditional, Not Absolute

An important property of nearly all moral law (all moral law according to some authorities) is that it is relative or conditional rather than absolute. The laws are subject to exceptions. Statements enjoining a respect for truth are found in all the major religions, but there are also specific acknowledgments of the occasional need for noncompliance. Confucius noted this in the dictum, "Frankness uncontrolled is effrontery." A Hindu moralist put it even more specifically: "Speak the truth; speak the pleasant; but never speak an unpleasant truth" (10,15). Chinese moralists, according to Hummell, go even further and generalize the need for exceptions. They regard as a reasonable ethical choice one that not only takes into account the abstract principle of right, but that also gives due weight to any extenuating circumstances.

It is easy to find other specific exceptions that are acknowledged. Nearly all religions condemn killing, but permit it in wartime. The rule, "Honor thy father and thy mother" might seem an unlikely place for exceptions, but it is possible to imagine situations in which most people probably would regard them as justified. What, for example, should an honest son do if he finds that his father is guilty of crimes and dishonest dealings as yet unknown to the law? Unless the son can talk his father into giving himself up, it seems to me that any ethical solution requires bringing dishonor to the father.

The relativity of moral law raises an important question: How do we determine when exceptions are to be made? Is it all a matter of subjective judgment or is there some master rule to which we can appeal? This brings us back again to the problem of value systems, and more specifically, to the question of whether there is one ultimate value.

Value and Value Systems

At the beginning of Chapter 1, we noted the need for a value system if we are to guide our conduct by enlightened self-interest. In Chapter 3, we named and discussed several value systems: *money,* the value system of commerce; *wants or preferences,* the system used in economic theory; and the *general welfare,* a broad and widely used concept, which we did not, at that point, attempt to define. In Chapter 5, we named *biological survival* as the value used in the study of natural selection. In this chapter we have had something to say concerning *conscience,* and *approbation* and *disapprobation.*

From our examination so far, we may conclude that value systems, or at least those systems used in any of the sciences, are based on quantifiable variables and ones that have positive and negative aspects. Thus in the money system we have dollars and cents, and credits and debts. Wants can be moderate and intense, and they have their opposite, aversion or dislike. The opposites, approbation and disapprobation, also vary in degree, and an individual's genes, in succeeding generations, can in varying degrees increase or decrease. The general welfare is the nearest to being an exception. It is not easy to identify a negative aspect.

It may simplify our further analysis of value systems if we start out with an analogy. Post offices have used for weighing letters a balance consisting of a pan where the letters are placed, a graduated arm, and a weight that can be moved back and forth on the arm to find the point at which balance occurs. In this analogy, pounds and ounces are the value system, the letter the thing to be weighed (evaluated), and the reading on the scale where balance occurs the value that we want to determine. The moving of the weight back and forth to discover the balance point can be compared to the fact-finding and reasoning whereby we decide on the ethicality of acts. The analogy is incomplete in one respect—weight, at least in common usage, has no negative value.

We turn now to the definition of terms necessary in the discussion of value systems. Concrete terms are easier to define than abstract ones. We therefore start with something concrete—a valued thing.

A *valued thing* is a wanted thing. A little consideration will show, I believe, that this is the customary meaning. Perhaps the most objective test of this is provided by the things for which people spend money. People buy things they want or desire. They may buy something they do not want to please the seller, but this in itself is gratifying a want. They may buy something they do not want to give to someone they believe does want it and whom they want to please. They may buy on impulse and later regret the temporary urging of desire. Always, however, the test of money, which certainly is one of our measures of value, tells us that want and value go together.

From what we have just said, we may conclude that when we talk about wants, we are dealing with two basic entities: (1) *want*, a subjective state or feeling that varies in degree and has both positive and negative aspects; and (2) *wanted things*. Wanted things are not confined to material things; they also include services such as medical care, experiences such as travel abroad, and acts such as going to a concert.

There is one other property of value systems that we need to note before we start defining. We have said that valued things are wanted things. We may agree from this that if we are properly to call any system a value system, the universe that it embraces must be valued and hence either be identical with the wanted universe or extensively overlap it. As we will show, the value systems already considered meet this test.

With these considerations in mind, we are ready to start defining.

A *value gauge* (or *scale* or *standard*) is a variable entity, with both positive and negative aspects, that is applicable as a gauge or "yardstick" to a universe identical with or that extensively overlaps the universe of things wanted. *Value* is sometimes used alone with this same meaning, but it also has a more restricted meaning, as follows.

A *value* is a number assignable to a particular member of a value universe when tested against the associated scale. Thus in the case of the monetary system, $3.50 may be the value assignable to a pound of steak. In some cases it may be impossible to assign a specific number, but quantity can be implied by the use of expressions such as great and small.

A *value universe* is all those entities to which a particular value gauge is applicable. A *value system* is a value gauge plus its associated value universe.

We now turn to a characterization of six well established value systems, plus two others that have been widely advocated. We start with the happiness system.

THE HAPPINESS SYSTEM

Happiness and its opposite, unhappiness, like our awareness of the color "red" and of the taste "sweet" and of all other testimony of the senses, are primary states of consciousness. The words used to designate these primary sensations are not amenable to the usual form of definition. Dictionaries define them by the use of synonyms, which really is no definition at all. Despite this definitional problem, nearly all adults use *happiness* and *unhappiness* as part of their regular vocabulary. Nevertheless, if the terms are to be given a prominent place in ethical theory, it is important that any possible ambiguities be removed.

We may stipulate immediately that the terms will be used in a broad sense. *Happiness,* as here used, includes contentment, pleasure, joy, delight, satisfaction, glee, enjoyment, and any other forms or guises of happiness for which language has a name. In like manner, *unhappiness* includes pain, misery, suffering, anguish, grief, and sorrow.

Children learn the meaning of happiness by three possible routes. If they hear the word used when they themselves feel happy, or when they see someone else with a broad smile or other outward manifestations of happiness, or when they are shown a potential cause of happiness such as a new toy, an association between the word and the meaning is created. With adequate repetition, the meaning becomes fixed. The less inclusive synonyms of happiness are given their meanings in the same way.

Perhaps the commonest route to fixation is through the association of the word *happiness* with some one of its many causes. This may be one reason that philosophers, with surprising frequency, talk about the *causes* of happiness as if they were happiness itself.

While we cannot differentiate happiness and unhappiness from other conscious states by naming any inherent properties, we can differentiate and therefore define them by pointing to properties that are associated with their inherent nature. These properties are that we wish for happiness and wish to avoid unhappiness. *Happiness* thus can be said to mean those conscious states that, when once we have experienced them, we seek to experience again. *Unhappiness,* correspondingly, can be said to mean those conscious states which, when once we have experienced them, we seek thereafter not to experience again.

The reality of the tendency of sentient creatures to seek repetition of pleasurable states is indicated by experiments in the laboratory. Neurophysiologists have conducted extensive studies of the function of different areas of the brain by inserting electrodes and observing the consequences of localized electrical stimulation. A common response is the movement of some specific muscle group. Evidently a motor area of the brain has been activated. But there are certain brain-areas where the response suggests that the electrical impulse has caused either pain or pleasure. An animal stimulated in a "pain" area acts as if seeking avoidance, and an animal stimulated in a "pleasure" area appears to seek recurrence. In fact a cat or rat or monkey will quickly learn to pull a lever to stimulate a pleasure area of its brain. A monkey has been observed to do this repeatedly; three times a second for as many as 16 hours. The pleasurable stimulus also has a pronounced or soothing effect on animals, while the pain stimulus has the opposite effect (6,11).

Evidence of a quite different sort confirms the association between pain and avoidance. The aberrant nature of man's reactions to the presumed causes of pain when the red flag of pain itself is removed is shown by one of those odd experiments of nature that congenital abnormalities sometimes provide. There is a rare pathological condition in which the subjects, although normal in other respects, are almost or quite insensible to superficial causes of physical pain. Although a review of reports on this subject identified only slightly over two dozen cases, the nature of the condition is well established. Affected individuals can detect, identify, and localize pinpricks and can distinguish differences of heat and cold, but lacerations and burns produce no suffering. In keeping with the principle that pain and avoidance are linked, these injuries also produce no avoidance. One girl of seven often deliberately burned herself because "it felt good," and most subjects show scars from earlier traumas (7).

In the light of the foregoing observations, we can say that happiness has value because it is wanted. In fact, if happiness is defined as the sum of those subjective experiences that we seek to repeat, it can be said to have value by definition. Of itself, however, it does not constitute a value system because it is not inclusive. We need to add to it the causes of happiness, because we seek these as a means to happiness itself. The *happiness value system,* then, may be defined as happiness and the means thereto, and unhappiness and the means of avoidance thereof.

We have noted that values assume positive and negative forms and that they are viewed as varying in quantity or degree. There is no question as to the negative form of happiness—it is unhappiness or pain or suffering. The measuring of degrees of happiness and unhappiness presents more complex problems. While these problems are no more difficult than those raised by other potential value standards, with the one exception of the monetary standard, it is in order to consider them because of the emphasis we are placing on the happiness value system.

Everyone, I assume, would agree that the diverse agreeable or pleasurable states that we can experience are not all of equal intensity. They can range from mild contentment to the ecstasy of the sex act. They likewise can vary in duration, with the less intense forms of pleasure generally being the more durable. The same is true of the painful states, which can be brief or extended and can range from mild discomfort to the agony of the torture chamber.

Most people probably could give some kind of rough estimate of the degrees of pain and pleasure they experience, but precise quantitation is at present beyond our grasp. Possibly neurophysiologists, with their ability to explore and experiment with specific pain and pleasure areas of the brain, ultimately will be able to devise a quantitative measure. Now, however, if we are to use the happiness value system as the measure of welfare, we must be content with approximation.

One stipulation acceptable to most people, I think, is that the *quantity of happiness and unhappiness* should be taken as the product of intensity and duration. There are many precedents for this. Electricity is measured in terms of kilowatt hours, where the kilowatts equal the intensity and the hours the duration. A 75-watt bulb burning for one hour uses the same amount of electricity as a 25-watt bulb burning for three hours. It is because of this principle, and the relatively short duration of the more intense pleasures, that the durable contentments are so important in our total happiness.

We have already presented evidence that happiness is, above all else, what we want. We have, indeed, made it so by definition, but I think the definition we have given is in accord with acceptable usage. Introspection will show that happiness, in a broad sense, *is* what we want or wish for or desire.

While it is doubtful if all moral philosophers will accept this statement, it does have support from distinguished sources. Thus Aristotle, in *Nicomachean Ethics,* wrote, "the chief good is evidently something final . . . if there is only one final end, this is what we will be seeking . . . we call that which is itself worthy of pursuit more final than that which is worthy of pursuit for the sake of something else. . . .

"Now such a thing happiness, above all else, is held to be, for this we choose always for itself and never for the sake of something else. . . ." (1).

Similar views were held by the utilitarians, who were the dominant force in British moral philosophy for over a century (16). Thus Bentham wrote: "Nature has placed mankind under the governance of two sovereign masters, *pain* and *pleasure*" (3).

We have characterized the happiness value system; we now turn to the consideration of several others.

THE WANT VALUE SYSTEM

This we define as wants and all things wanted. We include in things wanted the means of avoiding those entities for which we feel aversion. This system was adequately characterized in Chapter 3.

THE SURVIVAL VALUE SYSTEM

This we define as survival (in the evolutionary sense defined in Chapter 5) and the means thereto and their opposites. It was adequately characterized in Chapter 5.

The three value systems (happiness, wishes, and survival) that we have so far identified are closely related. Happiness, as we have pointed out, is always and universally wanted. Because of this, its causes, insofar as they are known, are always wanted and hence comprise a universe very similar to that of the want value system. The two, however, are not identical. Thus while relief from suffering is always part of the things-wanted universe, it cannot, if there is no source of relief, be part of the means to happiness universe. The happiness and want systems, in turn, are linked to the survival system because evolution has tied those things and acts conducive to survival to happiness and those things and acts that are not conducive to survival, to pain or unhappiness. The cases of congenital absence of pain sensibility on the body surface show what happens when the link is absent; individuals without the link are hanidcapped today and under primitive conditions would have been poor prospects for survival. Other much lesser flaws in the linkage occur, but given the tendency of sentient creatures to seek happiness and its causes and to avoid pain and its causes, it follows inevitably that adaptive behavior has been linked by evolution to happiness and nonadaptive behavior, to pain.

The three value systems based, respectively, on happiness, wishes, and survival, thus have much in common. Each identifies a specific universe of entities, and these universes are either identical or overlap extensively. They differ in one major respect: Wishes and happiness are subjective states, while survival is an objective phenomenon. However, they all give similar though not identical instructions as to conduct. We need a name for these three value systems, and because they all have a base in human nature, we shall call them the biological value systems.

THE GENERAL WELFARE

The general welfare is something of a catchall expression. The definition in Webster's Dictionary includes "Condition of health, happiness, prosperity, or the like." The inclusion in the expression of the word "general" implies that we are talking about something applicable to a group rather than separate individuals. Within this one restriction, we have considerable leeway in selecting a specific definition. For our purposes I select the following. *The general welfare* is the means to happiness available to the particular group implied by the term general in the context in which the expression is used. The group may very well be all mankind. This is a practical definition, because anyone trying to improve the condition of people has to work through the means to their well-being. These means can be determined in a number of cases with some precision (9). Improved housing and improved health are two likely candidates for attention.

THE APPROVAL AND DISAPPROVAL VALUE SYSTEM

Conscience is closely related to this system, but is generally thought of as a monitor of our own acts, whereas approval and disapproval are generally directed at the acts of others. In any case, because approval and disapproval are directed only at acts or choices, thereby making the universe of the approval and disapproval the acts or choices approved or disapproved, this system is far more restricted than the three biological value systems. The universe also varies from individual to individual because of the considerable individual variations in approval and disapproval. This may be true to some extent of the universe of wanted things, but not to the same degree. Despite these qualifications, there is overlap between the approval system and the three biological systems because we typically approve of acts conducive to group happiness and disapprove of acts conducive to group unhappiness. The system is therefore properly called a value system.

In our evaluation of systems, we assign the approval system a minor place, but Joseph Butler (cited by Sahakian) is clearly raising it to the level of an ultimate standard when he states, "Every man is naturally a law to himself, that every man may find within himself the rule of right, and obligations to follow it" (21). Plato, in the statement, "Virtue is neither natural nor acquired, but an instinct given by God to the virtuous," appears to treat conscience as the agency by which we learn the will of God. He thereby links these two proposed sources of moral values (17).

MONEY

Since money is unquestionably a value system, we need at least to mention it even though it would find little if any support as the ultimate moral guide.

THE WILL OF GOD

Another court of appeal with many advocates is the will of God. St. Augustine, in the statement, "I hold virtue to be nothing else than the perfect love of God," appears to be expressing a variant of this (21). Plato acknowledged the role of happiness or pleasure in the guidance of man's actions, but gave a superior place to "an acquired opinion which aspires after the best" or an "Idea of the Good." He seems to view this "acquired opinion" or "idea" as being a divine quality, so in some sense at least, Plato is among those who appeal to the will of God (18,24). The will of God is often viewed as having been revealed to a prophet in some mystical manner. The revelation to Moses on Mount Sinai of the Ten Commandments is a familiar example.

We may question whether the will of God is actually a value system, but since many people use or profess to use it as a guide to conduct, it deserves our consideration.

Self-Realization

Bradley suggests, as "the most general expression of the end in itself, the ultimate practical 'why,' the concept of self-realization" (5). We may question whether self-realization, any more than the will of God, is truly a value system, but since Bradley has found a substantial following in recent years, his suggestion also requires our consideration.

Which of the eight value systems we have described is best suited as a rational guide to conduct? To help us answer this question, we apply to the systems five tests. For the sake of brevity, we shall generally limit ourselves to only the most serious flaw or flaws of each system, since we need no more than this to rule most of the systems out.

1. *Is the value on which the system is based wanted for its own sake?* Only the happiness system meets this test. Happiness is both a measure of value and a thing valued. If there is any one thing that all humankind and indeed all sentient creatures would like to see maximized, it is happiness. Such maximization is the ideal goal for a rational ethic.

2. *Is the value quantitative and hence capable of serving as a value scale or gauge?* The three biological value systems and the money system meet this test. The general welfare meets it, but only because it is tied to the happiness system. Our feelings of approbation and disapprobation may vary in degree, so the system based on them might qualify. Self-realization and the will of God do not, or at least it would take a good deal of stretching to say that they do.

3. *Does the value have both positive and negative aspects so that it can measure both right and wrong?* The answers here are the same as for question 2, with one exception. The will of God can be pictured as commanding not doing as well as doing, and hence as having the two needed aspects.

4. *Can the value system be defined unambiguously?* Again we find the will of God and self-realization to be the two proposed guides least able to qualify. Most people probably do have some concept of what self-realization is, but what do believers mean by the will of God? If they say, as evidence of understanding, that they know His will, are they not really saying that they personally favor or approve the behavior they believe to be commanded?

5. *Is the value well adapted to instructing unselfishness, consideration for others, and the search for the common goals so necessary for a functioning society?* No value system can really serve as a moral guide unless it meets this test, and an examination of the evidence will show, I believe, that only happiness qualifies.

It has often been contended that a search for happiness leads to selfish, self-centered behavior. This misses the point that the goal of a rational ethic must be the total happiness. Enlightened self-interest tells us that we cannot be a happy island in a selfish world. A dedication to personal gratification, moreover, leads to discontent, not happiness. Evolution has fitted us for a life weighted more toward exertion than toward ease. We need genuinely to love and to be loved. We need the durable satisfactions that can come only from a feeling of accomplishment. The happiest people are not those who devote themselves to pleasures, but those who are dedicated to useful ends. As a Hindu philosopher put it, "He who chooses the pleasant misses the end" (15).

The want system has real problems as a guide to a cooperative society. As was pointed out in Chapter 3, wants are subject to both intrapersonal and interpersonal conflict, which can be resolved only by appeal to some standard other than wants. The total happiness is the ideal court of last appeal.

The survival value system has some real claim as a competitor against the happiness system, but it, like other alternatives, fails to meet our fifth test. Establishing it as an ultimate standard would not only deny people with no expectation of descendants and no close relatives any purpose in life, a condition these people would almost unanimously dispute, but would also make it theoretically difficult or impossible to enlist them in civic actions. The group could offer them no reward in return, though in actual practice, these people, because of freedom from family responsibilities, might take particular satisfaction in participating. The happiness code would be their actual guide. The survival system does deserve an important place in ethics, however, because it can enlarge our understanding of our social proclivities.

The approval and disapproval value system displays some of the more serious flaws. Rather than being a source of consensus, it is a means of promoting one's prejudices. Approval and disapproval reflect upbringing at least as much as heredity. We have noted the contradictions in approval to which this can lead. Only by appeal to the happiness system can these be resolved.

The money system, the one value system with something material for its measure, is useful in economic and ethical theory because it permits a precise quantitation of wants in the markeplace. I do not see, however, any way in which it can, of itself, indicate the degree to which money's uses serve social ends. There is some truth in the saying that money is the root of all evil. It cannot serve as a guide to right conduct.

While belief in the will of God may serve as a reinforcer of conscience, it fails totally as a moral guide. It is quite as apt to encourage devastating divisiveness as social unity. God, if we want to believe the contestants, has fought on both sides of many, and in fact probably most, wars. The claim by people on one side of a territorial dispute that a land was promised to them by God is likely to be both an admission that they have no valid claim to the land and a source of bitterness and probably war.

Because self-realization implies concern for self and not the group, it is more likely to be conducive to selfishness than to social responsibility. Even though its sponsors may insist on the need for good social relations if it is to be achieved, it provides no firm standard that demands unity. As we noted in Chapter 2, its wide acceptance in recent years has lowered rather than elevated moral standards (27).

We are led, then, to conclude that a happiness standard is the most promising standard for inclusion in a rational ethic. To serve this purpose, it must be appropriately formulated into moral law. Our next area of discussion will lead us back to this problem.

Rational Ethics

Moral law evolved because, by helping people live and work together, it helped to satisfy life's basic needs. Whatever our standard of satisfaction, it remains true today that the logical reason for each and every one of us to espouse and to live by a moral system is the benefits it returns that we all can hope to share. Fortunately, this logical reason for moral behavior is reinforced to a considerable degree by natural tendencies to be helpful and considerate as well as self-seeking. These tendencies are essentially fixed, but while man's social nature is now changing very slowly if at all, the world around us is undergoing revolutionary change. We need the best possible ethic, a rational ethic, to deal with the complex problems we now face.

If a moral system is justified by its benefits, then it is logical to seek to maximize these benefits. It is almost a truism that different standards offer different returns. Small changes in our moral beliefs can influence both the gains from observance and the likelihood that observance will be the rule. It is my personal conviction that the ethical codes of all the great religions—codes that have a great deal in common—and perhaps especially of Christianity, are well attuned to human needs. In our changing world, however, the least we can do is to ask how we should interpret specific moral principles to meet the conditions of the world today.

This leads us to the concept of a rational ethic. A *rational ethic* may be defined as that ethic that best serves the needs of those who espouse it. The use in this definition of the expression "best serves" raises again the issue of value judgments, but its validity is not contingent on the use of any one specific interpretation. A rational ethic will be an ethic in the humanist tradition, founded on reason and dedicated to human welfare (23).

It should be noted that our stipulated definition appeals to the self-interest of those who support the ethic. This is clearly implied by the words, "best serves the needs of those who espouse it." This does not rule out the possibility that an ethic thus defined can also serve the general interest. Indeed, it is a major function of a rational ethic to reconcile the welfare of the individual with that of the group. An uninstructed self-interest, however, may not recognize the need for a concern with the general welfare. We are brought back again to the search for self-interest rightly understood.

Since the definition of a rational ethic tells us, at least in general terms, the end to be sought, our subsequent need is to concentrate on the means. If, in what follows, we use "ought" or "must," we are using these terms in the engineering sense to distinguish appropriate from inappropriate means.

What, we may ask, are some of the properties that a rational ethic must have? What are the specifications necessary for an ethic that is to conform to our definition, for an ethic that will, in actual fact, best serve the needs of those who espouse it?

First, the ethic must be compatible with the moral nature with which the evolutionary process has endowed us. This, however, is not enough for today's world. To meet today's needs, a rational ethic must at the same time demand more of us than we can expect from our moral impulses, unaided by reason and foresight (25). In Chapter 5, we examined some of the long history of social evolution in mammals. It seems clear from the evidence that man, though he has considerable capacity for aggression, also has his full share of loving tendencies. Any ethic that is to work in practice must start with this foundation. Our moral nature, however, evolved in a tribal world; today's world is far more complex. Correspondingly, we must broaden the horizon of our ethical thinking, and our guide in doing this must be our reason. Specifically, there are two areas in which a rational ethic must assume a universal character.

The first of these areas, which constitutes our second requirement for a rational ethic, is a reach toward internationalism in the political arena. Throughout the long history of tribal evolution, ethical systems were always centered on the tribe. Tribal mores served the interests of the in-group, not the out-group. If a tribe had an ethic particularly conducive to group cohesion, its chance of survival was thereby increased. The ethics of the tribe was carried over into early agricultural and pastoral societies, as the Old Testament makes manifest in the story of David and Goliath, in its accounts of Sampson's slaying of the Philistines and of King Amaziah's slaughter of 10,000 captive Edomites, and by the many other tales of conflicts and battles in which God was always on the side of the Israelites.

In-group psychology is deeply ingrained in human nature and never will be eliminated, but in the truly tribal form it is today totally inadmissible and totally incompatible with a rational ethic. Jesus, in his later years, but even more clearly the Apostle Paul, preached an ethic that knew neither tribal nor national boundaries. Their followers carried Christianity to many lands, but they also at times made it an excuse for wars and persecutions. This perpetuation of tribalism had mixed results and is not our concern here. Beginning with the First World War, however, the situation began to change, and with the advent of the atomic bomb, the potential consequences of warfare became so utterly devastating as to be intolerable.

A rational ethic, then, must seek a sufficient degree of world unity to insure the elimination of war. The obstacles are formidable. Our tribal tendencies will always lead to some degree of nationalistic thinking, and up to a point this may be desirable. Diversity is one of our great riches. But nationalism can also lead to distrust and hatred, and this is particularly true in a world divided into democratic and totalitarian powers. The route to a political system that can guarantee peace may lead us through years of stress and perhaps sorrow, but we must pursue it until the haven of rule by law and justice is reached.

As a third requirement, a rational ethic should enlarge our circle of consideration. In the tribal world, people's daily actions impacted, and their love and neighborliness embraced, only a small circle of individuals, all known personally. In today's world, the circle of impact of our actions, and especially of the actions of our leaders, can be much broader, including individuals relatively or quite unknown. A society in which, as a result of a restricted circle of consideration, many individuals feel cheated is bound to experience serious economic and political problems. We will all gain if we can create a world where one man's suffering is every man's concern and no man's satisfaction is another man's sorrow.

Our fourth requirement for a rational ethic moves us away from inclusiveness and toward selectivity. Our neighborliness must be founded on a cognizance of individual differences.

We saw in Chapter 4 that there is persuasive evidence for the existence of individual differences in moral endowment and in Chapter 5 that, from the evolutionary point of view, mutual aid is counterproductive unless guided by a recognition of these differences. The need for this recognition is quite as great today as it was during the long years of tribal history. It is men and women of goodwill who will benefit most from a rational ethic and who must join together to create it and to make it work.

This dichotomy in a rational ethic between the need for inclusiveness and the need for selectivity raises some of the most difficult questions in the whole field of morality.

Fifth, a rational ethic should enlist the political process to insure adequate enforcement. Ethics and politics should work together. And indeed both the purposes and the methods of applied ethics and of politics have much in common. Both, in theory at least, are directed to the general welfare, both are concerned with the creation or implementation of law, both depend on consensus, and both use methods of enforcement. The methods of enforcement are different, but they complement each other. Enforcement through the political process is particularly necessary in those areas where our native moral tendencies are weak. Where the innate moral component is minimal, the cultural component, of which the political system is a major part, must carry the load.

A final requirement for a rational ethic is that it must be directed at the possible. It must be accepted, and be accepted by enough people and with enough conviction so that it will work. People must view it as an instrument appropriate to self-interest rightly understood.

If a rational ethic is to come into being, it will be through a process akin to the political process, but much less formalized. It will arise through prolonged discussion and debate, gradually moving in the direction of consensus.

Actually, this process is going on at the present time. In Congress, in newspapers, in churches, at scientific gatherings (some of them specifically devoted to ethical problems), and perhaps most usefully in magazines and journals (both general and ethically oriented), ethical issues of major concern today are being debated. The issues are wide-ranging, with topics under discussion including business ethics, management-labor relations, human applications of genetic engineering, abortion and birth control, the artificial prolongation of life of the terminally ill and of seriously defective infants, and the relations of democratic and totalitarian regimes. Any considerable consensus is a long way off, but we are moving in the right direction. "An ethical way will develop in which verifiable truth is more important than ethical fervor and persuasion more efficacious than guilt. *We can pioneer such an ethic*" (2).

Selecting a Philosophical Basis for a Rational Ethic

To arrive at a rational ethic, the great need is consensus, not acrimony, and discussions of value systems seem to produce more of the latter than the former. There is, therefore an element of risk in raising the value system issue when my major objective in this chapter and the next is to make at least some contribution to the greater use of reason in our ethical judgments. I take this risk because it is generally agreed that value systems are basic to any ethical philosophy. I hasten to add, however, that it is far more important for the achievement of a rational ethic that we establish a consensus concerning the working imperatives of morality than on the ultimate imperative defined in terms of a specific value system. I assume that philosophically inclined readers will welcome a discussion of value systems, even if they do not end up in agreement with the views here advanced. I add the hope that dissent in this area will not foreclose consideration of what we have to say about the working rules of a rational ethic.

We have already examined various alternative value systems and have concluded that a system based on happiness is the one most appropriate as a basis for moral choice. It remains to arrive at the precise formulation of the happiness standard most appropriate, in the context of a search for a rational ethic, for measuring the general welfare. With such a standard at hand, we should be able to propose precise definitions of right and wrong.

Philosophers have long debated what manner of formulation of happiness into moral law will best serve the welfare of the group. Bentham, in his early writings, used the expression "the greatest happiness of the greatest number" to indicate the end sought, but this is ambiguous. A more precise formula is necessary, as Bentham himself came to realize (3). There are at least two sets of alternatives on which such a formula might be based. We might stipulate that we are seeking the greatest total happiness, or that we are seeking the greatest average happiness for each individual. We also might stipulate that the population we are concerned with is mankind or that it is all sentient creatures. The choice between these alternatives could be debated endlessly—I know of no entirely convincing argument for preferring one to another. I may also add that I believe a good case can be made for the thesis that the different possible formulas, in actual practice, would dictate very similar moral choices. Thus, basing a formula on "all sentient creatures" rather than on "mankind" would specifically mandate kindness to animals; but because fondness for both pets and wild creatures is a common human trait, human happiness, to a considerable extent at least, requires the happiness of other living things. Singer, in his excellent book on ethics and sociobiology, chooses this larger circle. Perhaps my own preference as a working rule is the measure of the general welfare in terms of the total happiness of mankind, including all future generations as well as persons now living. This rule seems to me to be both simple and practical, and I shall use it in what follows even though the broader rule is philosophically more acceptable.

We are now in a position to define *right*. A *right act,* we may stipulate, is an act that, in the light of the best available evidence, will make a greater contribution to human happiness than any available alternatives. Armed with this definition, we may also state as the one moral law that is subject to no exceptions, the dictum, "Do right." We may also combine the law and the definition, and state: so act, in the light of the best available evidence, as to make the greatest possible contribution to the total happiness of mankind.

A few comments on the above definition and dictum are in order. Truth and falsehood are properties only of declarative statements. It is meaningless to ask whether definitions and imperatives are ture. The relevant question in regard to these two grammatical forms is whether or not they serve some intended purpose—whether they are useful. I submit that the above definition and dictum are useful because they are appropriate for a rational ethic.

The principal reason why an ethic founded on these definitions of right and wrong will serve well in the creation of a rational ethic is that an ethic founded thereon, under appropriate and attainable conditions, offers the best prospect for the individual happiness of its supporters. An ethic can increase the total happiness in either of two ways: (1) by increasing the average level of happiness, and (2) by increasing significantly the number of happy people. Since population increase can soon become overpopulation, any benefits thereof soon reach the point of diminishing returns. If we choose, however, to seek a higher average level of happiness, the chance that any one individual will find a satisfying life is increased. There can be no absolute guarantee that any one supporter of a rational ethic will be one of the gainers, but the odds are in his or her favor. Some form of martyrdom is always a possibility, but an otherwise excellent ethic that led to the martyrdom of a minority for the benefit of the majority would be counterproductive. A rational ethic must have safeguards against such an outcome, and indeed, in discussing mutual aid, we have specified such safeguards. An ethic, therefore, directed at the greatest total happiness of mankind is the best possible route to the happiness of its individual supporters.

A second point to be noted in regard to the definition and the dictum is that they require the estimate of future consequences. This is implied by the inclusion in the definition of a right act of the phrase, "best available evidence." Because of the inclusion of this phrase, the dictum "do right," although an absolute in the sense that it is not subject to exceptions, usually does not yield absolute answers. Because in many situations we have to choose from among possible alternative acts, and because these choices require the estimation of probabilities, our choices are based on probability, not certainty. And it cannot be overemphasized that the choices in some cases are extraordinarily complex and difficult.

While the above definition and dictum can give us a philosophically satisfying basis for a rational ethic, it is not these to which we are likely to turn when faced with most moral choices, but rather the moral principles inherited from the great moral teachers and confirmed by generations of actual practice. In the next chapter, we examine these working rules of a rational ethic.

7

The Rules We Live By

Introduction

Laws, in one or another of their forms, are an important part of the culture of all societies. Even in the most primitive tribes, they are found as rules and customs, although they are often more implicit than explicit (31). Following the introduction of agriculture and the accompanying growth of social complexity, the complexity of law increased. As writing evolved, laws began to appear in recorded form, and they comprise a significant part of the writings of all the great religions.* In modern society, the various levels of government turn out innumerable volumes of laws, rules and regulations.

Custom and moral law hold a special place in the total body of law. They are certainly the oldest form of law, appearing long before the development of writing, and these centuries of tradition lend them weight. They are reinforced by strong feelings of approbation and disapprobation to a degree not seen with society's other rules, and they are often the inspiration and the guide in the drafting of many of these rules. Custom can vary

---

*The antiquity of moral law and the considerable extent to which its content is universal is indicated by a translation of an Egyptian papyrus which antedates the birth of Moses by hundreds of years. Part of it reads:

> Hail to thee, great god, lord of the right and truth . . . I have not plundered god . . . I have never cursed god . . . I have not minished oblations . . . I have not committed murder . . . I have not committed adultery . . . I have not stolen . . . I have told no lies . . . I have not lusted . . .

The similarity to the Ten Commandments of these verses, which were recited by Egyptians at funerary rites, is obvious. (Farancyzk, Tom. 1985. Are the Ten Commandments Original and Unique? Free Inquiry (Fall), p. 43.)

a great deal from one society to another, but moral law has a universal quality. Its most basic principles are remarkably alike in all cultures.

Custom and moral law are potent modifiers of human behavior. Christians, following custom, take off their hats and keep on their shoes in their places of worship; Muslims do the reverse. To the true believers, any departures from these practices are a major offense. Christians have taken very seriously the precepts of the Old Testament. I think it is safe to say that the tragic case of the child executed for striking its parents that occurred in Geneva under the rule of Calvin would not have taken place were it not for the injunction in Exodus 21:15, "And he that smiteth his father, or his mother, shall be surely put to death." Likewise, the hanging of a number of presumed witches in Salem in 1692 and the earlier hanging and burning of hundreds on the same grounds in Europe (6), would, at the least, have met more resistance were it not for another biblical injunction, "Thou shalt not suffer a witch to live" (Exodus 22:18).

With the erosion of belief in the divine origin of moral law, there has been some weakening in the moral authority of even the most widely accepted principles. The pluralism of modern democratic societies may also have had a weakening influence. Nevertheless, moral law is still an important force in today's society. I suspect that most people who call themselves humanists take moral law very seriously. School children generally acquire a considerable understanding of and respect for law. Research by a team of psychologists confirms this view.

The authors conducted a study of 5000 school children, ages 10–14, in Greece, Denmark, India, Italy, Japan, and the United States. In the United States, blacks and whites were treated separately. All the children were given a battery of tests and 406 of them, in interviews, were asked 79 questions designed to reveal their understanding of and the degree to which they were influenced by rules and laws. The results are well summarized in the following quotation from a description of the study by one of the participants, June Tapp: "Considering the range of countries and the diversity of political, religious, and economic styles represented in our survey, it is remarkable that there should be such similarities across the seven cultures. The children see rules and laws as performing equivalent functions in the ordering of human conduct. They recognize the need for order in human affairs, and the role that rules and laws play in providing that order. They want a fair system—one that emphasizes equality and consensus. And they agree that with good reason or moral justification, rules could legitimately be violated." Especially significant in the context of this chapter were answers to a question concerning the existence of grounds for breaking a rule. In six of the seven tested groups, a majority of the students indicated that violations could be justified on moral grounds (29).

We can, I think, conclude that moral law, properly formulated and taught, can have a substantial and beneficent effect on the life of any community or nation. An examination, therefore, of the moral doctrines that are the basis of ethical thinking in the United States and other successful democracies is an appropriate part of our search for facts to help us recognize wherein lies self-interest rightly understood.

Moral laws fall into rather natural groups, each of which may conveniently be referred to as a *code*. Five of these codes can be distinguished: the Code of Family, the Code of Amity, the Code of Equity or Justice, the Code of Enmity, and the Code of Commonwealth. The Code of Equity can be subdivided, and if we count the two subdivisions there are six codes in all. We shall consider these in order.

The Code of Family

The family, in at least some form, is part of all human societies. No comparable unit is part of the life-style of most of our living primate relatives, and the family was lacking in our primate ancestors, but it probably appeared rather early in the human line. As we saw in Chapter 5, the family has contributed substantially to our potential for cooperative living. Because close family ties can reduce some of the sources of human conflict, rules that can strengthen these ties are important elements of all ethical systems.

The books of Exodus, Leviticus, and Deuteronomy in the Old Testament contain specific rules in regard to the marriage relationship, incest, and the obligations of children to parents.

"Thou shalt not commit adultery" is the seventh commandment, which, the Bible tells us, was communicated by the Lord to Moses. In Deuteronomy 22:22 we find the command, "If a man be found lying with a woman married to an husband, then they shall both of them die." On the other hand, a man may marry a maidservant he has purchased from his father and then take a second wife, provided the rights of the first wife are not thereby diminished. Divorce also is permitted (Deuteronomy 24:1). Rape of a "betrothed damsel" is punishable with death (Deuteronomy 22:25).

Incest is banned in all societies, and Hebrew society was no exception. It is specifically forbidden in Leviticus, Chapter 18, as is also a sexual relationship with close relatives by marriage.

The outlawing of incest has a biological foundation. Inbreeding increases the chance that deleterious recessive genes will be expressed. Hence throughout the long period of tribal evolution, tribes that practiced outbreeding were less apt to be burdened with handicapped individuals than were more inbred tribes. This conferred a selective advantage on the outbreeding tribes and apparently resulted in the evolution of an innate tendency to avoid marriage at least with siblings. Evidence for this comes from the Israeli cooperatives or kibbutzim. In these Israeli communities, all the children are reared together like members of one big family, spending only an hour or two at the end of the day with their parents. It was assumed that marriages among these children would not be uncommon. However, this has not been the case. A study has shown that among 2769 recorded marriages, none were between children reared together. Apparently close association during childhood inhibits romance, and hence becomes a bar to inbreeding (27). Incest is thus inhibited by the double barrier of both nature and nurture. It is interesting to note that at least some degree of aversion to close inbreeding has also been demonstrated in mice and several other species (32).

Besides its rules concerning relations between the sexes, the Old Testament also contains rules on parent-child relationships. The most familiar of these is the fifth commandment of Moses, "Honor thy father and thy mother." In keeping with the patriarchal nature of the pastoral society over which Moses presided, there are, however, other rules stating in much stronger terms the obligations of children. Deuteronomy 21:18–21, instructs that if a father can say to the elders of the city that his son "is stubborn and rebellious, he will not obey our voice, *he is* a glutton, and a drunkard," then the son shall be stoned to death. We also find in Deuteronomy 21 a rule concerning primogeniture and, specifically, instructions that a double portion of the father's inheritance is to go to the firstborn, whether or not he be the son of the favorite of two wives.

With the passage of time, modifications of the early Jewish rules regulating the family were made, at least in the teachings of some sects. Thus one of the Dead Sea Scrolls, found in 1947 in a cave near the Dead Sea and attributed to the community of Qumran, gives prescriptions regarding marriage that are similar to those later adopted by the Roman Catholic church. Divorce and polygamy are forbidden, and celibacy is given an important place in that sex is banned anywhere within the walls of Jerusalem (21). Jesus probably was influenced by this or by other similar sects that sought reform in Jewish religion as then practiced.

The following two precepts from the Koran illustrate the rules regarding family relations found in the teachings of religions other than Judaism. Honor towards parents is enjoined in the passage, "If either of thy parents or both should approach old age say not to them 'Oh,' reprovingly, nor rebuke them but always speak kindly to them." And with respect to the relations of husband and wife, we find, "Do not adopt a critical attitude toward your mate on account of any quality that you do not approve, for there must be many qualities in your mate that you do like and approve" (17).

These two rules from the Koran, because of their similarity to Judaeo-Christian standards, serve to emphasize the uniformity of religious teaching concerning family relations, but significant differences do exist as well, especially with respect to whether the marriage bond should be ploygamous or monogamous.

Just as there have been changes in the past, we can expect that there will be some changes in the future. In fact, in industrialized societies, the decrease in the number of children and the increase in the employment of women have already produced some changes and are likely to produce more. It is to be hoped, however, that we will remember the lessons from human evolution and develop a code of family that will strengthen, not weaken, family bonds. The family has served mankind well for over a million years. It is unlikely that we can develop a rational ethic in which it does not have a major place.

The Code of Amity

The Code of Amity is that group of moral principles, both implicit and explicit, that enjoins helpful and loving behavior varying from simple honesty and truthfulness to acts of genuine altruism. It is the centerpiece of all ethical systems. In keeping with its central role in ethics, amity finds its major biological counterpart in mutualism, which is the most characteristic form of biologically motivated social behavior.

In the following examination of the rules of amity, we shall confine ourselves to the Old and the New Testaments.*

Four of Moses' Ten Commandments (Exodus 20)—"Thou shalt not kill," "Thou shalt not steal," "Thou shalt not bear false witness against thy neighbor," and "Thou shalt not covet"—can all be assigned to the Code of Amity. In addition to these rules, however, there are many others in the Old Testament books of Exodus, Leviticus, and Deuteronomy that belong within this group. The following is a more or less random sampling:

"Thou shalt rise up before the hoary head, and honor the face of the old man" (Leviticus 19:32).

"*But* the stranger that dwelleth with you shall be unto you as one from among you, and thou shalt love him as thyself; for ye were strangers in the land of Egypt: I am the Lord your God" (Leviticus 19:32).

"If thou meet thine enemy's ox or his ass going astray, thou shalt surely bring it back to him again" (Exodus 23:4).

If there are any two of Jesus' sayings that are commonly viewed as the

---

*Because the various books of the New Testament were written after Jesus' death and may have been written by persons who had not personally known him, there is some uncertainty as to how accurately they convey his actual teachings (14). I nevertheless shall treat the ethics of the New Testament as reflecting Jesus' own views. Since I am not concerned with such strictly religious portions of the New Testament as the working of miracles and the divinity of Jesus, the degree to which these portions accurately convey Jesus' own views is irrelevant in this context.

essence of Christian teaching, they are the Golden Rule and the injunction, "Love thy neighbor as thyself."

The Golden Rule is found both in Matthew 7:12 and in Luke 6:31. In the King James version of Luke, it appears in the form: "And as ye would that men should do to you, do ye also to them likewise." This is followed in Luke but not in Matthew by several verses that clearly enjoin giving where we expect no return or at least no earthly return—in other words, that command the practice of altruism. Thus Luke 6:35 reads, "But love ye your enemies, and do good, and lend, hoping for nothing again; and your reward shall be great, and ye shall be the children of the Highest: for he is kind unto the unthankful and to the evil." It also is in this chapter of Luke that we are told to turn the other cheek.

The injunction "Love thy neighbor as thyself" occurs in the Old Testament in the nineteenth chapter of Leviticus and in at least six of the books of the New Testament. In Luke, Chapter 10, it is accompanied by the parable of the Good Samaritan, which in effect gives a definition of "neighbor." When the rule of neighborliness is viewed in the light of this definition, it takes on a meaning not, I suspect, generally appreciated, but that has great interest because of the parallel it presents to the lessons concerning mutualism that we have learned from evolutionary studies.

Saint Luke, in Chapter 10, tells of an interchange between Jesus and "a certain lawyer." The lawyer, having first repeated the injunction, "Thou shalt love the Lord thy God with all thy heart, and with all thy soul, and with all thy strength, and with all thy mind; and thy neighbor as thyself," subsequently asked, "And who is my neighbor?" It was in response to this question that Jesus told the parable of the Good Samaritan. The Samaritan, it will be remembered, aided a man attacked and wounded by thieves after a priest and a Levite had "passed by on the other side." At the end of the account, Jesus asked the lawyer, "Which now of these three, thinkest thou, was neighbor unto him that fell among thieves?" To which the lawyer replied, "He that showed mercy on him." Thus the parable, through the example of the Good Samaritan, defines neighbors as those people who are generous with their help, even to strangers. It is people such as these to whom the injunction "Love thy neighbor as thyself" applies. The selfish and the self-serving are excluded.

The clear implication of this parable is that amity must be selective. In the terminology of the evolutionary theory of social behavior employed in Chapter 5, Jesus is advocating a cognizant mutualism. It is interesting to note that in the Dead Sea Scriptures, written by the Essene Brotherhood whose writings are presumed to have influenced Christian thinking, there is a passage that specifically recommends cognizant mutualism. The passage reads: "Everyone is to be judged by the standard of his spirituality. Intercourse with him is to be determined by the purity of his deeds, and consort with him by his intelligence. This alone is to determine the degree to which a man is loved or hated" (21). The validity of this doctrine is borne out by ages of human experience.

The parable of the Good Samaritan teaches something else. The Samaritans were a fringe sect of Judaism and were looked down on by the orthodox Jews. Thus the Good Samaritan, by Jewish standards, was something of an outcast. The priest and the Levite, on the other hand, were representative of the orthodox majority. The inference clearly is that we are to judge neighborliness on an individual basis, not on the basis of sect, clan, or ethnic group.

Because of this aspect of the Parable of the Good Samaritan, it might be concluded that Jesus pictured his mission as being universal rather than tribal in its scope. Perhaps he did so view it in the later years of his ministry, but as McKown points out, there are a number of passages in the New Testament indicating that his message was intended only for the Jews. Thus in the tenth chapter of Matthew, which is devoted to the instructions he gave to the Twelve Apostles, we read:

"These twelve Jesus sent forth, and commanded them, saying, Go not into the way of the Gentiles, and into *any* city of the Samaritans enter ye not:

"But go ye rather to the lost sheep of the house of Israel." (Verses 5 and 6)

There are also passages in which Jesus seems to be referring to Gentiles as "dogs" and "swine" (20). (Mark 7:25–29; Matthew 7:6)

It was Paul, years after Jesus' death, who found a receptive audience among non-Jews in cities such as Corinth and who thereafter carried Jesus' message far beyond the confines of Israel.

There is, on the surface, a contradiction between the selective love of neighbors enjoined by the parable of the Good Samaritan and injunctions such as "Love ye your enemies" that recommend a wider charity. This apparent duality, it seems to me, actually reflects the wisdom of New Testament teaching. We all are guilty of mistakes and transgressions, and it would be a sad society indeed in which there was not a great deal of forgiveness by friends and neighbors. Harsh judgments are all too easy. If we require any urging, it is to love more, not less. That Jesus stated the need for selective love in a parable, rather than in a direct injunction, may be taken to mean that we should withhold our neighborliness only after thoroughly informed and careful judgment.

Another area of ethics that is part of amity is our obligation to the group. We referred to this in Chapter 5 as group mutualism. There is abundant evidence that this particular sense of obligation is part of our ethical thinking. Thus Linton notes in his paper *Universal Ethical Principles,* ". . . loyalty to any social unit to which an individual belongs is always regarded as a virtue, disloyalty as a vice" (18). In some of its manifestations, group loyalty can be pure tribalism, a dangerously competitive type of thinking that goes beyond a healthy patriotism. It can also take the constructive form of a general helpfulness in recognition of past, unrecompensed help from the group and its members. I suspect that many older people are glad to help youngsters without expecting a return because they remember occasions from their own youth when they were similarly helped. Philanthropy is another manifestation of this form of amity. Although more often implicit than explicit, this part of ethical doctrine can make important contributions to social living.

The Code of Equity or Justice

The ethic of virtually all societies includes two codes of Equity or Justice, not one. The first, since it is founded on the calculated self-interest of a substantial proportion of all members of the typical community, may be called head equity. In its simplest, unqualified form, it commands: Reward in proportion to service rendered and punish in proportion to disservice rendered. The second code is founded on sympathy and is designed to correct the injustices of nature. We may call it heart equity. It is the injunction: Work to alleviate the suffering of the unfortunate who suffer through no fault of their own. We will sometimes refer to head equity simply as equity.

These two ways of looking at justice effectively differentiate the philosophies of the two major political parties in the United States, the Republican and the Democratic. The Republicans are champions of head justice. They place considerable emphasis on rewarding our leaders and punishing our criminals. The grounds on which they distinguish between service and disservice may be open to question, but they are firm believers in the principle that success, which they tend to equate with service, should be well rewarded. They are not champions of equality. The Democrats, on the other hand, emphasize help to the unfortunate, thus championing a rule of heart equity. In the words of Jay Rockefeller, now U.S. Senator from West Virginia, "The Democratic Party does, and I think should, continue to symbolize an interest in and compassion for those who have not been able to make it" (25). This difference between the parties is, of course, primarily a matter of degree, but it is sufficiently fundamental in their thinking so that we may appropriately use as alternative names for the two kinds of equity the expression Republican equity and Democratic equity.

Actually, heart justice has much in common with the principles that make up the Code of Amity, and we might have elected to place it there if our views concerning it had not been colored by the political process. Democrats like to refer to its application as a "right," or to speak of it as "social justice." In the words of Fritz Mondale, 1984 Democratic candidate for President, "My faith unmistakably has taught me that social justice is part of a Christian's responsibility." It is easier to get votes for something that is a "right" or is an application of "justice" than for something that can be advocated only as an application of brotherly love. From the philosophical point of view, a more valid reason for treating help to the unfortunate as a manifestation of justice is that it can be regarded as man's answer to the injustices of nature.

Head equity or head justice actually is more complex than we have so far indicated. It is time that we examined some of these complexities.

Justice, in this coldly rational form, has two aspects: reward for services rendered and punishment for disservice rendered. These are sometimes referred to as *distributive justice* and *retributive justice*. While the purpose of both aspects is to induce right or socially useful behavior, they are quite different in their nature. We shall consider them separately, taking up first negative or retributive justice.

Retributive justice, in the simple form in which we have stated it, is a characteristic part of early moral doctrine. Thus in the Koran we find the injunction, "If a man commits a trespass his punishment shall be proportional thereto." But we also find in the Koran, "The recompense of evil is a penalty proportional thereto, but he who forgives the trespass of another intending thereby to effect a reformation (in the offender) shall have his reward with God" (17).

This passage from the Koran makes an important point. The purpose of punishment is not revenge, but to encourage socially beneficial conduct. If there are situations in which strict proportionality between crime and punishment is not the best way to deter crime, then we should depart from the rule of proportionality. As Bentham makes clear in his *Principles of Morals and Legislation,* there are many factors that should be taken into consideration (2). We cannot examine these here. It is sufficient to note that the sole purpose of justice is to benefit society, and it should be administered accordingly.

Most of the judgments made in the execution of retributive justice have been formalized by law and custom into definite patterns and placed in the hands of judges and other members of the legal apparatus. Many of the judgments of distributive justice, on the other hand, are not guided by fixed rules. These judgments, however, are commonly and widely made and are an important part of our social system.

There are a number of quite different situations in which we make distributive judgments. Every promotion in the business and academic worlds involves a judgment of this variety. Officials weighing the promotion must decide whether the past record of service and the probable capacity for future service of any candidate indicate an ability to handle the responsibilities of the job and merit the accompanying financial rewards. In democratic nations, but not in nations under absolute rule, citizens have the privilege and responsibility, through their votes, of passing judgments on their leaders and on candidates for leadership positions.

Recipients of the numerous honors and prizes awarded annually in most modern societies are chosen by processes involving distributive judgments. The award of scholarships to students also involves such judgments, but there is much more emphasis on the capacity for future service and less on service rendered. Distributions of property under our legal system are an even more special case, since the patterns of distribution may be determined much more by law and precedent than by any considerations of past service rendered.

Justice, in both its aspects—the justice of cool reason and the justice of compassion—has a well-established place in our society. Since our concern is ethics, however, it is appropriate to ask how successfully justice provides the sort of service to society that ethics demands. Does it, in its existing form, have imperfections? Can it be improved, and if so, how? Are the two aspects of equity necessarily separate, or can they in some way be united? We now turn to a consideration of these questions.

Retributive justice is a part of all civilized societies and, at an early stage in the history of these societies, was formalized and placed on a legalistic basis. The principle of punishment in proportion to disservice is universally recognized, as also is the need for qualifications. As Morrison notes, the Puritans in the early days of this country aimed to make the punishment fit the criminal rather than the crime (23). Even today, however, there are great differences in the way disservice is determined. In democratic societies, the standards applied are, in general, ethically valid. In many nations under absolute governments, and even in some under less despotic authoritarian governments, this is not the case. Disservice in these societies is measured, not in terms of the welfare of the society as a whole, but in terms of the welfare of the ruling clique. The tragic consequence is the incarceration or confinement in slave labor camps and often the torture or death of thousands of prisoners who have been arrested solely for political reasons.

In democratic societies, debate over the fairness and efficacy of the criminal justice system has been going on for a long time. We examined a few of the problems in Chapter 4. The system on the whole works as well as can be expected and as long as public concern remains alive, we can reasonably hope for improvement where improvement is possible.

If there are serious flaws in our practice of head equity, it is in the distributive component and in its inseparable companion, the choice of leaders. Our applications of these companion functions of head equity, in both the business and political systems of free societies certainly are by no means unrelated to valid measures of service to society as a whole. Nevertheless, they fall short of the standards we should seek, and curiously, despite the vital importance of equitable reward to the welfare of the community, these shortcomings seem sometimes to arouse less popular concern than the way we treat our criminals.

Because our businesses are typically authoritarian in structure, promotions are determined by a few people at the top. Thus there is an inevitable tendency to base the promotions, which in turn determine rewards, on loyalty and service to the management. Since profits are a chief goal of management, these are important in gauging merit by business standards. Under our economic system, profits and social benefit are related, but the departures from a strict correlation are sufficient to make promotions on the basis of ability to turn a profit an ethically flawed measure for determining rewards. The flaws are exaggerated if, as sometimes happens, the top executives are more interested in lining their own pockets than in the health of the business they run.

We examined some of the problems in this area in Chapters 2 and 3 and cannot go further into this subject here. It is, I think, clear that this is an area where improvement should be sought. Lest this statement be misinterpreted, perhaps I should add that I believe improvements will come through refinements in our present system, not through its overthrow.

With respect to our political system, it is sufficient to say that the criticisms we often hear in the United States today of the way our elections work are more than justified. Our population and our way of life have changed over the years, and it is not surprising if a system that was at least moderately successful in the past should now display unacceptable flaws.

Our practice of heart equity also has its flaws. In early societies, aid for the unfortunate was the responsibility of the individual; now this responsibility has in large part been delegated to the government. In our complex society, measures such as government-sponsored medical care, unemployment insurance, and old age pensions are a necessity, but leaving the applications of compassion primarily to the government also has its risks. It invites the adoption of poorly conceived but politically appealing measures; it opens the door to the cheat; and it risks making the needy who could learn to stand alone permanently dependent on a government-provided crutch. It also, as Hoff notes, can lower the ethical standards of the fortunate. People who have voted for social programs can easily convince themselves that their duty has been done and hence lead them to neglect thereafter the possibility of helping the unfortunate directly. The aged and the poor need the cheer that only personal kindness can bring (13).

A search for improvements in the practice of equity is an appropriate area of ethical concern, and we now turn to a consideration of possible approaches. We draw substantially on our examination of the biological basis of social behavior in Chapters 4 and 5 and of moral law in Chapter 6.

As we saw in Chapter 5, the practice of head justice is favorable to the evolution of strong social bonds. Genes conducive to service are favored in a social environment in which service is rewarded and disservice punished. Evolutionary forces thus have given us some natural tendency to apply justice in this form.

The evolutionary origin of heart justice has very different roots. Its manifestation in the form of help to the unfortunate is less easily explained than the manifestations of head justice, but we found evidence in Chapter 5 that it does occur and not only in primitive man, but also in the chimpanzee and the highly social African wild dog. Only in man, however, is it a common practice to extend help not only to near kin who are injured or handicapped but also to unrelated and often unknown sufferers. In this form, it may at least in part be an accident of evolution, but it does have evolutionary roots.

In addition to its evolutionary roots, the practice of justice in all its aspects has long-established roots in our ethical thinking. The great religious teachers have generally recognized the existence of differences in our moral natures. Thus Buddha, in speaking of the caste system incorporated in the Vedas, said, "But that is not true to the First Law of Life. People are only divided into good people and bad people. They who are good, are good; and they who are bad, are bad. And it does not make any difference in what family they are born" (7). And Jesus, in speaking to his disciples, said, "So it shall be at the end of the world: the angels shall come forth, and sever the wicked from among the just" (Matthew 13:49). The great religious teachers also generally recognized the need for justice, though with a tendency to emphasize its punitive or retributive side.

We have already noted that the Koran enjoins the practice of justice, and also recognizes the need for qualifications. One of the clearest statements concerning the value of justice was made by Confucius, who was more concerned with promoting good government than with personal salvation. At one time, Confucius served as Chief Magistrate of the city government of Chung-tu. Subsequently, in speaking with the Duke of Lu, he said of his practice of justice: "I rewarded those who were good, and I punished those who were bad. The people saw that is was good to be good and bad to be bad, and they became good. And good people are loyal to each other and to the government ." (7)

Justice has a prominent place in the ethical doctrine of the Old Testament. Generally, it is a stern justice. Thus in Exodus 21 we find, "Eye for eye, tooth for tooth, hand for hand, foot for foot, burning for burning, wound for wound, stripe for stripe." In the New Testament, on the other hand, there appears the oft-quoted dictum that seems to ban all judgment, "Judge not that ye be not judged" (Matthew 7:1). Taken in context, however, this appears to be a warning against prejudiced judgment of neighbors rather than a repudiation of judgment in general. Indeed, in John 7, Jesus is quoted as saying "Judge not according to appearance, but judge righteous judgment." And the whole of Matthew, Chapter 23, is devoted to an indictment by Jesus of the scribes and Pharisees. Perhaps as a reflection of this position of Jesus and his disciples as antiestablishment critics and outsiders, there is a tendency to leave punishment of wrongdoing to the hereafter. Thus the passage from Matthew already quoted predicting the separation by angels of "the wicked from among the just," is followed by, "And they shall cast them into the furnace of fire: there shall be wailing and gnashing of teeth." But there is also the well-known incident, recounted in several of the Gospels, in which Jesus, armed with "a scourge of small cords," drove the moneychangers from the temple.

Heart justice as well as head justice has a clear place in traditional ethical thinking. In primitive societies, according to Linton, care of the poor and unfounate is always approved, though there is also a tendency for charity to be bestowed ostentatiously as a means of acquiring prestige. In Proverbs, Chapter 31, which extols the virtuous woman, we find as one of her attributes that "she stretcheth out her hand to the poor; yea, she reacheth forth her hands to the needy." And in the eighteenth chapter of Luke we find Jesus instructing the rich man who inquired after the route to eternal life, to "sell all that thou hast, and distribute unto the poor, and thou shalt have treasure in heaven: and come, follow me."

There is an element of conflict in our two forms of justice in that head justice is a divider of mankind and heart justice an equalizer. Heart justice is egalitarian; head justice can be so interpreted as to lead to an extreme form of elitism. Is there any way in which the two can be united in a common standard? I venture to propose three rules that point the way to a practical solution.

1. Insofar as possible, we should make widely and easily available a diversity of opportunity appropriate for the full development of the constructive talents represented in the population. There is an element of elitism in this rule, but talented people are a resource too valuable to be neglected.

2. Access to the various opportunities for jobs, education, or training should be based strictly on ability. Such factors as sex, race, age, family, and financial circumstances, and, insofar as they are not incompatible with the performance of useful work, physical and mental handicaps, should be irrelevant.

3. While opportunities may separate people, attitudes should unite them. Managers should treat workers as equals. They should feel that employees work with them, not under them. An example of the wrong approach is the provision of separate dining rooms for workers and management, a common practice in factories in the United States and Britain (19). To make Rule 3 work, we will need leaders whose attitudes are wholeheartedly democratic.

Japan has tended to operate its industries in conformity with these rules, and that has been a reason for its industrial success. In this country, our intellectual leaders and to some extent our business leaders are moving in the direction indicated. Two American writers who have discussed these problems are Rawls, in his *A Theory of Justice* (24), and Lodge in *The American Disease* (19). While their approaches to the problems are very different from mine, their proposed solutions are basically similar.

There is one problem area relevant to the administration of justice that will not be solved by the application of these rules alone. There appears to be a tendency, even in developed countries, for poverty and big families to go together. There are also reports of a rather high frequency of childbearing by women dependent for support on government aid. There is an injustice in these reproductive patterns since they create a need for financial rewards quite out of proportion to service rendered. I suspect that the application of our three rules would contribute to the solution of these problems, but other and even more difficult measures may be necessary.

We have treated equity as a moral principle without any defense of so doing, when in fact, at least in the case of head equity, some defense may be necessary. One valid reason for treating equity as part of moral law is that the great religions command it, but it also has three of the four properties that we have assigned to moral law. It is widely approved, it has generality, and the group benefits from its observance. As at present practiced, however, it is typically all benefit and no sacrifice for most of its practitioners. Putting a criminal in prison is, except for the cost, which is paid by way of taxes, essentially pure gain for the honest citizen. And distributive justice likewise, can bring generous rewards to the successful. These rewards may come at the price of long hours of hard work, but the real question for ethics is whether they also come at the price of placing the general welfare above personal gain. Insofar as our leaders enlarge their circle of consideration and think of the welfare of the many and not the few, then to that extent our practice of head equity will have become truly moral.

The Code of Enmity

In Chapter 5, we saw that it is characteristic of social species that they display cooperation within the social group, and competition, sometimes of a violent nature, between social groups. Man is a social species and a tendency to this dual pattern of behavior is part of his nature.

In this chapter, we are concerned not with the genetic or nature components of our social behavior, but with the cultural or nurture components. We are thus led to the question: Are there, as part of our culture, ethical principles that correspond to our intragroup cooperativeness and our intergroup competitiveness?

The answer, I suggest, is yes. We have already examined, in our discussion of the Code of Amity, cultural factors conducive to cooperation. But there are also elements in our culture that can call forth our combativeness, and these elements constitute the *Code of Enmity*. It was

Herbert Spencer who coined this expression, followed by Sir Arthur Keith and S. J. Holmes. I also have adopted from these authors the expression, the "ethics of amity" (or the Code of Amity). I differ from all three authors in that I have found it appropriate, for my purposes, to recognize several codes of ethics in addition to those of amity and enmity.

While the ethics of amity is present in our culture in a very explicit form, the teaching of enmity tends to be more implicit. It may, for example, appear as a glorification of military victories, or of acts of unusual bravery. It is common in this form in the Old Testament, but it occurs also as explicit exhortations to engage the enemy. Thus in Leviticus 26:7 we find "And ye shall chase your enemies, and they shall fall before you by the sword." Likewise in Deuteronomy, Chapter 20, verses 16–18 read, "But of the cities of these people, which the Lord thy God doth give thee for an inheritance, thou shalt save alive nothing that breatheth. But thou shalt utterly destroy them; namely, the Hittites, and the Amorites, the Canaanites, and the Perizzites, the Hivites, and the Jebusites; as the Lord thy God hath commanded thee: That they teach you not to do after all their abominations, which they have done unto their gods; so should ye sin against the Lord your God." Other nations of the Middle East during this period of history doubtless inculcated in their citizens a similar, extreme tribal enmity. Unfortunately, too much of it still persists today.

In the New Testament, there are no exhortations to war comparable to those in the Old Testament. On the contrary, we find phrases such as the well-known verse in the Beatitudes, "Blessed are the peacemakers: for they shall be called the children of God" (Matthew 5:9). But we also find Jesus saying: "Suppose ye that I am come to give peace on earth? I tell you, Nay; but rather division: For from thenceforth there shall be five in one house divided, three against two, and two against three" (Luke 12:51,52). Tragically, Christian behavior through the centuries has conformed more closely to this prediction of Jesus than it has to his praise of peacemaking. Christian sects and nations have fought many wars on religious grounds, or at least with religion as an excuse. The strain of enmity in human nature runs deep.

Amity is the opposite of enmity and, indeed, in one situation may be said to be the obverse—the opposite side of the coin. In warfare, aggression toward the enemy is accompanied by the sacrifice of lives for the community. Sometimes lives are given intentionally for a comrade: the ultimate act of altruism.

Fortunately, this capacity for altruism, which was born at least in part in the crucible of tribal warfare, need not be lost in peace. As we noted in Chapter 5, acts of heroic sacrifice are not altogether uncommon in peacetime.

Since enmity is so different from amity, it is appropriate to ask if the ethics of enmity really should be regarded as part of moral law. The answer is yes—the Code of Enmity, in at least some of its applications, does conform to our definition. Our basic teachings of enmity have generality; they are widely approved (though the degree of approval may fluctuate with circumstances); they demand sacrifice of the individual; and their observance can benefit the group. I use the phrase "can benefit" rather than "do benefit" advisedly. The one great difference between enmity and the other moral codes is that, whereas the observance of the other codes almost always benefits society, the adherence to enmity can, in some circumstance, prove to be a tragic mistake. Throughout the tribal stage of human evolution, war quite generally benefited the victor, and probably in most instances mankind as a whole. Nuclear war in this age, on the other hand, would be the ultimate disaster for all mankind. We cannot yet say categorically that all war is bad, but the risks have increased enormously.

The Code of Commonwealth

The *Code of Commonwealth* is that group of principles, both implicit and explicit, that is concerned with people consulting with one another to resolve conflicts and to create rules, compacts, alliances, and associations designed to advance the common good. It demands positive action by the group as a whole. Because of this it particularly depends on leadership. In situations in which the other moral codes fall short, either in content or observance, the Code of Commonwealth provides needed guidance and enforcement. It is the essence of democratic society.

The Code of Commonwealth has ancient roots. But of all the codes that we have considered, it is the one least ingrained in human nature, yet the one whose use has been subject to the greatest expansion in civilized, and especially in democratic, societies. As a general principle, it is an implicit rather than an explicit part of our culture. But while not explicitly stated in any formal injunction, many of its manifestations, such as constitutions, do provide formalized rules of consultation and law-making.

In primitive tribes the practice of Commonwealth took two forms: a gathering of tribal members for consultation and the formation of alliances. The word *powwow,* now part of the English language, attests the occurrence of American Indian tribal meetings. The formation of alliances between tribes appears to have been rather common. Chagnon has described such associations in the South American Yanomamö despite the relative aggressiveness of these people (4).

Perhaps the most remarkable alliance among primitive tribes that has been recorded was the Iroquois League. The Iroquois were a group of related tribes sharing a common language and living in central New York state. The League was formed in the late fifteenth century after the Iroquois had developed a well-established trade with white settlers, but there is no indication that the concept of the League was influenced by the whites. Its formation may, however, have been stimulated by the near extinction of beavers, the major source of pelts for trade in the area that the Iroquois occupied. Since the Iroquois had become very dependent on trade, the decline in the population of beavers in their territory became a source of pressure to conquer adjacent unexploited territories. The League at first included five tribes or "nations"; a sixth was added later. The structure of the League was remarkably sophisticated, with a provision for an annual meeting of delegates whose number and powers were precisely defined. Benjamin Franklin is said to have admired the political skill displayed in the design.

A major function of the League was to insure internal peace, but this security at home, in turn, was intended to permit aggressive action by the members against neighbors. And, indeed, over a period of decades, the Iroquois built by conquest an extensive empire. Their power was recognized by the English, who courted their friendship (10,22). Hoebel has described similar well organized and thoroughly democratic organizations in the Cheyenne Indians of the Great Plains (12).

The formation of associations is characteristic of, and indeed is the very basis of, all democratic societies. Whereas the formation of associations is generally repressed under absolute governments, it is the life blood of democracies. De Tocqueville, in his *Democracy in America,* testified to the number and diversity of these structures in the United States in the 1830s. He writes:

> "The political associations that exist in the United States are only a single feature in the midst of the immense assemblage of associations in that country. Americans of all ages, all conditions, and all dispositions constantly form associations. They have not only commercial and manufacturing companies, in which all take part, but associations of a thousand other kinds, religious, moral, serious, futile, general, or restrictive. . . . I have often admired the extreme skill with which the inhabitants of the United States succeed in proposing a common object for the exertions of a great many men and in inducing them voluntarily to pursue it" (5).

Of course many associations, both political and nonpolitical, are designed with some purpose other than advancing the common good. In fact some, such as the PACS concerned with raising money to aid the election of candidates whose political stand will profit the organizing group, may be primarily selfish. Others, such as those typical of organized crime, are wholly criminal. These, clearly, are not applications of the principle of commonwealth. Many others, however, are representative of the principle of commonwealth. The American Cancer Society and Ralph Nader's Public Citizen, to take two examples, certainly are intended to advance the general welfare. And always in democracies, it is the presumed function of governments at all levels to benefit the citizens they are designed to serve.

While there is no semblance of a "Law of Commonwealth" in the teachings of the great religions, there are two philosophical doctrines going back to the seventeenth and eighteenth centuries that are relevant to this ethical code. These are the concepts of *compact* and of *natural rights* both of which played an important role in the development of democracy in the United States.

Thomas Hobbes stated the principal of *compact* as follows: "That a man be willing, when others are so too, as farforth, as for peace, and defense of himself he shall think necessary, to lay down this right to all things; and be contented with so much liberty against other men as he would allow other men against himself" (11). Hobbes exaggerated the role of compact in the development of social groups, but the concept was esteemed and used by the Founding Fathers of the United States. Indeed, it was used in 1620 by the Pilgrims before their landing at Plymouth. At that time they drew up the Mayflower Compact, in which they bound themselves together "into a civil body politick, for (their) better ordering and preservation. . ." (3).

The concept of *rights* found its way both into the American Declaration of Independence and into the Constitution. In the Declaration, it appears in the statement, "We hold these truths to be self-evident, that all men are created equal, that they are endowed by their creator with certain unalienable Rights, that among these are Life, Liberty, and the pursuit of Happiness." The concept appears in the Constitution in the form of the Bill of Rights. One ethical philosopher, Virginia Held, places great emphasis on rights (9), and in the political arena, the concept is very much alive today. Thus we find Senator Edward Kennedy referring, in 1978, to health care as a "basic human right."

The concept of rights has a valid place in the Code of Commonwealth. Unfortunately it has been abused because of the tendency of politicians to give politically popular measures a legitimacy they do not necessarily deserve by referring to them as "rights." I know of no evidence that there is any such thing as a "natural" right, or a right with which people "are endowed by their Creator." Rights are strictly a *human* creation. Insofar as we possess them, it is because we have, through our government, established them as a fixed policy. The Bill of Rights, which places specific restrictions on the powers of government to interfere with individual liberty, represents a legitimate use of the concept. However, presumed rights, which place a demand on the government rather than a restriction (although in some cases they may be sufficiently broad in their benefits to warrant the title), certainly need to be scrutinized with great care and, if adopted, probably continually monitored.

Freedom was proclaimed by James Otis and Samuel Adams as a God-given right (26) and indeed it is one of our great blessings. However, one of the weaknesses of Commonwealth today is the extent to which freedom is abused. Russian novelist and expatriate Alexander Solzhenitsyn offered an interesting commentary on this in the acceptance speech given on his receipt in 1976 of the Amerian Friendship Award of the Freedom Foundation. In the speech, he listed fifteen abuses of freedom that he has observed in the United States (30). The full list deserves reading; I quote only two of his items.

"Freedom! to spit in the eye and in the soul of the passerby with advertising."

"Freedom! of indifference to a distant alien's trampled freedom."

A concept appropriate for inclusion in the Code of Commonwealth, and one that has not received the attention it deserves, is the need for finding the appropriate *mean* so greatly emphasized by Aristotle. We need to find the best balance—the "Golden Mean"—between freedom and planned order, between government-provided welfare and individual responsibility, between equality and the concentration of wealth coexisting with poverty. The concept of the mean has an appropriate place in any personal philosophy, but many of its most important though difficult applications are in government affairs.

Are There Other Moral Codes?

The books of Exodus, Leviticus, and Deuteronomy of the Old Testament contain many injunctions that clearly belong in one of the first four codes—the codes of Family, Amity, Equity, and Enmity—in our list. They also contain many other injuctions that do not belong in any of these four codes or in the Code of Commonwealth. Three examples are:

"Thou shalt have no other gods before me." (Exodus 20:3)

"And thou shalt make an altar to burn incense upon: of shitten wood shalt thou make it." (Exodus 30:1)

"For every meat offering for the priest shall be wholly burnt: it shall not be eaten." (Leviticus 6:23)

These rules, and the many others like them, concern either presumed supreme beings or the priesthood and the manner of worship of a supreme being. Their observance may have helped give the Israelites coherence and hence have returned a certain benefit to this one group of people, but they are not designed to benefit people in general. They are tribal law, not moral law. We might place them in a *Code of Diety* and a *Code of Piety,* but these two codes do not belong in the same ethical category as our other five.

While we can find no additional moral codes in religious teachings, there is potentially one other of such recent origin that its moral import is only beginning to be recognized. This is what we may call the *Code of Conservation.* In Chapter 3, when we discussed waste, we gave some indication of the potential importance of this code. Its development and implementation must be a major theme of any rational ethic.

Ethics and the Survival of Democracy

While the Code of Commonwealth is particularly relevent to the functioning of democracy, the successful application of all the codes is essential for democracy's success. Indeed, rational ethics and true democracy are dedicated to the same end—the general welfare—and the two are totally interdependent. In the long run, only the moral can be free. This is no novel concept. It was voiced by Samuel Adams more than 200 years ago when he wrote, "We may look up to Armies for our Defense, but Virtue is our best Security. It is not possible that any State should long remain free, where Virtue is not supremely honored" (1). According to Rossiter, this was the view of many Americans in the early years of this country (26). It is this principle above all others that must lead us to believe that self-interest rightly understood dictates the adoption and practice of high moral standards.

8

Looking to the Future

Introduction

The major thesis of this volume is that self-interest rightly understood leads us to socially responsible behavior. Moral action—action which serves the best interests of the group—follows naturally from an informed and reasoned self-interest.

In this final chapter, we summarize our conclusions in regard to the need for a more ethical society, and likewise, some conclusions in regard to the form a rationally designed ethic must take. We conclude by examining some of the conditions necessary if this ethic is to be put to work.

The Need for a More Ethical Society

Why should self-interest dictate ethical behavior? Can it be shown, on rational grounds, that people will be better off if they do what is right? The answer, I believe, is yes.

In Chapter 3, we examined seven patterns of behavior most people would agree are wrong because, in the long run, they hurt society and its members, and yet which many people condone and practice. These are patterns of behavior whose ill effect is worked, not on friends and neighbors, but on strangers. This is why people tolerate them. As an example, one of the seven wrongs we discuss is the encouragement by bankers, credit agencies, merchants, and manufacturers of buying on credit and often of buying the unneeded or the unaffordable. The result is a growth of personal debts, sometimes beyond the ability of the debtor to repay. Not only does the debtor suffer; the growth of debt is like the use of a drug— the whole social organism is hurt. Taken together, these seven wrongs account for many of the ills of our economy, and the magnitude of these ills may not yet be fully apparent.

Further evidence of the need for morality can be found in the thesis that democracy can survive only in an ethical society. This thesis, as we noted in Chapter 7, was widely esteemed in Colonial America. I believe that it can be documented by substantial evidence, but this is outside the scope of this volume. Here we merely can suggest some signs of its validity.

I probably need not argue the advantages of living under a democratic as contrasted with an autocratic government. The steady outflow of refugees from nations under absolute rule should be sufficient to convince anyone of this point. Millions of people have given a clear answer by "voting with their feet." We must preserve democracy, and to preserve democracy, we must be moral.

The Ethical Potential of Our Species

Nature has endowed the human species with a considerable but imperfect capacity for ethical conduct. This is the conclusion of Chapter 5, in which we examined the evolution of man's social nature. Nearly all this evolution took place when man and his ancestors were living in small social groups or tribes. Under these conditions, acts affected only friends or near neighbors—the circle of impact was limited.

Now all this is changed. In modern society, our acts, and especially the acts of our leaders, can influence, and influence profoundly, people remote and unknown. The circle of impact, and indeed also the potential magnitude of impact, has enlarged profoundly. The growth of technology is increasing the problem. Under these circumstances, our natural ethical tendencies, although a vital part of ethical behavior, no longer provide a sufficient basis for successful social living.

To solve this problem, we need to invoke the cultural component in our prosocial behavior—the moral principles that are a traditional part of all cultures and embedded in all the world religions. We need also to refine and expand these principles to make them adequate for today's world. The role to be assumed by the cultural as compared with the innate component in our social conduct will have to be enlarged to meet today's needs. However, we fortunately are endowed with two faculties that give us reason to hope that the task of making moral law work will not be beyond us. These faculties are reason and the moral sense or conscience.

While most primitive societies and most religions attribute moral law to some divine source (8), we have to believe today that this law was man-made. Reason has certainly always played a part in its design; our goal today should be to use our reason and our wealth of knowledge to design a truly rational ethic.

In Chapter 6, we discuss some of the properties of a rational ethic. One essential property is that it be widely accepted and applied. We must have pioneers in the realm of moral thinking. Martyrdom also has sometimes served moral ends. Nevertheless, for a small fraction of the population to attempt to carry the whole moral load would be counterproductive. To insure a substantial sharing of moral behavior, we must appeal to the human conscience. Conscience can be instructed; our approbations and disapprobations, at least in considerable part, are learned. And since people are, although in widely varying degrees, sensitive to group approval and disapproval, moral instruction can spread the load of moral observance.

Men and Women of Goodwill

In Chapter 4 we examined the genetic basis of social behavior. As we have already indicated, evolutionary studies indicate that we possess a natural endowment of social propensities. The importance of the genetic studies is the evidence they provide that, like all genetically determined traits, our moral propensities are variable. People are not either good or bad. Most of us are somewhere in the middle. However, at one extreme there are sociopaths who seem totally lacking in the moral sense and at the other extreme, people truly dedicated to good works.

This diversity in our moral tendencies makes it necessary to ask if all people would choose a moral society as representing the optimum for them personally. The answer clearly is no. For a gangster, the most unrewarding society would be a genuinely moral one. In fact, gangsters tend to move to areas where law enforcement is lax and where officials are easily corrupted.

We can conclude, then, that it is people with natural moral tendencies—men and women of goodwill—who benefit from a moral society. They are the ones who need kindness and honesty and who can be expected to collaborate to make moral law work.

There is an interesting parallel here with the role, noted in Chapter 5, of the practice of cognizant mutualism in the evolution of man's social nature. Cognizant mutualism was defined as mutual aid or cooperation that is selective; the cheat is intentionally excluded. The participants take note of the responsiveness of their associates and extend aid only to those who have demonstrated that they will extend aid in return. The practice of cognizant mutualism is found in all societies.

It is interesting to note that Jesus, in the parable of the Good Samaritan, recognizes the need for selectiveness in the choice of those to whom we apply the injunction, "Love thy neighbor as thyself." This subject is discussed in Chapter 7.

While a rational ethic must have a universal quality, it must also be restrictive. It is men and women of goodwill who need a rational ethic, and it is they who must make it work.

An Era of Renewal

We are living in a revolutionary age. Technology is transforming the physical aspects of our living, and there is a widely felt need for change in our education, our business practices, and our politics. Actual changes, however, have been modest. Can we reasonably hope for an era of genuine ethical renewal?

In Chapter 3 we cited the opinion of authorities that a major economic collapse may occur before the end of the 1980s. This is perhaps a minority view, and there can be no certainty in predictions of this sort, but we did find substantial evidence that supports it. History shows that it is in times of crisis that major changes in public thinking and major institutional developments occur. If we are prepared, the misfortune of hard times can be turned into an opportunity for social and moral advance.

Getting Ethics Applied

Ethical theory is meaningless unless it is put to work. What is required for the institution of a rational ethic?

There are, I suggest, three requirements that must be met if we are to have major ethical reform. The first is a crisis that will create a broad demand for reform. Such a crisis, as we have noted, may be at hand. The second is sufficient degree of consensus, especially among people with leadership potential, so that concerted action can be taken. The third is a program of education and institutional change. Consensus can establish the ends, but only education and appropriate institutions can provide the means.

The obstacles are substantial. There is no likelihood that the totalitarian powers will participate in any general consensus. I write, therefore, as in the rest of this volume, with the successful democracies and especially the United States in mind, though a broader consensus should be our ultimate goal. In this country, we are far from agreement on the need for a rational ethic, let alone on its philosophical background or the detailed instructions it should convey. Some people take a religious approach to ethical problems, some a humanistic approach. Each of these groups in turn is itself divided. There are many religions and religious sects to which people seeking religious guidance can turn. The humanists, likewise, appeal to different value systems.

Some confluence of thinking among the philosophically inclined is not impossible, but probably the best goal we can hope for as a general consensus is the welfare of mankind, including generations yet to come as well as those now living. If we are to get an agreement on working principles it will be, as suggested in Chapter 1, through the gathering of facts and the application of reason. It is by way of this route that men and women of goodwill the world around can hope ultimately to find broad common ground amidst the diversity of their self-interests.

While it is to the human reasoning faculty that we must appeal for a moral plan, it is the human conscience that must be invoked as the first step in its implementation. How, then, do we instruct our consciences? The best opportunity for reaching many people is through the schools. The media, the churches, and above all the home also form our moral standards for good or ill, but these molders of the public conscience are less easily attuned to a felt public need than are the schools.

The New England Puritans of the seventeenth century believed that "The purpose of education . . . was to prepare children to lead a moral and virtuous life" (4). This conception of education was still alive two centuries later. Douglas Sloan, in an article on this subject, writes: "Throughout most of the nineteenth century, the most important course in the college curriculum was moral philosophy taught usually by the college president, and required of all senior students. . . . The task was nothing less than to show that religion, science, and the human mind, all, *if rightly understood* [italics mine], revealed and contributed to the highest values of the individual and society" (6).

All this has now changed. In Chapter 2 we cited Benson and Engeman's *Amoral America,* which documents the decline in the teaching of morality in American schoolrooms (1). Sloan has noted the same trend and so also has the Thomas Jefferson Research Center (2). It should be possible to reverse this process and, indeed, there seems to be a growing feeling that this is a real necessity today.

There are three general approaches to moral instruction:

1. the formulation and presentation of clear and specific moral values;
2. the absorption of ethical concepts from reading matter that indirectly teaches moral standards; and
3. the encouragement of students to think ethically by the presentation and guided discussion of selected situations that raise moral issues.

Good example is of course also profoundly important, but although related to our second approach it is less a matter of instruction than are the three listed. The precise effectiveness of each of these methods probably depends on both the age and the personality of each student. How and when each should be used is a problem for the educational psychologist. I will not attempt to discuss it, but I believe that all three approaches have a place. What we can usefully do here is to briefly consider some steps that might be taken to put all three to use.

The teaching in the schools of ethical doctrines associated with a particular religion is limited in the United States by the First Amendment to the Constitution, which stipulates: "Congress shall make no law respecting an established religion, or prohibiting the free exercise thereof. . . ." The Supreme Court has interpreted this to mean that religion cannot be taught in public schools. Presumably this would not bar the teaching of ethical principles independently derived, even if they resembled recognized religious precepts (and in fact, some state universities offer courses in world religions).

To what source, then, should we turn for a group of laws or commandments that embody the best of today's ethical thinking? I am not aware of any humanist writings that provide the answer. The two Humanist "Manifestos" prepared by members of the organized humanist movement represent a start, but are widely controverted as they stand (7). Perhaps the best source for a specific formulation of moral law suitable for school use would be a convocation of respected citizens interested in ethics and its teaching. The group should be as broadly representative as would be compatible with reaching a consensus. It seems likely that such a group could draw up a list of simply stated moral principles that the courts would find acceptable.

Ethics can be taught not only through the use of explicit moral injunctions, but also implicitly through the use of appropriate reading matter. This was a common practice in American schools in the early days of the republic. Most children now get little or none of this sort of reading matter. Children's television programs also are mostly devoid of any moral content. Indeed there is evidence that the frequent scenes of violence incite some youngsters to crime (2), and that the general effect of the programs is amoral (3).

The third method of teaching ethics—the planned stimulation of ethical analysis by the students themselves—has been put to use recently in a number of schools and seems to have achieved considerable success. It is reported to have reduced disciplinary problems ranging from vandalism to neglect of studies. The material used in some recent applications of this method has been prepared by the American Institute for Character Education in San Antonio, TX and is distributed by the Thomas Jefferson Research Center (1143 N. Lake Ave., Pasadena, CA 91104).

This last method of teaching ethics is the most intellectual of the three methods I have listed. If any method is appropriate to older and more advanced students, this should be it. It has been used by professors of moral philosophy in colleges; it has also been tried out on dental students, though in this preliminary work the intent was more to gauge than to influence moral thinking (5). The author found that when cases that raise ethical issues are presented, "some students are so preoccupied with the technical aspects of the case (prescribing the correct bridgework, for instance) that the problem of values is hardly recognized." Perhaps we need to introduce this form of teaching in schools of business, law, and medicine.

Rest, in the same article in which this study of dental students is summarized, looks at some of the broader aspects of teaching ethics (5). He concludes that education can have a powerful effect on the development of moral judgment, but that our knowledge of how this effect is produced is inadequate. He argues for more collaboration between ethics instructors and psychologists. He also notes the variability in our moral natures and the presence of differences between individuals both in the capacity for reasoned moral judgment and in the spontaneous tendency to weigh the possible consequences of acts on the welfare of others. This suggests that, ideally at least, the teaching of moral principles should not conform to a rigid pattern, but should be adjusted to the individual or group being taught.

While moral instruction, working through conscience as the enforcer, is an important part of the socialization process, it alone is not enough. Reward and punishment based on appropriate rules and regulations are also essential.

To expect all this of our teachers would be asking a great deal. We already ask that they not only maintain discipline and import the familiar subjects of the school curriculum, but that they help deal with some of our social problems. The teaching profession deserves high respect in any case, but if teachers are to be asked also to instill moral standards, they should be respected and rewarded accordingly. This is the situation of teachers in Japan, where the educational process is eminently successful by any standard and where moral instruction has been an accepted part of education for many years. In Japan, "teachers enjoy respect and high status, job security, and good pay. (They) are hired for life, at starting salaries equivalent to starting salaries for college graduates in the corporate world" (9).

The major function of moral law and moral education is to instruct the conscience, which then becomes an enforcing agent. In all societies, however, punishment or the threat of punishment is also used as an enforcer (8). In modern societies, the state acts both in this role and as a creator of juridical laws, many of which profoundly affect our lives. Under these circumstances the nature of state institutions assumes great significance. While this makes governmental design an appropriate concern of the ethicist, it is outside the scope of this volume.

This volume has been dedicated to the assembling of facts appropriate to the development of an informed and well reasoned self-interest. What does self-interest, thus instructed, tell us? To each of us, there is one basic message that it delivers.

If you love liberty, prefer honesty to deceit, value gentleness and kindness in your fellow man, cherish family and friends, and glory in the beauty of the world and the wonderful diversity of its creatures, then for the sake of yourself and your children and all that is good in the human spirit, join with like-thinking men and women everywhere to build, on the foundations of knowledge and reason, a future where goodness and love shall prevail.

References

The shortened forms of publishers names used in our list of references are the forms given in *Books in Print*. The reader who wishes a publisher's full name and address should consult the key in *Books in Print*.

Chapter 1—Introduction

1. Dawkins, Richard. 1976. The selfish gene. Oxford U Pr, New York.
2. de Tocqueville, Alexis. 1835. Democracy in America, Vol. II. (Translated from the French.) 1945 edition, Knopf, New York.
3. Lippmann, Walter. 1929. A preface to morals. Macmillan, New York.
4. Singer, Peter. 1981. The expanding circle: Ethics and sociobiology. FS & G, New York.

Chapter 2—A Revolutional Age

In addition to the references given below, in discussing current events I have used material from newspapers and magazines that have not been specifically cited.

1. Adams, James T. 1936. The living Jefferson. Scribner, New York.
2. Adams, Thomas B. 1981. A new nation. Globe Pequot, Chester, CT.
3. Bailyn, Bernard. 1967. The ideological origins of the American Revolution. Belknap Pr, Cambridge, MA.
4. Barzun, Jacques. 1941. Darwin, Marx, Wagner. Little, Boston.
5. Bell, Daniel. 1973. The coming of post-industrial society. Basic, New York.
6. Benson, George C.S., and T.S. Engeman. 1975. Amoral America. Hoover Inst. Pr, Stanford, CA.
7. Berlin, Isiah. 1963. Karl Marx, his life and environment, 3rd ed. Oxford U Pr, New York.
8. Bredin, H. 1982. Unmanned manufacturing. Mechanical Engineering (Feb.), pp. 20–25.
9. Carson, Rachel. 1962. Silent Spring. HM, Boston.

10. Chandler, Alfred D., Jr. 1977. The visible hand: The managerial revolution in American business. Belknap Pr, Cambridge, MA.

11. Darlington, Cyril D. 1969. The evolution of man and society. Allen Unwin, London.

12. Darwin, Charles. 1874. The descent of man, 2nd ed. 1898 printing, Appleton, New York.

13. de Tocqueville, Alexis. 1856. The old régime and the French Revolution. (Translated from the French) Doubleday, New York.

14. Durant, Will, and A. Durant. 1975. The story of civilization, XI. The age of Napoleon. S & S, New York.

15. Duus, P. 1982. Japan—taking off. Wilson Quart. (Winter), pp. 109–121.

16. Fisher, George P. 1873. The Reformation. Scribner, New York.

17. Fisher, Herbert A.L. 1939. A history of Europe. HM, Boston.

18. Fromm, Erich. 1961. Marx's concept of man. 1961. Ungar, New York.

19. Heer, Friedrich. 1962. The medieval world. Europe 1100–1350. (Translated from the German) World Publishing Co., Cleveland.

20. Hexter, J.H. 1984. One humanist at the Humanities Center. Natl. Humanities Center Yearbook 6 (No. 1):1–6.

21. Hobbes, Thomas. 1651. Leviathan (selections). 1955 edition, Great Bks Foundation, Chicago.

22. Hobsbawn, Eric J. 1962. The age of revolution. Europe 1789–1848. Weidenfeld and Nicolson, London.

23. Hoffstadter Robert. 1944. Social Darwinism in American thought, 1860–1915. U of Pa Pr, Philadelphia.

24. Landes, David. 1969. The unbound Prometheus: Technological change and industrial development in western Europe from 1750 to the present. Cambridge U Pr, New York.

25. Laszlo, Ervin, et al. 1877. Goals for mankind. Dutton, New York.

26. Lewontin, R.C., et al. 1984. Not in our genes: Biology, ideology, and human nature. Pantheon, New York.

27. Lodge, George C. 1984. The American disease. Knopf, New York.

28. Maccoby, Michael. 1976. The gamesman. S & S, New York.

29. Morley, John. 1900. Oliver Cromwell. The Century Co., New York.

30. Nader, Ralph. 1965. Unsafe at any speed. Grossman, New York.

31. Needham, Joseph. 1981. Science in traditional China. Harvard U Pr, Cambridge, MA.

32. Nef, John V. 1964. The conquest of the material world. U Of Chicago Pr, Chicago.

33. Packard, Vance. 1960. The wastemakers. McKay, New York.

34. Palmer, Robert R. 1959. The age of the Democratic Revolution. A political history of Europe and America, 1760–1800. Princeton U Pr, NJ.

35. Quillian, William F., Jr. 1945. The moral theory of evolutionary naturalism. Yale U Pr, New Haven, CT.

36. Rossiter, Clinton L. 1953. Seedtime of the Republic. HarBrace, New York.

37. Sarton, George. 1948. The life of science: Essays on the history of civilization. Henry Schuman, New York.

38. Schevill, Ferdinand. 1940. A history of Europe. HarBrace, New York.

39. Singer, Peter. 1981. The expanding circle: Ethics and sociobiology. FS & G, New York.

40. Spencer, Herbert. 1892. Principles of ethics. Williams and Norgate, London.
41. Stephenson, Carl. 1935. Mediaeval history. Europe from the fourth to the sixteenth century. Har-Row, New York.
42. Tarkunde, V. M. 1980. Towards a fuller consensus in humanist ethics. Pages 154–171 *in* Morris B. Storer, Ed. Humanist ethics: Dialogue on basics. Prometheus Bks, Buffalo, NY.
43. Weber, Max. 1959. The Protestant ethic and the spirit of capitalism. (Translated from the German) Scribner, New York.
44. Webster, Charles. 1976. The greast instauration: Science, medicine and reform, 1626–1660. Holmes and Meier, New York.
45. Wertime, Theodore A., and J.D. Nuckly, Eds. 1981. The coming of the Age of Iron. Yale U Pr, New Haven, CT.
46. Yankelovich, Daniel. 1981. New rules. Random, New York.

Chapter 3—What's Gone Wrong

A. INTRODUCTION

1. Etzioni, A. 1986. Founding a new socioeconomics. Challenge (Nov.–Dec.), pp. 13–17. See also article by Eicher and interview with Simon in the same issue of Challenge.
2. Hirsch, Fred. 1976. Social limits to growth. Harvard U Pr, Cambridge, MA.
3. Lodge, George C. 1984. The American disease. Knopf, New York.
4. Magnum, G.L., et al. 1987. The three (at least) worlds of economic theory. Challenge (March-April), pp. 57–59.
5. McPherson, M.S. 1983. Want formation, morality and some "interpretive" aspects of economic inquiry. *In* Normal Haan, et al., Eds. Social science as moral inquiry. Columbia U Pr, New York, pp. 96–124.
6. Rhodes, Steven E. 1985. The economist's view of the world: Government, markets, and public policy. Cambridge U Pr, New York.
7. Smith, Adam. 1759. The theory of moral sentiments. 1982 edition, Liberty Classics, Indianapolis, IN.
8. Steppacher, Rolf, et al., Eds. 1977. Economics in institutional perspective. Lexington Bks, Winchester, MA.
9. U.S. News and World Report, Feb. 21, 1983. A conversation with Derek Bok, p. 83.
10. Walker, Amasa. 1969. The science of wealth: A manual of political economy. Kraus Reprint Co., New York. (Reprint of seventh edition, Little, Boston. 1874.)
11. Wash. Post Natl. Weekly, June 23, 1986, pp. 6–7.
12. Wash. Post Natl. Weekly, May 11, 1987, pp. 6–7.

B. WASTE

Material from a few newspaper and magazine articles has been used in this section in addition to material from the sources cited.

1. Barnet, Richard J. 1980. The lean years. Politics in an age of scarcity. Institute for Policy Studies, Washington, DC.
2. Boslough, J. 1981. Rationing a river. Science 81 (June), pp. 26–37.
3. Brody, H. 1987. Energy-wise buildings. High Tech. (Jan.), pp. 36–39.
4. Brown, L.R. 1978. The global economic prospect: New sources of economic stress. Worldwatch Paper 20, Worldwatch Inst., Washington, DC.
5. Brown, L.W. 1981. Interview in US News World Report (Nov. 2), p. 58.
6. Brown, Lester R. 1981. Building a sustainable society. Norton, New York.
7. Brown, Lester R., et al. 1987. State of the world 1987. Norton, New York.
8. Canby, T.Y. 1980. Our most precious resource; water. Natl. Geogr. 158 (no. 2), pp. 144–179.
9. Ceron, J.P. 1977. Accelerated obsolesence or increased durability: The need for a public policy. *In* Rolf Steppacher et al., Eds. Economics in institutional perspective. Lexington Bks, Winchester, MA.
10 Coffin, T. 1982. Looking ahead: Some answers on food and fuel. Washington Spectator (May 15).
11. DuPont Today Vol. 3, No. 1 (1982).
12. Epstein, Samuel, et al. 1982. Hazardous waste in America. Sierra Club, San Francisco.
13. Fields, Shirley F., et al. 1979. Where have the farmlands gone? National Agricultural Lands Study. U.S. Government Printing Office, Washington, DC.
14. Fuller, R. 1980. Inflation: the rising cost of living on a small planet. Worldwatch Paper 34, Worldwatch Inst., Washington, DC.
15. Gabor, Dennis and U. Colombo. 1978. Beyond the age of waste. A report of the Club of Rome. Pergamon Pr, New York.
16. Georgiou, G. 1987. Oil market instability and a new OPEC. World Policy J. (Spring), pp. 195–311.
17. Hibbard, W.R., Jr. 1968. Mineral resources: Challenge or threat? Science 160:143–149.
18. Hirst, E., and J.C. Moyers. 1973. Efficiency of energy use in the United States. Science 179:1299–1304.
19. Kapp, Karl W. 1950. The social costs of private enterprise. Harvard U Pr, Cambridge, MA.
20. Landsberg, H.H. 1985. A time for every purpose: The case of energy. Issues in Sci. and Tech. (Winter), pp. 14–27.
21. Lang, J.S. 1987. The disposable society. US News World Report (June 1), p.68.
22. Laszlo, E., et al. See Chapter 2.
23. Lubow, V. (Cited by Packard).
24. MacDonald. 1980. Detroit 1985. Doubleday, New York.
25. Murdoch, William W., Ed. 1971. Environment; resources, pollution, and society. Sinauer Assocs, Stanford, CT.
26. National Georgraphic. 1981. Energy. Nat. Geogr. Special Report (Feb. 1981).
27. Neary, J. 1981. Pickleweed, Palmer's grass, and Saltwort. Science 81 (June), pp. 39–43.
28. Packard, Vance. 1960. See Chapter 2.
29. Page, Talbot. 1977. Conservation and economic efficiency. Johns Hopkins U Pr. Baltimore.

30. Peipert, J.R. 1984. Denuding of land plays part in African famine. AP in Bangor Daily News, Nov. 28.
31. Pimental, D., et al. 1976. Land degredation: Effects on food and energy resources. Science 194:149–155.
32. Pimental, D. 1982. The vanishing land. *In* Issues and studies, 1981–1982. The National Research Council, Natl. Acad. Sci., Washington, DC, pp. 104–105.
33. Porter, G. 1983. A watchful eye is best tactic as CO_2 rises over next century. News Report (March), Natl. Acad. Sci., Washington, DC, pp. 8–13.
34. Rapaport, R. 1981. Garbaeology. Science 81 (May), pp. 76–78.
35. Reisner, M. 1981. Are we headed for another "Dust Bowl"? Read. Dig. (May), pp. 87–92.
36. Rogers, P. 1986. Water. Not as cheap as you think. 1986. Tech. Rev. (Nov.-Dec.), pp. 31–43.
37. Royston, M.G. 1980. Making pollution prevention pay. Harvard Bus. Rev. (Nov.–Dec.), pp. 6–22.
38. Seisler, J. 1987. Home-heating advances save energy dollars. High Tech. (March), pp. 57–59.
39. Sheets, K.R. 1981. Is U.S. paving over too much farmland? US News World Report (Feb. 2), pp.47–48.
40. Sheets, K.R. 1981. Water: Will we have enough to go around? US News World Report (June 29), pp. 34–37.
41. Spurr, S.H. 1979. Silviculture. Sci. Am. 240 (No. 2):76–91.
42. Stobaugh, R., and D. Yergin, Eds. 1979, Energy future. Random, New York.
43. U.S. Bureau of Mines. Nov. 1976. Natural gas annuals. Salient statistics of national gas in the U. S. (Section XIII, Table 5).
44. Walter, S. 1982. Oil from rock. Eng. News (Feb.), pp. 32–41.
45. Weisskopf, M. 1986. Plastic: The waterborne plague. Wash. Post Natl. Weekly (Dec. 29), p. 7.
46. Zweibel, K. 1987. Photovoltaics: A worthy investment. High Tech. (May), p. 8.

C. Misguidance in the Marketplace

Material from some newspaper and magazine articles has been used in this section in addition to material from the sources cited.

1. Comptroller General. 1981. Consumer products advertised to save energy—let the buyer beware. U.S. General Accounting Office, Washington, DC. Document No. HRD-81-85-115899.
2. Elkind, David. 1984. All grown up and no place to go. Addison-Wesley, Reading, MA.
3. Gibney, Frank. 1960. The operators. Har-Row, New York.
4. Hiaasen, C. (Knight-Ridder Newspapers). 1987. Why newspapers display cigarette advertising. Bangor Daily News, Jan. 15.
5. Kapp, Karl W. 1950. See Chapter 3, Section B.
6. Laszlo, Ervin, et al. 1977. See Chapter 2.
7. Lichty, L.W., et al. 1981. Television in America. Wilson Quart. (Winter), pp. 53–101.

8. MacDonald, Donald. 1980. See Chapter 3 under Waste.

9. Mayer, Martin. 1958. Madison Avenue, U.S.A. Har-Row, New York.

10. Megin, Elliot. 1985. Just a pinch between your cheek and your gum. Public Citizen (Spring) pp. 28–32.

11. Reeves, R. Reality in advertising. 1961. Knopf, New York.

12. Schrank, Jeffrey. 1975. Deception detection. Beacon Pr, Boston.

13. Stellwagon, Lindsey D. 1982. Consumer fraud. Natl. Inst. of Justice, U.S. Dept. of Justice.

14. US News World Report, Sept. 22, 1980. Interview with Stephen A. Newman, expert on consumer affairs. Consumer ripoffs—how to protect yourself.

D. DEBT ENCOURAGED

Material from a number of newspaper and magazine articles has been used in this section in addition to material from the sources cited. The use of uncited sources is greatest in the discussion of the encouragement of debt.

1. Alm, R., and R.J. DeLouise. 1985. For some real inflation look at the third world. US News World Report (July 15), pp. 49–51.

2. Anderson, J. 1986. Helms hearings expose corruption in Mexico. Bangor Daily News, June 18.

3. Associated Press. 1980. Family finances reported grim. Bangor Daily News, Feb. 23.

4. Associated Press. 1985. U.S. posts worst trading year ever. Bangor Daily News, Jan. 31.

5. Associated Press. 1985. U.S. owes more to foreigners than they owe us. Bangor Daily News, March 18.

6. Associated Press. 1985. Argentina building an undersea navy. Portland Press Herald, Sept. 11.

7. Associated Press. 1986. Charges of capital flight put Mexico on defensive. Bangor Daily News, June 13.

8. Associated Press. 1987. United Nations report warns Latin American nations face dire deterioration. Bangor Daily News, Jan. 17.

9. Associated Press. 1987. Banks show net loss because of foreign loans. Bangor Daily News, June 16.

10. The Bank Credit Analyst, May 1985. BCA Publications Ltd., Montreal, Canada.

11. Bartlett, Sara, et al. 1987. The home equity gold rush. Business Week (Feb. 9), pp. 64–70.

12. Bork, Robert H., Jr., et al. 1987. Latin America's new dance of debt. US News World Report (March 16), p. 55.

13. Bremmer, A.F. 1987. Dim prospects for the federal budget deficit. Challenge (May–June), pp. 58–59.

14. Brinner, R. E., and K.J. Kline. 1985. What caused inflation to collapse? A new market realism: Wage moderation. Challenge (Sept.–Oct.), pp. 27–29.

15. Brzoska, M. 1987. The accumulation of military debt. Pugwash Newsletter (April), p. 99.

16. Bureau of the Census, U.S. Dept. of Commerce. 1979. Statistical Abstracts of the U.S., 1978, p. 543.

17. Campbell, Colin D. 1978. An introduction to money and banking, third edition. Dryden Pr, Hinsdale, IL (The material I have used appears on pp. 42, 100, 255, 261, 266, 278.)

18. Cardoso, E.A. 1987. Latin American debt: Which way now? Challenge (May–June), pp. 11–17.

19. Chicago Tribune. 1985. Congress takes note of credit card costs. Bangor Daily News, Oct. 2.

20. Congressional Budget Office. 1985. The economic and budget outlook: Fiscal years 1986–1990.

21. Dale, R.S. 1983. International banking is out of control. Challenge (Jan.–Feb.), pp. 14–19.

22. Durkin, T.A., and G.E. Elliehausen. 1978. 1977 consumer credit survey, Board of Governors, Fed. Reserve Sys.

23. Foweraker, J. 1987. What's good for Citicorp . . . Challenge (Jan.–Feb.), pp. 47–50.

24. Galbraith, John K. 1958. The affluent society. HM, Boston.

25. General Motors Acceptance Corp. 1979. Annual Report.

26. Greenspan, A. 1985. Talk at annual meeting of Governors, Feb. 1985, cited by David Broder, Sarasota Herald, Feb. 27.

27. Guenther, R., et al. 1985. Commercial properties encounter hardships, hurting many S&Ls. Wall Street J., Aug. 8.

28. Heinemann, H.E. 1983. Issue in Polish debt: How to borrow again. New York Times (June 17).

29. Hurley, E.M. 1976. Survey of finance companies, 1975. Fed. Reserve Bull. (March), pp. 197–207.

30. Internatl. Monetary Fund. 1981. Annual report. Internatl. Monetary Fund, Washington, DC.

31. Jarvis, P. 1985. Big consumer debt may be future problem. Bangor Daily News, Aug. 6. (Jarvis cites as his source the Bank Credit Analyst (9). A comparison of a chart on consumer installment debt in this publication with figures on the gross national product shown in the Economic Report of the President for 1982 confirm Jarvis's statement.)

32. Kraft, J. 1985. The mystery of the mighty dollar is fraught with unknowns. Tampa Tribune-Times, Feb. 17.

33. Mayer, Martin. 1974. The bankers. Weybright and Tally, New York.

34. MacDonald, D. 1980. See Chapter 3 under Waste.

35. McClain, D. 1985. What caused inflation to collapse? Stabilizing oil and farm prices holds the key. Challenge (Sept.–Oct.), pp. 23–26.

36. Migdail, C.J. 1985. After Mexico's vote, more headaches for the U.S. US News World Report (July 22), pp. 37–38.

37. Minsky, H.P. 1985. Money and the lenders of last resort. Challenge (March–April), pp. 12–18.

38. Moore, B.J. 1981. Is the money stock really a control variable? Challenge (July–Aug.), pp. 43–46.

39. Neikirk, W. 1985. Debt is most important U.S. commodity. Bangor Daily News, Aug. 3, citing the Chicago Tribune.

40. Packard, Vance 1960. See Chapter 2.

41. Porter, S. 1978. Saving too little? Bangor Daily News, Dec. 22.

42. Pugwash Symposium. 1987. Foreign debts and international stability. Pugwash Newsletter (April), pp. 91–105.

43. Ross, N.L. 1986. A man's home may become his credit card. Wash. Post Natl. Weekly (Aug. 4), p. 18.

44. Rowen, H. 1987. Facing facts. Wash. Post Natl. Weekly (June 8), p. 5.

45. Sanders, J.W. and S.R. Schwenninger. 1986–87. The democrats and a new grand strategy. Part II. World Policy J. (Winter), pp. 1–50.

46. Scherschel, P.M. 1984. Global financiers: The jet set of banking fraternity. US News World Report (May 7), pp. 62–63.

47. Silk, L. 1980. Economic scene. What causes inflation? New York Times, Dec. 3.

48. Smith, Adam. 1759. See Chapter 3 under Introduction.

49. Thornton, J. 1981. Behind the surge of personal bankruptcies. US News World Report (May 25), pp. 87–88.

50. Time, Oct. 30, 1964. The importance of being in debt.

51. US News World Report, July 29, 1985. Foreigners put their money on America.

52. Von Hoffman, N. 1986. Who really pays for those bad foreign loans? Bangor Daily News, April 11.

53. World Bank. 1987. Developing country debt. The World Bank, Washington DC.

E. Unneeded Wealth Misused

Material from a few newspaper and magazine articles has been used in this section in addition to material from the sources cited.

1. Adams, W. 1982. Mega-mergers spell danger. Challenge (March–April), pp. 12–17.

2. Alm, R. 1984. Impact of the new surge in research dollars. US News World Report (Oct. 8), pp. 45–48.

3. Associated Press. 1979. Lunch is loser to booze, drugs. Bangor Daily News, Nov. 24.

4. Associated Press. 1984. Tax cuts not used for investment, study finds. Bangor Daily News, Jan. 28.

5. Associated Press. 1986. Japanese corporations and research at U.S. universities. Bangor Daily News, March 22.

6. Associated Press. 1987. Corporations accused of mishandling billions of dollars in retirement funds. Bangor Daily News, March 25.

7. Baldridge, Malcolm. 1986. Remarks before the Center for Strategic and International Studies, Nov. 5. Dept. of Commerce.

8. Behar, B., and J. Block. 1984. The 400 richest people in America. Forbes (Oct.), pp. 69–186.

9. Break, George F., and J.A. Pechman. 1975. Federal tax reform: The impossible dream? Brookings Inst., Washington, DC.

10. Broder, D.S. 1987. The high road to lower finance? Wash. Post Natl. Weekly (June 21).

11. Brophy, Beth, et al. 1986. Middle-class squeeze. US News World Report (Aug. 18), pp. 36–41.
12. Brown, A.L., and G.A. Dancke. 1987. The rising electronic sun: Japan's photovoltaics industry. Issues in Sci. and Tech. (Spring), pp. 69–77.
13. Buchele, R. 1984. Reaganomics and the fairness issue. Challenge (Sept.–Oct.), pp. 25–31.
14. Bucy, J.F. 1985. Meeting the competitive challenge: The case for R&D tax credits. Issues in Sci. and Tech. (Summer), pp. 69–78.
15. Centocor. 1984, 1986. Annual Reports. Also interview with Michael Wall, Board Chairman.
16. Common Cause. 1979. How money talks in Congress. Common Cause, Washington, DC. See also Federal Election Commission, Release of Jan. 17, 1983.
17. Coburn, Don. 1986. The millions without health insurance. Wash. Post Natl. Weekly (July 21).
18. Congressional Budget Office. 1987. The economic and budget outlook: Fiscal years 1988–1992.
19. Currier, C. 1982. Why the bulls took over Wall Street. Bangor Daily News, Nov. 5.
20. Doan, M. 1981. There's no recession in the luxury market. US News World Report (Nov. 23), pp. 53–54.
21. Drumtra, J. 1985. Taxes push poor below the poverty line. Public Citizen (Oct.), pp. 6–7.
22. Edmondson, Brad. 1987. Who gives what to charity? Am. Demographics (Nov.).
23. Field, T.F. 1981. Estate and gift taxes come under heavy attack. Taxation with Representation Newsletter (May).
24. Forbes, Special Issue. 1985. The Forbes four hundred.
25. Gabor, A. 1987. Here come the venture capitalists—again. US News World Report (May 18), pp. 52–53.
26. Gest, T., and P.M. Scherschel. 1985. Stealing $200 billion "the respectable way." US News World Report (May 20), pp. 83–85.
27. Guttmann, P.M. 1979. Statistical illusions, mistaken policies. Challenge (Nov.-Dec.), pp. 14–17.
28. Hale, J. 1985. Speaker extols "venture capital clubs." Bangor Daily News, Oct. 18.
29. Kilborn, P.T. (New York Times News Service). 1983. Americans now saving less. Bangor Daily News, Sept. 7.
30. Lear, N. 1987. Our Babylon is a tower of greed and gratification. Wash. Post Natl. Weekly (April 20).
31. MacNeil/Lehrer News Hour, Public Broadcasting System, April 8, 1987.
32. McIntyre, R.S. 1987. More tax cuts for the rich? Wash. Post Natl. Weekly (Feb. 23).
33. Mashek, J.W., and K. Johnson. 1985. Will money sway fight on tax reform? US News World Report (Sept. 23), p. 24.
34. Meyer, Richard. 1985. Running for shelter. Tax shelters and the American economy. Public Citizen, Washington, DC.
35. Michel, R.C., et al. 1984. Are we better off in 1984? Challenge (Sept.–Oct.), pp. 10–17.

36. Moynihan, D.P. 1985. We can't avoid family policy much longer. Challenge (Sept.–Oct.), pp. 9–17.

37. National Science Foundation. 1987. Human talent for competitiveness. N.S.F.

38. Pechman, J.A., and M.J. Mazur. 1984. The rich, the poor, and the taxes they pay: An update. The Public Interest (Fall), pp. 28–36.

39. Pechman, J.A., and B.A. Okner. 1972. Individual income tax erosion, by income classes. *In* The economics of federal subsidy programs. Brookings Institute (reprint 230), pp. 13–40.

40. Peters, Thomas J., and R.W. Waterman, Jr. 1982. In search of excellence. Warner, New York.

41. Pifer, A. 1987. Philanthropy, voluntarism, and changing times. Daedalus (Winter), pp. 119–131.

42. Porter, M.P. 1987. From competitive advantage to corporate strategy. Harvard Bus. Rev. (May–June), pp. 43–59.

43. Rich, S. 1986. Are you really better off than you were 13 years ago? Wash. Post Natl. Weekly (Sept. 8).

44. Rosenberg, C. 1984. Joe Kennedy's energy firm mixes idealism, business savy. UPI article in Bangor Daily News, Dec. 15.

45. Rottenberg, D. 1986. The most generous living Americans. Town and Country (Dec.).

46. Ross, M. 1979. Giving and getting. Dartmouth Alumni Mag. (Dec.).

47. Rowen, H. 1987. Why Darman bailed out. Wash. Post Natl. Weekly (May 4).

48. Scherschel, P.M. 1984. The comeback of risk takers: They're reshaping business. US News World Report (Sept. 24), pp. 60–62.

49. Scherschel, P.M., and R.F. Black. 1985. Helping fund a takeover? You could get burned. US News World Report (May 13), pp. 75–76.

50. Schmitt, R.W. 1985. Engineering research and international competitiveness. High Tech. (Nov.), pp. 13–14.

51. Schorsch, L.L. 1987. Can big steel change bad habits? Challenge (July-Aug.), pp. 32–40.

52. Scripps Howard News Service. 1987. Quality of U.S. products scorned. Bangor Daily News, Aug. 20.

53. Seaberry, J. 1987. Been up so long it looks like down to me. Wash. Post Natl. Weekly (Jan 19).

54. Senator George Mitchell. 1987. Analysis of federal tax burdens prepared by Congressional Budget Office at Mitchell's request.

55. Sherrid, P. 1986. Millions for megadeal makers. US News World Report (Jan. 20), pp. 45–47.

56. Sherrid, P. 1986. Making the best deals their own. US News World Report (June 30), p. 47.

57. Sherrid, P. 1986. Wall Street's junkyard jitters. US News World Report (Dec. 8), pp. 52–53.

58. Sloan, A. 1984. Luring banks overboard. Forbes (April), pp. 39–43.

59. Statistical Abstracts of the U.S. 1986. Bureau of the Census, Washington, DC.

60. Stern, Philip M. 1973. The rape of the taxpayers. Random, New York.

61. US News World Report, June 3, 1978. Middle managers taste the bitter fruit of inflation, pp. 49–51.

62. US News World Report, Dec. 24, 1979. Wallets open up for risky ventures, pp. 75–76.
63. US News World Report, Aug. 13, 1984. New poverty count stirs political pot, p.8.
64. US News World Report, July 29, 1985. Foreigners put their money on America, p. 45.
65. US News World Report, March 2, 1987. GM's bonus babies.
66. Vickrey, W. 1975. Private philanthropy and public finance. *In* Edmund S. helps, Ed. Altruism, morality, and economic theory. Russel Sage Foundation, New York, pp. 149–169.
67. Vise, D., and S. Coll. 1987. The scandal won't be over till the traders stop singing. Wash. Post Natl. Weekly (May 25).
68. Ward, B. 1982. The greatest American dream. Sky (June), pp. 84–89.
69. Wash. Post Natl. Weekly, March 2, 1987. A paid-for Congress.
70. Wiener, L., and R.J. Morse. 1985. The swelling ranks of U.S. millionaires. US News World Report (March 18), p. 53.
71. Wingo, W.S., and R.J. Morse. 1982. Executives' pay goes up, up and away. US News World Report (May 24), pp. 59–61.
72. Work, C.P., and G. Bronson. 1985. After the takeover: How raiders perform. US News World Report (Sept. 9), pp. 72–73.
73. Work, C.P., and R.J. Morse. 1985. Executive pay goes sky-high. US News World Report (April 22), pp. 60–65.
74. Work, C.P., and S. Peterson. 1985. The raider barons: Boon or bane for business? US News World Report (April 8), pp. 51–54.
75. Work, C.P., and J. Seamonds. 1985. What are mergers doing to America? US News World Report (July 22), pp. 48–50.

F. Security Sought through Power or Privilege

Most of my sources for this section have been newspaper and magazine articles, e.g., US News World Report, July 7, 1980, p. 45; Oct. 6, 1980, p. 96; Aug. 17, 1981, p. 33; March 2, 1981, p. 59; Nov. 1, 1982, pp. 51–55; Dec. 13, 1982, p. 85; Dec. 20, 1982, pp. 58–61; June 20, 1983, pp. 74–77. I have also, with respect to lawyers and law suits, used the publications of HALT (Help Abolish Legal Tyranny), e.g.: Talentam, V. 1982. ABA, Harvard hold conference on dispute resolution, Americans for Legal Reform, Vol. 3, Number 3.

1. Bok, D.C. 1983. A flawed system. Harvard Mag. (May–June), pp. 38–71.
2. Burger, W. 1984. Address at the Midyear Meeting of the American Bar Association. Americans for Legal Reform 4 (Spring), p. 5 (brief summary)
3. Heller, W.H. 1981. Shadow and substance of inflation policy. Challenge (Jan.–Feb.), pp. 5–13.
4. United Press International. 1984. Court Mediation Service a success. Bangor Daily News, April 3.

G. Dishonesty Condoned

Material from many newspaper and magazine articles has been used in this section in addition to material from the sources cited.

1. Arieff, I., and D. Tarr. 1980. Hill ethics image hurt again despite many rule changes. Congresssional Quart. Weekly Report (Feb. 9), pp. 329–331.
2. Arkes, H. 1981. Morality and the law. Wilson Quart. (Spring), pp. 100–111.
3. Bayh, B. 1977. Challenge for the third century: Education in a safe environment. Report of the Subcommittee to Investigate Juvenile Delinquency. U.S. Govt. Printing Office, Washington, DC.
4. Bell, Daniel, and I. Kristol, Eds. 1981. The crisis in economic theory. Basic, New York.
5. Block, A. 1978. Combat neurosis in inner-city schools. Am. J. Psychiatry (Oct.), 135:1189-1192.
6. Parke, R.D. et al. 1977. Some effects of violent and nonviolent movies on the behavior of juvenile delinquents. Advances Exper. Social Psychology 10:135-172.
7. Read (study by the editors). 1981-1982. Do schools need more discipline? Read 31:(No. 1), pp. 4-7; (No. 9), pp. 3-5.
8. Rickover, H.G. 1982. Economics of defense policy. Hearing before the Joint Economic Committee of Congress, Jan. 28. (Quoted in Washington Spectator, April 1, 1982).
9. US News World Report, July 23, 1979. In hot pursuit of business criminals.

M. Major Crime Unpunished

Material from many newspaper and magazine articles has been used in Chapter 3, Section H, in addition to material from the sources cited.

1. Anderson, J. 1985. Drug kingpins get only a wrist slap by courts. Bangor Daily News, May 2.
2. Bangor Daily News, Sept. 4, 1980. Maine native's testimony helped convict Bonanno.
3. Bayh, B. 1977. See Chapter 3 under Dishonesty Condoned.
4. Chiles, L. 1981. How the IRA helps the mob. Read. Dig. (March), pp. 135–138.
5. Cressey, Donald R. 1969. Theft of the nation: The structure and operation of organized crime in America. Har-Row, New York.
6. Encyclopedia Brittanica, 1911. Mafia.
7. Freeman, E. A., and T. Ashby. 1911. Sicily. Encylo. Brit. (11th Edit.).
8. Kwitny, Jonathan, 1979. The Mafia in the marketplace. Norton, New York.
9. Riesel, V. 1967. Why did Senator Long not probe Cosa Nostra? Bangor Daily News, Nov. 14.
10. Smith, D.C., Jr. 1976. Mafia: The prototypical alien conspiracy. Ann. Am. Acad. Political and Social Sci. 423:75–88.
11. US News World Report, Feb. 25, 1980. FBI's war on Mafia—Better luck this time?
12. US News World Report. Beating the Mob (Feb. 3, 1986). The Mafia gets an unhealthy dose of sunlight (Oct. 6, 1986). Convictions and conflict: A top cop's tale (March 23, 1987).

I. CONCLUSIONS

1. Alm, R. 1985. It's back to the doghouse for economists. US News World Report (Feb. 4), p. 55.
2. Batra, Ravi. 1985. The Great Depression of 1990. Venus Bks, Dallas.
3. Brock, J.W. 1987. Bigness is the problem, not the solution. Challenge (July–Aug.), pp. 11–16.
4. Dean, J.W. 1982. Why economists disagree. Wilson Quart. (Autumn), pp. 86–97.
5. Etzioni, A. 1986. Founding a new socioeconomics. Challenge (Nov.–Dec.), pp. 13–17.
6. Galbraith, J.K. 1987. The 1929 parallel. The Atlantic (Jan.), pp. 62–66.
7. Islam, S. 1987. What's causing America's capital imports? Challenge (Sept.–Oct.), pp. 4–11.
8. Kaufman, H. 1986. In the shadow of financial exhilaration. Challenge (July–Aug.), pp. 4–10.
9, Moffit, M. 1987. Shocks, deadlocks, and scorched earth: Reaganomics and the decline of U.S. hegemony. World Policy J. (Fall), pp. 553–582.
10. Scitovsky, T. 1974. Are men rational or economists wrong? *In* Paul A. David and M.W. Reder, Eds. Nations and households in economic growth. Academic Pr, New York, pp. 223–235.
11. Simon, W.E. 1976. A challenge to free enterprise. *In* Ivan Hill, Ed. The ethical basis of economic freedom. Am. Viewpoint, Inc., Chapel Hill, NC.
12. Walker, Amasa. 1866. See Chapter 3 under Waste.
13. Wilbur, Charles K., and K.P. Jameson. 1983. See Chapter 3 under Waste.
14. Wilson, J.Q. 1985. The rediscovery of character: Private virtue and public policy. The Public Interest (Fall), pp. 3–16.

Chapter 4—The Biological Background of Social Behavior: Genetic Factors

Besides the references cited below, I have, in discussing crime, drawn on some studies described in newspaper and magazine articles based on government or government supported studies.

1. Alper, J. 1985. The roots of morality. Science 85 (March), pp. 70–76.
2. Batchelder, P., et al. 1982. The effects of age and experience on strain differences for nesting in *Mus musculus*. Behav. Genet. 12:149–159.
3. Bauer, F.J. 1956. Genetic and experimental factors affecting social reactions in male mice. J. Comp. Physiol. Psych. 49:359–364.
4. Bayh, Birch. 1977. See Chapter 3 under Dishonesty Condoned.
5. Bouchard, T. 1981. Personal communication.
6. Chaiken, J.M., and M.R. Chaiken. 1983. Crime: trends and targets. Wilson Quart. (Spring), pp. 103–115.
7. Chaudhari, N., and W.E. Hahn. 1983. Genetic expression in the developing brain. Science 220:924–928.
8. Checkley, Hervey M. 1964. The mask of sanity: An attempt to clarify some

issues about the so-called psychopathic personality. 4th edit. Mosby, St. Louis, MO.

9. Chorover, Stephen L. 1979. From genesis to genocide. MIT Pr, Cambridge, MA.

10. Christiansen, K.O. 1977a. A review of studies of criminality among twins. *In* Sarnoff A. Mednick and K.O. Christiansen, Eds. Biosocial basis of criminal behavior. Gardner Pr, New York, pp. 45–88.

11. Christiansen, K.O. 1977b. A preliminary study of criminality among twins. *In* op. cit, pp. 89–108.

12. Coleman, D. 1955. Personal communication.

13. Conrad, John P., and S. Dintz, Eds. 1977. In fear of each other. Lexington Bks, Winchester, MA,

14. Dalgard, O.S., and E. Kringlen. 1976. A Norwegian twin study of criminality. Brit. J. Criminology 16:213–231.

15. DeFries, J.C. 1980. Genetics of animal and human behavior. *In* George W. Barlow and J. Silverberg, Eds. Sociobiology: Beyond nature/nurture. Westview, Boulder, CO, pp. 273–294.

16. deGrouchy, Jean, and C. Turleau. 1977. Clinical atlas of human chromosomes. Wiley, New York.

17. Dryja, T.P., et al. 1984. Homozygosity of chromosome 13 in retinoblastoma. N.E. J. Med. 310:550–553.

18. Eleftheriou, Basil E., and J.P. Scott, Eds. 1971. The physiology of aggression and defeat. Plenum Pr, New York.

19. Elliott, F.A. 1982. Neurological findings in adult minimal brain dysfunction and the dyscontrol syndrome. J. Neur. Mental Disease 170:680–687.

20. Fredericson, E. 1952. Reciprocal fostering of two inbred mouse strains and its effect on the modification of inherited aggressive behavior. Am. Psychologist 7:241–242.

21. Fredericson, E., and E.A. Birnbaum. 1954. Competitive fighting between mice with different hereditary backgrounds. J. Genet. Psychol. 84:271–280.

22. Fredericson, E., et al. 1955. The relationship between heredity, sex, and aggression in two inbred mouse strains. J. Genet. Psychol. 887:121–130.

23. Fristrom, J.W., and P.T. Spieth. 1980. Principles of genetics. Chiron Pr, Concord, MA.

24. Fuller, J.L. 1979. Genetic analysis of deviant behavior. *In* D.J. Keehn, Ed. Psychopathology in animals. Research and clinical implications. Academic Pr, New York, pp. 61–79.

25. Fuller, J.L. 1982. Psychology and genetics: A happy marriage? Canadian Psychol. 23:11–21.

26. Gilbert, G.M. 1950. The psychology of dictatorship. Ronald Pr, New York.

27. Ginsburg, B., and W.C. Allee. 1942. Some effects of conditioning on social dominance and subordination in inbred strains of mice. Physiol. Zool. 15:485–506.

28. Glueck, Sheldon, and E. Glueck. 1943. Criminal careers in retrospect. The Commonwealth Fund, New York.

29. Green, Margaret C., Ed. 1981. Genetic variants and strains of the laboratory mouse. Gustav Fisher Verlag, New York.

30. Hamparian, Donna, et al. 1978. The violent few: A study of the dangerous juvenile offender. Lexington Bks, Winchester, MA.

31. Hancock, R., and T. Boulikas. 1982. Functional organization in the nucleus. Internat. Rev. Cytol. 79:165–214.

32. Harlow, Harry F. 1971. Learning to love. Albion, San Francisco.

33. Hassold, T., et al. 1980. A cytogenetic study of 1000 spontaneous abortions. Ann. Hum. Genet. 44:151–178.

34. Haug, M., and P. Pallaud. 1981. Effect of reciprocal cross-fostering on aggression of female mice toward lactating strangers. Developmental Psychobiol. 14:177–180.

35. Hauser-Urfr, I.H., and J. Stauffer. 1985. Comparative chromosome analysis of nine squamous cell carcinoma lines from tumors of the head and neck. Cytogenet. Cell Genet. 39:35–39.

36. Hearnshaw, Leslie S. 1979. Cyril Burt, psychologist. Cornell U Pr, Ithaca.

37. Hirschi, T. 1983. Families and crime. Wilson Quart. (Spring) pp. 132–139.

38. Hirschi, T., and M.J. Hindelang. 1977. Intelligence and delinquency: A revisionist review. Am. Sociol. Rev. 42:571–587.

39. Holden, C. 1980. Twins reunited. Science 80 (Nov.), pp. 55–59.

40. Hsu, T.C. 1979. Human and mammalian cytogenetics: A historical perspective. Springer, New York.

41. Hubel, D.H. 1979. The brain. Sci. Am. 241 (Sept.), pp. 45–53.

42. Hutchings, B., and S.A. Mednick. 1977. Criminality in adoptees and their adoptive and biological parents. *In* Sarnoff A. Mednick and K. O. Christiansen, Eds., Biosocial basis of criminal behavior. Gardner Pr, New York, pp. 127–141.

43. Hyde, J.S. 1983. The genetics of agonistic and sexual behavior. *In* J.L. Fuller and E.C. Simmel, Eds., Behavior genetics. L. Erlbaum Assocs, Hillsdale, NJ, pp. 409–434.

44. Jeffery, C.R., Ed. 1979. Biology and crime. Sage Research Progress Series in Criminology, Vol. 10. Sage, Beverly Hills, CA.

45. Kagan, Jerome. 1984. The nature of the child. Basic, New York.

46. Kessler, S., et al. 1975. The genetics of pheromonally mediated aggression in mice. I. Strain differences in the capacity of male urinary odors to elicit aggression. Behav. Genet. 5:233–238.

47. Lagerspitz, K., and S. Hautojarvi. 1967. The effect of prior aggressive or sexual arousal on subsequent aggressiveness or sexual reactions in male mice. Scand. J. Psychol. 8:1–6.

48. Langan, P., and L. Greenfield. 1983. Career patterns in crime. Bureau of Justice Statistics. (Summarized in article by United Press, International, July 23, 1983).

49. Lejeune, J., et al. 1959. Les chromosomes humains en culture de tissus. C.R. Acad. Sci. (Paris) 248:1721.

50. Levine, L., et al. 1979. Functional relationships between genotypes and environments in behavior. Effects of different kinds of early social experience on interstrain fighting in male mice. J. Hered. 70:317–320.

51. Loehlin, J.C., and R.C. Nichols. 1976. Heredity, environment and personality: A study of 850 sets of twins. Texas U Pr, Austin.

52. Lundin, L.G. 1979. Evolutionary conservation of large chromosome segments reflected in mammalian gene maps. Clin. Genet. 16:72–81.

53. Marler, P., and S. Peters. 1981. Sparrows learn adult songs and more from memory. Science 213:780–782.

54. Maxson, S.C., and A. Trattner. 1981. Interaction of genotype and fostering in the development of behavior of DBA and C57 mice. Behav. Genet. 11:153–165.

55. McCracken, A.A., et al. 1978. Twins and Q-banded chromosome polymorphisms. Human Genet. 45:253–258.

56. McKay, R.D.G., and S.H. Hockfield. 1982. Monoclonal antibodies distinguish antigenically discrete neuronal types in the vertebrate central nervous system. Proc. Natl. Acad. Sci. 79:6747–6751.

57. McKusick, V.A. 1985. The human gene map. 1 December 1984. Clin. Genet. 27:207–239.

58. McKusick, V. A. Personal communication.

59. Mednick, Sarnoff A., and Karl O. Christiansen, Eds. 1977. Biosocial basis of criminal behavior. Gardner Pr, New York.

60. Mednick, S.A., et al. 1982. Biology and violence. *In* Marvin E. Wolfgang and N.A. Weiner, Eds. Criminal violence. Sage, Beverly Hills, CA, pp. 21–80.

61. Miller, Stuart J., et al. 1982. Careers of the violent. Lexington Bks, Winchester, MA.

62. Nagle, James J. 1979. Heredity and human affairs, 2nd ed. Mosby, St. Louis, MO.

63. Newman, H.H., et al. 1937. Twins: A study of heredity and environment. U of Chicago Pr, Chicago.

64. Nielsen, J., and I. Sillesen. 1975. Incidence of chromosome aberrations among 11,148 newborn children. Human Genet. 30:1–12.

65. Nissen, H.W. 1956. Individuality in the behavior of chimpanzees. Am. Anthrop. 58:407–417.

66. Obrien, S.J. 1973. On estimating functional gene numbers in eukaryotes. Nature 242:52–54.

67. Office of Juvenile Justice and Delinquent Prevention, U.S. Dept. of Justice. 1981. Analysis of national crime victimization survey data to study serious delinquent behavior. Washington, DC. [Cited in Wilson Quart. (Winter '82), pp. 47–48.]

68. Omenn, G.S. 1983. Medical genetics, genetic counseling, and behavior genetics. *In* John L. Fuller and E.C. Simmel, Eds. Behavior genetics. L. Erlbaum, Hillsdale, NJ, pp. 155–187.

69 Panel on Research on Rehabilitative Techniques. 1981. The rehabilitation of criminal offenders: Problems and prospects. Natl. Acad. Pr, Washington, T.C.

70. Plomin, Robert, et al. 1980. Behavior genetics. A primer. WH Freeman, San Francisco.

71. Prentice, N.M., and F.J. Kelly. 1983. Intelligence and delinquency: A reconsideration. J. Social Psych. 60:327–337.

72. Radke-Yarrow, M., et al. Children's prosocial dispositions and behavior. *In* Paul H. Mussen, Ed., Charmichael's manual of child psychology (Vol. IV, 4th ed.). Wiley, New York, pp. 469–545.

73. Radke-Yarrow, M., and C. Zahn-Waxler. 1984. Roots, motives and patterns in children's prosocial behavior. *In* J. Reykowski, et al., Eds. The development and maintenance of prosocial behavior: International perspectives. Plenum Pr, New York, pp. 81–99.

74. Raeburn, Paul. 1983. An uncommon chimp. Science 83 (June), pp. 40–48.

75. Reed, Sheldon. 1980. Counseling in medical genetics, 3rd ed. A. R. Liss, New York.

76. Rennie, Ysabel. 1978. The search for criminal man: A conceptual history of the dangerous offender. Lexington Bks, Winchester, MA.

77. Rensberger, B. 1983. The nature-nurture debate. I. Margaret Mead; II. On becoming human. Science 83 (April), pp. 28–46.

78. Rest, J. A. 1982. A psychologist looks at the teaching of ethics. Hastings Center Report (Feb.), pp. 29–36.

79. Restak, R.M. 1982. Newborn knowledge. Science 82 (Jan.–Feb.), pp. 58–65.

80. Roderick, T., and M. Davisson. 1985. Personal communication.

81. Rushton, J.P., et al. 1984. Altruism and genetics. Acta Genet. Med. Gemellol. 33:265–271.

82. Samenow, Stanton E. 1984. Inside the criminal mind. Times Bks, New York.

83. Scarr, S. 1968. Environmental bias in twin studies. Eugenics Quart. 15:34–40.

84. Schulsinger, F. 1977. Psychopathy: Heredity and environment. *IN* Sarnoff A. Mednick and K.O. Christiansen, Eds. Biosocial bases of criminal behavior. Gardner Pr, New York. pp. 109–125.

85. Scott, J.P. 1966. Agonistic behavior in mice and rats: a review. Am. Zoologist 6:683–701.

86. Scott, J.P. 1979. Critical periods in organizational processes. *In* F. Falkner and J.M. Tanner, Eds. Human growth, vol. 3. Plenum Pr, New York, pp. 223–241.

87. Scott, J.P., and E. Fredericson. 1951. The causes of fighting in mice and rats. Physiol. Zool. 24:273–309.

88. Scott, J.P., and M.-V. Marston. 1953. Nonadaptive behavior resulting from a series of defeats in fighting mice. J. Abnorm. Soc. Psych. 48:417–428.

89. Scriver, C.R., et al. 1978. Genetics and medicine: An evolving relationship. Science 200:946–952.

90. Sondern, Frederic, Jr. 1959. Brotherhood of evil: The Mafia. Farrar, Straus and Cudahy, New York.

91. Sorokin, Pitirim A. 1950. Altrustic love. A study of American "good neighbors" and Christian saints. Beacon Pr, Boston.

92. Southwick, C.H. 1968. Effect of maternal environment of aggressive behavior of inbred mice. Communications in Behavioral Biol. 1:129–132.

93. Sperry, Roger. 1982. Some effects of disconnecting the cerebral hemispheres. Les Prix Nobel, 1981, pp. 209–219.

94. Staub, Ervin. 1978. Positive social behavior and morality, Vol. 2. Academic Pr, New York.

95. Stene, J., et al. 1981. Paternal age and Down's syndrome. Human Genet. 59:119–124.

96. Therman, Eeva. 1980. Human chromosomes: Structure, behavior, effects. Springer-Verlag, New York.

97. Valzelli, L., and S. Bernasconi. 1979. Aggressiveness by isolation and brain scrotonin turnover changes in different strains of mice. Neuropsychobiology 5:129–135.

98. Vandenberg, S.G. 1967. Hereditary factors in psychological variables in man, with special emphasis on cognition. *In* J.N. Spuhler, Ed. Genetic diversity and human behavior. Aldine, Chicago, pp. 99–133.

99. vom Saal, F.S., and F.H. Bronson. 1980. Sexual characteristics of adult

female mice are correlated with their blood testosterone levels during prenatal development. Science 208:597–599.

100. Welch, A.S., and B.L. Welch. 1981. Isolation, reactivity and aggression: Evidence for involvement of brain catecholamines and serotonins. *In* Basil E. Eleftheriou and J. P. Scott, Eds. The physiology of aggression and defeat. Plenum Pr, New York, pp. 91–142.

101. Weltman, A.S., et al. 1967. Effects of isolation on maternal aggressiveness and body growth rates of offspring. Experientia 23:782–784.

102. Wilson, James Q. 1983. Thinking about crime, 2nd ed. Basic, New York.

103. Wilson, J.Q., and P.J. Cook. 1985. Unemployment and crime—what is the connection? The Public Interest, No. 79 (Spring), pp. 3–8.

104. Witkin, H.A., et al. 1977. XYY and XXY men: Criminality and aggresssion. *In* Sarnoff A. Mednick and K.O. Christiansen, Eds. Biosocial basis of criminal behavior. Gardner Pr, New York. pp. 165–187.

105. Wolfgang, M.E., et al. 1972. Delinquency in a birth cohort. U Of Chicago Pr, Chicago.

106. Yamazaki, Y., et al. 1983. Sensory distinction between $H\text{-}2^b$ and $H\text{-}2^{bml}$ mutant mice. Proc. Natl. Acad. Sci. 80:5685–5688.

107. Yunis, J.J., et al. 1980. The striking resemblance of high-resolution G-banded chromosomes of man and chimpanzee. Science 208:1145–1148.

108. Yunis, J.J., et al. 1984. High-resolution chromosomes as an independent prognostic indicator in adult acute nonlymphocytic leukemia. N.E. J. Med. 311:812–818.

Chapter 5—The Biological Background of Social Behavior: Evolutionary Factors

1. Altmann, S.A. 1962. A field study of the sociobiology of rhesus monkeys, Macaca mulatta. Ann. N.Y. Acad. Sci. 102:338–435.

2. Axelrod, R., and W.D. Hamilton. 1981. The evolution of cooperation. Science 811:1390–1396.

3. Boorman, S.A., and P.R. Leavitt. 1973. A frequency-dependent natural selection model for the evolution of social cooperation networks. Proc. Natl. Acad. Sci. 70:187–189.

4. Burroughs, C. A. 1959. Cliff dwellers' secrets. Natl. Geogr. 106:619–625.

5. Bushnell, David I., Jr. 1934. Tribal migration east of the Mississippi. Smithsonian Miscellaneous Collections, Vol. 89, No. 2. Washington, D.C.

6. Bygott, J.D. 1972. Cannibalism among wild chimpanzees. Nature 238:410–411.

7. Carpenter, C.R. 1942. Societies of monkeys and apes. Biol. Symp. 13:177–204.

8. Cashdan, E. 1983. Territoriality among human foragers: Ecological models and an application to four bushmen groups. Current Anthrop. 24:47–55.

9. Chagnon, Napoleon A. 1977. Yanomamö: the fierce people, 2nd ed. HR & W, New York.

10. Chagnon, N.A. 1980. Kin-selection theory, kinship, marriage and fitness among the Yanomamö indians. *In* George W. Barlow and J. Silverberg,

Eds. Sociobiology: Beyond nature/nuture? Westview, Boulder, CO, pp. 545–571.

11. Chagnon, N.A., and R.B. Hames. 1979. Protein deficiency and tribal warfare in Amazonia: New data. Science 203:910–913.

12. Cohen, J. E. 1969. Natural primate troops and a stochastic population model. Am. Nat. 103:455–477.

13. Darwin, Charles. 1859. The origin of species. (1963 printng, WSP, New York.)

14. Darwin, Charles. 1874. See Chapter 2.

15. Davies, O. 1956. African Pleistocene pluvials and European glaciations. Nature 178:757–759.

16. de Waal, Frans. 1982. Chimpanzee politics. Har-Row, New York.

17. de Waal, F., and Roosmalen. 1979. Chimpanzees. Reviewed in Charles T. Snowden. 1983. Ethology, comparative psychology, and animal behavior. Ann. Rev. Psychol. 34:63–94.

18. Dixson, A. F. 1981. The natural history of the gorilla. Weidenfeld and Nicolson, London.

19. Eibl-Eibesfeldt, I. 1975. Aggression in the !Ko-Bushmen. *In* Martin A. Nettleship, et al., Eds. War, its causes and correlates. Mouton, The Hague, pp. 281–295.

20. Errington, P.L. 1939. Reactions of muskrat populations to drought. Ecology 20:168–186.

21. Eshel, I., and L.L. Cavalli-Sforza. 1982. Assortment of encounters and evolution of cooperativeness. Proc. Natl. Acad. Sci. 79:1331–1335.

22. Fleagle, J.G., et al. 1980. Sexual dimorphism in early anthropoids. Nature 287:328–329.

23. Ford, E.B. 1955. Rapid evolution and the conditions which make it possible. Symp. Quant. Biol. 20:230–238.

24. Fossey, D. 1970. Making friends with mountain gorillas. Natl. Geogr. 137:48–67.

25. Fossey, D. 1981. The imperial mountain gorilla. Nat. Geogr. 159:501–523.

26. Franklin, W.L. 1981. Living with guanacos: Wild camels of South America. Natl. Geogr. 160:63–75.

27. Givens, R.D. 1975. Aggression in nonhuman primates: Implications for understanding human behavior. *In* Martin A. Nettleship, et al., Eds. War, its causes and consequences. Mouton, The Hague, pp. 263–280.

28. Goodall, J. 1970. Tool using in primates and other vertebrates. Adv. Study Behavior 3:195–249.

29. Goodall, J. 1979. Life and death at Gombe. Natl. Geogr. 155:592–621.

30. Hamburg, David A., and E.R. McCown. 1979. The great apes. Benjamin Cummings, Menlo Park, CA.

31. Hamilton, W.D. 1963. The evolution of altruistic behavior. Am. Nat. 97:354–356.

32. Hamilton, W.D. 1964. The genetical theory of social behavior, I, II. J. Theoret. Biol. 7:1–52.

33. Hammond, A.L. 1983. Tales of an elusive ancestor. Science 83 (Nov.), pp. 36–43.

34. Harding, R.S.O. 1981. An order of omnivores: Nonhuman primate diets in the wild. *In* Robert S.O. Harding and G. Teleki, Eds. Omnivorous pri-

mates: Gathering and hunting in human evolution. Columbia U. Pr, New York, pp. 191–214.

35. Hayden, B. 1981. Subsistence and ecological adaptations of modern hunter-gathers. *In* Robert S.O. Harding and G. Teleki, Eds. Omnivorous primates: Gathering and hunting in human evolution. Columbia U. Pr, New York, pp. 344–421.

36. Hoebel, E. Adamson. 1960. The Cheyennes, Indians of the Great Plains. HR&W, New York.

37. Hoese, H.D. 1971. Bottle-nosed dolphins. Review in E.O. Wilson. 1975. Sociobiology: The new synthesis. Belknap Pr, Cambridge, MA, p. 475.

38. Holloway, Ralph. Casts of inside of skulls of early man. Review in Richard E. Leaky and R. Lewin. 1977. Origins. MacDonald and Jane's, London, p. 197.

39. Hrdy, S.B. 1977. Infanticide as a primate reproductive strategy. Am. Sci. 65:40–47.

40. Hrdy, Sarah B. 1981. The woman that never evolved. Harvard U Pr, Cambridge, MA.

41. Hrdy, S.B. 1983. Heat loss. The absence of estrus reflects a change in sexual strategy. Science 83 (Oct.), pp. 73–78.

42. Hrdy, S.B., and W. Bennett. 1981. Lucy's husband: What did he stand for? Harvard Mag. (July–Aug.), pp. 7–9, 46.

43. Isaac, G., and D. Crader. 1981. To what extent were early hominids carnivores? An archaeological perspective. *In* Robert S.O. Harding and G. Teleki, Eds. Onmivorous primates: Gathering and hunting in human evolution. Columbia U. Pr, New York, pp. 37–103.

44. Jarvis, J.U.M. 1981. Eusociality in a mammal: Cooperative breeding in naked mole-rat colonies. Science 212:571–573.

45. Jenness, D. 1970. Eskimo. *In* Encyclopaedia Britannica. William Benton, Chicago.

46. Johanson, Donald, and M. Edey. 1981. Lucy: The beginnings of humankind. S&S, New York.

47. Jolly, A. 1966. Lemur social behavior and primate intelligence. Science 153:501–506.

48. Jouventin, P., and A. Cornet. 1980. The sociobiology of pinnipeds. Adv. Study Behavior 11:121–141.

49. Kay, R.F. 1982. Sexual dimorphism in Ramapithecus. Proc. Natl. Acad. Sci. 79:209–212.

50. Kennedy, G.E. 1978. Hominoid habitat shifts in the Miocene. Nature 271:11-12.

51. Kolata, G.B. 1977. Human evolution: Hominoids of the Miocene. Science 197:244–245, 294.

52. Konner, M.J. 1972. Aspects of the developmental ethology of a foraging people. *In* N.G. Blurton Jones, Ed. Ethological studies of child behavior. Cambridge U Pr, New York, pp. 285–304.

53. Krebs, C.J., et al. 1973. Population cycles in small rodents. Science 179:35–41.

54. Kropotkin, Prince. 1922. Mutual aid: A factor in evolution, 2nd ed. Knopf, New York.

55. Leaky, Richard E., and R. Lewin. 1977. Origins. MacDonald and Jane's, London.

56. LeBoeuf, B.J., and R.S. Peterson. 1969. Social status and mating activity in elephant seals. Science 163:91–93.

57. Leigh, E.G., Jr. 1977. How does selection reconcile individual advantage with the good of the group? Proc. Natl. Acad. Sci. 74:4542–4546.

58. Lewin, R. 1984. Man the scavenger. Science 224:861–862.

59. Lorenz, Konrad. 1966. On aggression. (Translated from the German) Bantam, New York.

60. Lovejoy, C.O., et al. 1977. Paleodemography of the Libben site, Ottawa County, Ohio. Science 198:291–293.

61. Loy, T. 1983. Prehistoric blood residues: Detection on tool surfaces and identification of species of origin. Science 220:1269–1270.

62. MacLeish, K., and J. Launois. 1972. The Tasadays: Stone Age cavemen of Mindanao. Natl. Geogr. 142:219–242.

63. Meehan, W.P., and J.P. Henry. 1981. Social stress and the role of attachment behavior in modifying aggression. *In* Paul E. Brain and D. Denton, Eds. Multidisciplinary approaches to aggression research. Elsevier/North-Holland, Amsterdam, pp. 209–223.

64. Mellen, Sydney L.W. 1981. The evolution of love. WH Freeman, San Francisco.

65. Murie, A. 1944. Hunting by wolves. Review in E.O. Wilson. 1975. Sociobiology: The new synthesis. Belknap Pr, Cambridge, MA, p. 54.

66. Neel, J.V. 1970. Lessons from a "primitive" people. Science 170:815–822.

67. Neel, J.V. 1972. The genetic structure of a tribal population, the Yanomama Indians. Ann. Hum. Genet. 35:255–259.

68. Neel, J.V. 1980. On being headman. Perspectives in Biol. and Med. 23:277–294.

69. Neel, J. V., and K.M. Weiss. 1975. The genetic structure of a tribal population, the Yanomama Indians. XII. Biodemographic studies. Am. J. Phys. Anthrop. 42:25–52.

70. Packer, C., and A.E. Pusey. 1982. Cooperation and competition within coalitions of male lions: Kin selection or game theory? Nature 296:740–742.

71. Packer, C., and A.E. Pusey. Adaptations of female lions to infanticide by incoming males. Amer. Nat. 121:716–728.

72. Parkman, F. 1900. Pioneers of France in the New World. Little, Boston.

73. Pearl, M.C., and S.R. Schulman. 1983. Techniques for the analysis of social structure in animal societies. Adv. Study Behavior 13:107–146.

74. Plomin, Robert J., et al. 1980. Behavioral genetics. A primer. WH Freeman, San Francisco.

75. Point, Father Nicolas. 1967. Wilderness kingdom. Indian life in the Rocky Mountains, 1840–1847. (Translated from the French.) HR&W, New York.

76. Raeburn, P. 1983. An uncommon chimp. Science 83 (June), pp. 40–48.

77. Rensberger, B. 1984. A new ape in our family tree. Science 84 (Jan./Feb.), p. 16.

78. Reynolds, V., and F. Reynolds. 1965. Chimpanzees of the Bundongo Forest. *In* Irven DeVore, Ed. Primate behavior. HR&W, New York, pp. 368–424.

79. Rodman, P., and H. McHenry. Why man became biped. Summary in

Sarah B. Hardy and W. Bennett. 1981. Lucy's husband: What did he stand for? Harvard Mag. (July–Aug.), pp. 7–9, 46.

80. Roper, M.K. 1969. A survey of evidence for intrahuman killing in the Pleistocene. Current Anthrop. 10:427–459.

81. Rushton, J.P., et al. 1984. Genetic similarity theory: Beyond kin selection. Behav. Genet. 14:179–193.

82. Sahlins, M.D. 1959. The social life of monkeys, apes, and men. Human Biol. 31:54–73.

83. Sahlins, M.D. 1960. The origins of society. Sci. Amer. (Sept.), pp. 76–87.

84. Schaffer, W.M. 1968. Character displacement and the evolution of the Hominidae. Am. Nat. 102:559–571.

85. Schaller, G.B. 1965. The behavior of the mountain gorilla. *In* Irven DeVore, Ed. Primate behavior. HR&W, New York, pp. 324–367.

86. Schaller, G.B. 1969. Life with the king of beasts. Natl. Geogr. 135:496–519.

87. Schaller, George B. 1972. The Serengeti lion. U of Chicago Pr, Chicago.

88. Scott, John P. 1958. Animal behavior. U of Chicago Pr, Chicago.

89. Shipman, R., and R. Potts. Toolmarks on bones as evidence of scavenging. Summarized in Roger Lewin. 1984. Man the scavenger. Science 224:861–862.

90. Short, Roger. Testis weight in chimpanzees. Summarized in Sarah B. Hrdy. 1983. Heat loss. The absence of estrus reflects a change in sexual strategy. Science 83 (Oct.), pp. 73–78.

91. Sibley, C.G., and J.E. Ahlquist. 1984. The phylogeny of the hominoid primates, as indicated by DNA-DNA hybridization. J. Molecular Evol. 20:2–15.

92. Simonds, P.E. 1965. The bonnet macaque of South India. *In* Irven DeVore, Ed. Primate behavior. HR&W, New York, pp. 175–196.

93. Snell, G.D. 1932. The role of male parthenogenesis in the evolution of the social hymenoptera. Am. Nat. 381–384.

94. Snowdon, C.T. 1983. Ethology, comparative psychology, and animal behavior. Ann. Rev. Psychol. 34:63–94.

95. Southwick, C.H., et al. 1965. Rhesus monkeys in North India. *In* Irven DeVore, Ed. Primate behavior. HR&W, New York, pp. 111–159.

96. Spencer, Herbert. 1892. Principles of ethics. Williams and Norgate, London (Printing of 1904).

97. Storer, M.B. 1971. Toward a theory of moral debt. Inquiry 14:355–385.

98. Sugiyama, Y. 1976. Life history of male Japanese monkeys. Adv. Study Behavior 7:255–284.

99. Susman, R.L., and J.T. Stern. 1982. [Summary of work, under title "Handy hominids," in Science 82 (Nov.), p. 7].

100. Thornton, Francis B. 1953. Sea of glory. The magnificient story of the four chaplains. P-H, Englewood Cliffs, NJ.

101. Trivers, R.L. 1971. The evolution of reciprocal altruism. Quart. Rev. Biol. 46:35–57.

102. van Lawick-Goodall, Hugo, and J. van Lawick-Goodall. 1971. Innocent killers. HM, Boston.

103. van Lawick-Goodall, J. 1974. Infanticide in African wild dogs. Reiview in E.O. Wilson. 1985. Sociobiology: The new synthesis. Belknap Pr, Cambridge, MA, p. 512.

104. Wade, M.J. 1978. A critical review of the models of group selection. Quart. Rev. Biol. 53:101–114.
105. Washburn, S. 1982. Fifty years of human evolution. Bull. Am. Acad. Arts Sci. 35:25–39.
106. Weitkamp, L.R., and J.V. Neel. 1972. The genetic structure of a tribal population, the Yanomama Indians. Ann. Hum. Genet. 35:433–444.
107. West-Eberhard, M. J. 1975. The evolution of social behavior by kin selection. Quart. Rev. Biol. 50:1–33.
108. Williams, B. J. 1980. Kin selection, fitness and cultural evolution. *In* George W. Barlow and J. Silverberg, Eds. Sociobiology: Beyond nature/nuture? Westview Pr, Boulder, CO, pp. 573–587.
109. Wilson, D.S. 1980. The natural selection of populations and communities. Benjamin Cummings, Menlo Park, CA.
110. Wilson, Edward O. 1971. The insect societies. Belknap Pr, Cambridge, MA.
111. Wilson, E.O. 1973. Group selection and its significance for ecology. Bio. Sci. 23:631–638.
112. Wilson, Edward O. 1975. Sociobiology: The new synthesis. Belknap Pr, Cambridge, MA.
113. Wilson, Edward O. 1978. On human nature. Harvard U Pr, Cambridge, MA.
114. Wilson, E.O. 1979. Sociobiology: Sex and human nature. Wilson Quart. (Autumn), pp. 95–105.
115. Woolfenden, G. E. 1973, 1974. Florida scrub jay. Review in E.O. Wilson. 1975. Sociobiology: The new synthesis. Belknap Pr, Cambridge, MA.
116. Wright, S. 1931. Evolution of Mendelian populations. Genetics 16:97–159.
117. Wright, S. 1955. Classification of the factors of evolution. Cold Spring Harbor Symp. Quant. Biol. 20:16–24.
118. Zimmerman, L.J., and R. Alex. 1981. The Crow Creek experience. Early Man (Autumn), pp. 3–10.
119. Zimmerman, L.J., and R.G. Whitten. 1980. Prehistoric bones tell a grim tale of Indian V. Indian. Smithsonian (Sept.), pp. 100–107.

Chapter 6—What Is Ethics?

1. Aristotle. Nicomachean ethics. In Richard M. McKeon, Ed. Introduction to Aristotle. Random House, New York.
2. Beattie, Paul H. 1985. Twenty years in a Unitarian pulpit. First Unitarian Church, Pittsburgh, PA.
3. Bentham, Jeremy. 1789. The principles of morals and legislation. Hafner, New York (1948 edition). See Introduction by L.J. LaFleur.
4. Blanshard, B. 1966. Morality and politics. *In* Richard T. DeGeorge, Ed. Ethics and society. Doubleday, New York, pp. 1–23.
5. Bradley, Francis H. 1876. Ethical studies. Oxford U Pr, London.
6. Caggiula, A.R., and B.G. Hoebel. 1966. "Copulation reward site" in the posterior hypothalamus. Science 153:1284–1285.
7. Critchley, M. 1956. Congential indifference to pain. Ann. Inst. Med. 45:737–747.
8. Darwin, Charles. 1874. See Chapter 2.

9. Easterlin, Richard A. 1974. Does economic growth improve the human lot? Some empirical evidence. *In* Paul A. David and M.H. Reder, Eds. Nations and households in economic growth. Academic Pr, New York, pp. 89–125.

10. Hummell, A.W. 1952. Some basic moral principles in Chinese culture. *In* Ruth N. Anshen, Ed. Moral principles of action. Har-Row, New York, pp. 598–605.

11. Lilly, John C. 1961. Man and dolphin. Doubleday, New York.

12. Linton, R. 1952. Universal ethical principles: An anthropological point of view. 1952. *In* Ruth N. Anshen, Ed. Moral principles of action. Har-Row, New York, pp. 645–660.

13. MacKaye, James. 1939. The logic of language. Dartmouth College Publications, Hanover, N.H.

14. Muller, H.J. 1961. Survival. AIBS Bullentin 11:15–24.

15. Nikhilananda, Swami. 1952. Hindu ethics. *In* Ruth N. Anshen, Ed. Moral principles of action. Har-Row, New York, pp. 616–644.

16. Plamenatz, John. 1949. The English utilitarians. Basil Blackwell, Oxford.

17. Plato. Meno. 1957 edition. The Great Books Foundation, Chicago.

18. Plato. Phaedrus. *In* Irwin Edman, Ed. 1928. The works of Plato. S&S, New York, pp. 263–329.

19. Robertson, A. 1945. Morals in world history. C.A. Watts and Co., London.

20. Ruse, M., and E.O. Wilson. 1986. Moral philosophy as an applied science: A Darwinian approach to the foundations of ethics. Philosophy 61:173–192.

21. Sahakian, William S. 1964. Systems of ethics and value theory. Littlefield, Totowa, NJ.

22. Steiner, Franz. 1956. Taboo. Cohen and West, London.

23. Storer, Morris B., Ed. 1980. Humanist ethics. Prometheus Bks, Buffalo, NY.

24. von Fritz, Kurt. 1952. Relative and absolute values. *In* Ruth N. Anshen, Ed. Moral principles of action. Har-Row, New York, pp. 94–121.

25. Wilson, E.O. 1980. Comparative social theory. The Tanner Lectures on Human Values, 1980, Vol. 1. Cambridge U Pr, New York, pp. 49–73.

26. Wilson, Edward O. 1984. Biophilia. Harvard U Pr, Cambridge, MA.

27. Yankelovich, David. 1981. See Chapter 2.

Chapter 7—The Rules We Live by

1. Adams, Samuel. Quoted by Rossiter, Clinton L. 1953. See Chapter 2.

2. Bentham, Jeremy. 1789. See Chapter 6.

3. Bradford, William. Of Plimoth Plantation, 1620–1647. Modern Lib, New York (1967 edition).

4. Chagnon, Napoleon. 1977. See Chapter 5.

5. de Tocqueville, Alexis. 1835. See Chapter 1.

6. Fiske, John. 1902. New France and New England. HM, Boston.

7. Gaer, Joseph. 1951. How the great religions began. Dodd, New York.

8. Gaster, Theodor H. 1956. The Dead Sea Scriptures. Doubleday, New York.

9. Held, Virginia. 1984. Rights and goods: Justifying social action. Free Pr, New York.

10. Hewitt, J. N. B. 1918. A constitutional league of peace in the Stone Age of

America. Smithsonian Institution, Ann. Report, Washington, DC, pp. 527–545.
11. Hobbes, Thomas. 1651. See Chapter 2.
12. Hoebel, E. Adamson. 1960. See Chapter 5.
13. Hoff, Christina. 1982. When public policy replaces private ethics. Hastings Center Report (Aug.), pp. 13–14.
14. Hoffmann, R.J. 1985. The origins of Christianity: Free Inquiry (Spring), pp. 50–56.
15. Holmes, S.J. 1948. Life and morals. Macmillian, New York.
16. Keith, Sir Arthur. 1946. Essays on human evolution. Watts, London.
17. Khan, Sir M.Z. 1952. Moral principles as the basis of Islamic culture. *In* Ruth N. Anshen, Ed. Moral principles of action. Har-Row, New York, pp. 559–577.
18. Linton, R. 1952. See Chapter 6.
19. Lodge, George C. 1984. See Chapter 2.
20. McKown, D.B. 1986. A humanist looks at the future of Unitarian Universalism. Religious Humanist (Spring), pp. 58–64, 70.
21. Milgrom, J. 1978. The Temple Scroll. Biblical Archeologist 41:105–120.
22. Morgan, Lewis H. 1901. League of the Ho-dé-no-sau-nee or Iroquois. New edition. Dodd, New York. (Originally published 1851.)
23. Morrison, Samuel E. 1930. Builders of the Bay Colony, HM, Boston.
24. Rawls, John. 1971. A theory of justice. Belknap Pr, Cambridge, MA.
25. Rockefeller, Jay. 1980. Interview with US News World Report, Dec. 1, 1980.
26. Rossiter, Clinton L. 1953. See Chapter 2.
27. Sheeper, J. 1971. Mate selection among second-generation kibbutz adolesents and adults: Incest avoidance and negative imprinting. Archives of Sexual Behavior 1:293–307. (Summarized in Wilson, Edward O. 1978. See Chapter 5.)
28. Spencer, Herbert. 1892. See Chapter 5.
29. Tapp, J.L. 1970. A child's garden of law and order. Psychology Today (Dec.), p. 29.
30. US News World Report, Sept. 13, 1976.
31. Westermark, Edward. 1912. The origin and development of moral ideas, 2nd ed. Macmillan, London.
32. Yanai, J., and G.E. McClearn. 1972. Assortative mating in mice and the incest taboo. Nature 238:281–282.

Chapter 8—Looking to the Future

1. Benson, George C.S. and T.S. Engeman. 1975. See Chapter 2.
2. Cannon, M.W. 1982. Crime and the decline of values. In Owen Peterson, Ed. Representative American Speeches, 1981–1982. Wilson Co., Bronx NY, pp. 62–78.
3. Elkind, David. 1984. See Chapter 3 under Misguidance in the Marketplace.
4. Graham, P.A. 1984. Teaching in America. Wanting it all. Wilson Quart. (New Years), pp. 47–58.
5. Rest, J.R. 1981. See Chapter 4.

6. Sloan, Douglas. 1980. The teaching of ethics in the American undergraduate curriculum, 1876–1976. In Daniel Callahan and Sissela Bok, Eds. Ethics teaching in higher education. Plenum Pr, New York, pp. 2–58.
7. Storer, M.B. 1971. See Chapter 5.
8. Westermark, Edward. 1912. See Chapter 7.
9. White, M.I. 1984. Japanese education: How do they do it? The Public Interest (Summer), pp. 87–101.

Author Index

Subject Index